D1071210

Ecological Paradigms Lost

Routes of Theory Change

A volume in the
Academic Press | THEORETICAL ECOLOGY SERIES

Published Books in the Series

J. M. Cushing, R. F. Costantino, Brian Dennis,
Robert A. Desharnais, Shandelle M. Henson,
Chaos in Ecology: Experimental Nonlinear Dynamics, 2003.

Ecological Paradigms Lost
Routes of Theory Change

Kim Cuddington
Ohio University

Beatrix E. Beisner
University of Quebec at Montreal

Amsterdam Boston Heidelberg London New York Oxford
Paris San Diego San Francisco Singapore Sydney Tokyo

Elsevier Academic Press
30 Corporate Drive, Suite 400, Burlington, MA 01803, USA
525 B Street, Suite 1900, San Diego, California 92101-4495, USA
84 Theobald's Road, London WC1X 8RR, UK

This book is printed on acid-free paper. ∞

Library of Congress Cataloging-in-Publication Data
Application submitted

British Library Cataloguing in Publication Data
A catalogue record for this book is available from the British Library

ISBN-13: 978-0-12-088459-3
ISBN-10: 0-12-088459-3

For all information on all Elsevier Academic Press publications visit our Web site at www.books.elsevier.com

Printed in the United States of America
05 06 07 08 09 10 9 8 7 6 5 4 3 2 1

CONTENTS

Part II EPIDEMIOLOGICAL ECOLOGY

Part III COMMUNITY ECOLOGY

Foreword
Robert Paine

It would be both arrogant and foolish for a non-historian but active ecologist to evaluate Thomas Kuhn's (1962) philosophical impact on science, especially ecology and evolution. Kuhn suggested that the *ordinary science* most of us do, testing research philosophies identified with Popper (1959), Platt (1984), or Lakatos (1978), failed to describe how disciplines evolve. Rather, research was seen to focus on fine-tuning central *paradigms*. Accumulating anomalous results and building skepticism eventually generate a *scientific revolution* and a new major paradigm. The term *paradigm* has stuck; but like *niche* and *community*, it defies strict, generally acceptable definition. Hilborn and Mangel (1997) and Gotelli and Ellison (2004) provide brief contemporary perspectives by ecologists. Ecology seems to have a surfeit of paradigms.

A related issue continues to perplex me: What exactly constitutes a *theory*? Dictionaries are of little help because they embrace wording from qualitative "contemplation" to "scientifically acceptable general principles offered to explain phenomena" (Websters 1947). MacArthur (1968) caught the term's sense as employed by most ecologists: "A theory must be falsifiable to be useful to a scientist, but it does not *in itself* have to be directly and easily verified by measurement. Most often it is the consequences of the theory that are verified or proved false." Hypotheses identify testable cause-and-effect mechanisms. Thus cosmologists continue their quest for a grand unified theory coupling gravity to other forces; that theory is a goal, approached by testing predictions or incorporating novel phenomena. But does ecology have a dynamic and quantitative theory of succession, arguably ecology's first organizing principle, or food web organization, or biodiversity, or ecosystem functioning? And if modelling has been applied to these verbal cornerstones of ecology, do they generate testable predictions? There can be no doubt that theoretical ecology, embracing both *theories* (e.g., island biogeography) and *models* (e.g., metapopulations, Ecosim/Ecopath), shows signs of continuing to mature into a sub-discipline in its own right, drawing from mathematics, awesome computational power, and statistical sophistication. But I retain the suspicion that we have just scratched the surface: of understanding an ecological world characterized by nonlinear dynamics, time lags, and disequilibria; of balancing the convenience of models based on often false assumptions (e.g., normal distributions) or estimated parameters (e.g., saturation densities, rates of increase) but with readily interpreted output; or of factoring out the true consequences of process or observation errors. No one I know has ever claimed that disentangling Darwin's metaphor of a "tangled bank" would be easy. Quantitative models or theories generating testable predictions and withstanding challenges to the necessary underlying assumptions will provide the litmus test for a future maturity.

I have suggested elsewhere (Paine 2002) that because our discipline is nonlinear and multi-causational, in McIntosh's (1987) phrase, pluralistic, it therefore does not map comfortably onto Kuhn's perspective based in the physical sciences. On the other hand, there has unquestionably been continued conceptual evolution in ecology, usually nurtured by technological advances (bioenergetic models, molecular genetics, and more powerful computational and analytical techniques spring to mind). "Continued" evolution doesn't necessarily imply a smooth continuum: "Fits and spurts" would be the best descriptor of the developmental trajectory as novel ideas were advanced and challenged. All but those on the lunatic fringe of reality have survived, often in an improved form. Our genre of biology clearly has benefited from an intuition based on observation and logic. Many of our discipline's early household names fit this description as a reading of Darwin, Wallace, or Bates illustrates. Foraging theory (Krebs & Davies 1997) provides an example, as do the multiple trade-offs by zooplankton forced to balance growth and safety (Gliwicz 2004). Is macroecology really new, or can it be traced comfortably to Preston (1948, 1962)? Equally, consideration of the mathematics of trade-offs between competitive prowess and dispersal probably originated with Skellam (1951); I doubt that the views were heretical at the time, and their eventual descendant, the metapopulation concept (Hanski 1999), has expanded in scope and continues to flourish. If there is an ecological connection to Kuhnian thought, it might be the MacArthur and Wilson (1963) theory of island biogeography. While it had (and still does have) its critics, it abruptly changed the way ecologists thought about site size and distance from source populations. Paradigm shift—?, revolution—no; progress—yes.

The participants in a Western Society of Naturalists' symposium (Ecology, 2002, 1479–1559) were asked to focus on "Paradigms in Ecology: Past, Present, and Future" (Graham & Dayton 2002). Paradigms were equated to fundamental principles; participants were also asked to develop a historical overview and identify "how ecological breakthroughs really happen". That event, like the book in hand, suggests a continuing recognition by biologists that ours is a historical science, that the present is not independent of either the deeper or recent past. Participants in both ventures had to confront "chicken or egg" style questions, or what led to what. Darwin's (1859) synthetic views were based on and developed from a massive amount of observation, a.k.a. natural history. The stage had been set in part by his grandfather, and the concept of natural selection catalyzed by Wallace's letter. Volterra's involvement with those infamous equations bearing his name was stimulated by a conversation with his father-in-law D'Ancona, a noted hydrobiologist.

So, as ecology has evolved from fascination with the intimate details of plants, animals and their interconnections to today's vastly more quantitative, and experimental endeavors, have paradigms been discarded? The Clements-Phillips (Phillips 1934, 1935) perspective of communities as *superorganisms* might be one, but even that survives in muted form as *Gaia* (Lovelock & Margulis 1974). And the early debates on superorganisms generated positive results; Tansley (1935) in a response to Phillips coined the term *ecosystem*, identifying the necessary and interactive connections between physical and biological processes. In evolution, blending vs. particulate inheritance could be another. But Darwin recognized that a blending mechanism had serious problems, and it was discarded in the early 1900s with the rediscovery of Mendel. Instead of losing paradigms, have we as environmental biologists rather gained perspective through time? Further, my impression is that ecological criticism usually (but not always; e.g., see Simberloff 1982) involves non-citation, that is, simply ignoring rather than attacking the study in question. On the other hand, the acceptance of a novel idea is often rapid and enthusiastic (e.g., island biogeography) regardless of the practitioner's age or status. Conceptual rejection is much harder to identify except with a historical perspective as a brainstorm simply disappears from the research landscape.

Ecology and evolutionary concepts certainly have an ontogeny that, with very few exceptions, involves collection of facts and appreciation of patterns, leading to attempts at synthesis or generalization with models. Gause's (1934) classic experiments were designed explicitly to amplify the Lotka-Volterra equations. A lesser appreciated example is Fisher-Piette's (1935) essay on the history of a mussel bed, with attention to qualitative but potentially cyclical changes in mussels, barnacles, and their mutual enemies. It inspired the theoretician Kostitzin (1939) to think about predator-prey cycles. Arguably, a notable exception to the biology-first proposition might be Lewontin's (1969) essay on alternative states in ecological systems. Although ecologists had already shown (Brooks & Dodson 1965, Paine 1966) that dramatic shifts in community composition could be forced by predator addition or deletion, there is little biology in Lewontin's seminal work. Most reviews of this ecologically interesting and important applied subject (e.g., Beisner *et al.* 2003) cite Lewontin as its intellectual founder. No paradigm was lost here. Rather, a mathematical framework was offered, gradually accepted and adapted to the peculiarities of our complex natural world.

This volume's authors were also challenged to both think about possible paradigm development and subsequent shifts and therefore to provide the underlying historical context. When and even where should such a history begin? Jackson *et al.* (2001) have demonstrated that ecological baselines are malleable, historically fragile entities. The same must be true for intellectual ones. Anyone willing to expend the time (and brave the dust) of pre–computer-accessible journals in a library can enter a surprisingly sophisticated world. One doesn't have to explore deeply, even excluding such pioneers as Darwin (1859) and Forbes (1887). Qualitatively, lots of today's concepts have deep roots. Limiting factors as they apply to population productivity were tested as early as 1843 (Tansley 1904). Trade-offs and their consequences are clear in the forgotten work of Baker (1910). Hatton (1938) terminates a remarkably modern treatise with a discussion of pair-wise interactions. Gillett's (1962) paper anticipated the increasingly quantitative plant biodiversity debate (Janzen 1970, Connell 1970, Hubbell 2002). Perhaps especially remarkably, Farrow (1917) in a footnote, but based on his previous experimental studies, provided a differential equation describing the dynamics of vegetation "luxuriance" as influenced by rabbit consumption. Of course, there are a host of other examples, and the aforementioned are simply personal favorites. Another approach to ecological paleontology, very different from that espoused in this volume, is the Leibold-Wootton (2001) examination of the impact on subsequent ecological thought of Elton's (1927) seminal book, *Animal Ecology.* Their retrospective represents an enlightening and comfortable way to see how some present concepts are hardly independent of the past. Allee *et al.* (1949) provide a now dated entrée. Reading Cole (1957) on the roots of demography can be humbling, as can reading Hutchinson (1978) on an amazing array of ecological and evolutionary history. More specialized books (and there are an increasing number), like Tobey (1981) on the early days of plant ecology or Golley (1993) on the history of the ecosystem concept, will also prove rewarding.

Whether our discipline has paradigms or not seems immaterial. We do have deep roots; rediscovery of some ecological wheel should be embarrassing. Exploration of our intellectual history can add a sense of completeness and a recognition that our field, indeed, is built on the shoulders of giants.

REFERENCES

Allee, W.C., Emerson, A.E., Park, O., Park, T., & Schmidt, K.P. (1949). "Principles of Animal Ecology." Saunders, Philadelphia.

Anonymous. (1947). "Webster's Collegiate Dictionary," 5th Edition. G. & C. Merriam Co., Springfield, Massachussetts.

Baker, S.M. (1910). On the causes of the zoning of brown seaweeds on the seashore. *New Phyt.* **9**, 54–67.

Beisner, B.E., Haydon, D.T., & Cuddington, K. (2003). Alternative stable states in ecology. *Front. Ecol. Environ.* **1**, 376–382.

Brooks, J.L., & Dodson, S.I. (1965). Predation, body size, and composition of plankton. *Science* **150**, 28–35.

Cole, L.C. (1957). Sketches of general and comparative demography. *Cold Spring Harb. Symp. Quant. Biol.* **22**, 1–15.

Connell, J.H. (1970). On the role of natural enemies in preventing competitive exclusion in some marine animals and rain forest trees. In "Dynamics of Populations" (P.J. Den Boer & G.R. Gradwell, Eds.). PUDOC, Wageningen, pp. 298–312.

Darwin, C. (1859). "The Origin of Species by Means of Natural Selection." Reprinted by Random House, Inc., New York.

Elton, C. (1927). "Animal Ecology." MacMillan, New York.

Farrow, E.P. (1917). On the ecology of Breckland: IV—Experiments mainly relating to the availability of water supply. *J. Ecol.* **5**, 1–18.

Fischer-Piette, E. (1935). Histoire d'une moulière: Observations sur une phase de déséquilibre faunique. *Bull. Biol.* **2**, 1–25.

Forbes, S.A. (1887). The lake as a microcosm. *Bull. Peoria Sci. Assoc.* 1887: 77–87. Reprinted in *Bull. Ill. Nat. Hist. Surv.* (1925), **15**, 537–550.

Gause, G.F. (1934). "The Struggle for Existence." Reprinted by Hafner Publishing, New York.

Gillett, J.B. (1962). Pest pressure, an underestimated factor in evolution. *Syst. Assoc. Publ.* **4**, 37–46.

Gliwicz, Z.M. (2004). "Between Hazards of Starvation and Risk of Predation: The Ecology of Offshore Animals." Excellence in Ecology, Book 12. (O. Kinne, Ed.). International Ecology Institute, Oldendorf/Luhe.

Golley, F.B. (1993). "A History of the Ecosystem Concept in Ecology." Yale University Press. New Haven.

Gotelli, N.J., & Ellison, A.M. (2004). "A Primer of Ecological Statistics." Sinauer Associates, Inc. Sunderland.

Graham, M.H., & Dayton, P.K. (2002). On the evolution of ecological ideas: Paradigms and scientific progress. *Ecology* **83**, 1481–1489.

Hanski, I. (1999). "Metapopulation Ecology." Oxford University Press, Oxford.

Hatton, H. (1938). Essais de bionomie explicative sur quelques especes intercotidales d'algues et d'animaux. *Ann. Inst. Oceanogr. Monaco* **17**, 241–348.

Hilborn, R., & Mangel, M. (1997). "The Ecological Detective: Confronting Models with Data." Princeton University Press, Princeton.

Hubbell, S.P. (2001). "The Unified Neutral Theory of Biodiversity and Biogeography." Princeton University Press, Princeton.

Hutchinson, G.E. (1978). "An Introduction to Population Biology." Yale University Press, New Haven.

Jackson, J.B.C., & 18 others. (2001). Historical overfishing and the recent collapse of coastal ecosystems. *Science* **293**, 629–638.

Janzen, D.H. (1970). Herbivores and the number of tree species in tropical forests. *Am. Nat.* **104**, 501–528.

Kostitzin, V.A. (1939). "Mathematical Biology." George G. Harrap & Co., London.

Krebs, J.R., & Davies, N.B. (1997). "Behavioural Ecology, an Evolutionary Approach." Blackwell Science, Oxford.

Kuhn, T.S. (1962). "The Structure of Scientific Revolutions." Chicago University Press, Chicago.

Lakatos, I. (1978). "The Methodology of Scientific Research Programs." Cambridge University Press, New York.

Leibold, M.A., & Wootton, J.T., Eds. (2001). "Animal Ecology." University of Chicago Press, Chicago.

Lewontin, R.C. (1969). The meaning of stability. *Brookhaven Symp. Quant. Biol.* **22**, 13–24.

Lovelock, J.E., & Margulis, L. (1974). Atmospheric homeostasis by and for the biosphere. *Tellus* **26**, 2–9.

MacArthur, R.H. (1968). "The Theory of the Niche." Syracuse University Press, Syracuse.

MacArthur, R.H., & Wilson, E.O. (1963). An equilibrium theory of insular zoogeography. *Evolution* **17**, 373–387.

McIntosh, R.P. (1987). Pluralism in ecology. *Annu. Rev. Ecol. Syst.* **18**, 321–341.

Paine, R.T. (1966). Food web complexity and species diversity. *Am. Nat.* **100**, 65–75.

Paine, R.T. (2002). Trophic control of production in a rocky intertidal community. *Science* **296**, 736–739.

Phillips, J. (1934–35). Succession, development, the climax, and the complex organism: An analysis of concepts. *J. Ecol.* **22**, 554–571; **23**, 210–243; **23**, 488–508.

Platt, J.R. (1964). Strong inference. *Science* **146**, 347–353.

Popper, K.R. (1959). "The Logic of Scientific Discovery." Harper & Row, New York.

Preston, F.W. (1948). The commonness, and rarity, of species. *Ecology* **29**, 254–283.

Preston, F.W. (1962). The canonical distribution of commonness and rarity. *Ecology* **43**, 185–215, 410–432.

Simberloff, D. (1982). The status of competition theory in ecology. *Ann. Zool. Fennici.* **19**, 241–253.

Skellam, J.G. (1951). Random dispersal in theoretical populations. *Biometrika* **38**, 196–218.

Tansley, A.G. (1904). The Rothamsted agricultural experiments. *New Phyt.* **3**, 171–176.

Tansley, A.G. (1935). The use and abuse of vegetational concepts and terms. *Ecology* **16**, 284–307.

Tobey, R.C. (1981). "Saving the Prairies: The Life Cycle of the Founding School of American Plant Ecology." University of California Press, Berkeley.

PREFACE

The idea for this book grew out of a symposium sponsored by the Theoretical Section at the annual meeting of the Ecological Society of America (ESA) in Tucson, AZ, that we organized in 2002. We were both bright-eyed postdocs when we decided it would be a great idea not only to host a symposium but also to put together a book on the same topic. We have learned a lot in the intervening 2 years.

We came upon the topic of theory change in ecology by several routes. During a conversation with a graduate student at a poster session at the 2001 ESA meeting, we realized that the student had limited knowledge of the very standard theory upon which their research was based. Similarly, discussions with others solidified in our minds that much of the older literature was being lost to new students as the switch to electronic databases became the norm. At the beginning of our graduate careers in the early 1990s, we still went to the library regularly with a long list of papers to dig from the stacks and drag to the photocopier, which was invariably placed far from the ecology journals. A graduate student's passage through the library could be traced by the trail of bits of paper that had been used to mark the treasured articles. By the time we were finishing our PhD's, and to this day, we and our fellow academics were well entrenched at our computers in our offices, reading the hottest *new* papers online by the dozen—realizing how easy it is to "not really need" that article that is only available at the library, in another building, buried in some stack somewhere. This, in addition to the electronic reference databases that go back only a limited number of years, depending on how rich one's university library is, tends to limit exploration of older literature.

After asking ourselves whether it *really* mattered if one focused on new material, we decided to do a more rigorous analysis of the issue. The question of whether an understanding of history is important to ecologists led us to the more philosophical question of how theory change in ecology proceeds. Is it an evolutionary, building-block type of change, or is it one that grows by paradigmatic shifts? We decided to get at this question by asking both practicing ecologists and philosophers what they thought. The result is this book.

The volume you now hold would not have existed without input from many people, including the original speakers of our ESA symposium and the Theoretical Ecology Chapter of the ESA. In particular, we are grateful to Gregg Hartvigsen, Lou Gross, Joan Roughgarden, Michael Rosenzweig, and David Tilman for their efforts towards making the symposium a reality. We thank Alan Hastings for suggesting that this book be a contribution to the Theoretical Ecology series at Elsevier Academic Press. We thank our authors for their dedication to the project, and we thank our slew of anonymous reviewers—many of whom came through in record time. Finally, we thank J.J. Hubert, statistics professor at the University of Guelph, whose long weekly assignments made us realize how well we could work together.

Kim Cuddington
Athens, Ohio
Beatrix E. Beisner
Montreal, Quebec
October 2004

LIST OF CONTRIBUTORS

T.F.H. Allen: University of Wisconsin–Madison, Madison, WI 53706-1381, United States of America.

Beatrix E. Beisner: Department of Biological Sciences, University of Quebec at Montreal, P.O. Box 8888, Downtown Station, Montreal, Quebec, H3C 3P8, Canada.

David Castle: Department of Philosophy, University of Guelph, Guelph, Ontario, N1G 2W1, Canada.

Kim Cuddington: Department of Biological Sciences/Quantitative Biology Institute, Ohio University, Athens, OH 45701, United States of America.

Troy Day: Departments of Mathematics and Biology, Jeffery Hall, Queen's University Kingston, ON, K7L 3N6, Canada.

Kevin de Laplante: Department of Philosophy and Religious Studies, Iowa State University, Carrie Chapman Catt Hall, Ames, IA 50011-1302, United States of America.

André M. De Roos: Institute for Biodiversity and Ecosystem Dynamics, University of Amsterdam, PB 94084, NL-1090 GB Amsterdam, The Netherlands.

Alan Hastings: Department of Environmental Science and Policy, University of California, Davis, CA 95616, United States of America.

Hans Heesterbeek: Faculty of Veterinary Medicine, University of Utrecht, P.O. Box 80.163, NL-3508TD Utrecht, The Netherlands.

Robert D. Holt: Department of Zoology, P.O. Box 118525, University of Florida, Gainesville, FL 32611-8525, United States of America.

Henry S. Horn: Department of Ecology and Evolutionary Biology, Princeton University, Princeton, NJ, 08544-1003, United States of America.

Anthony R. Ives: Department of Zoology, University of Wisconsin–Madison, Madison, WI 53706, United States of America.

Matt J. Keeling: Maths Institute & Department of Biological Sciences, University of Warwick, Gibbet Hill Road, Coventry, CV4 7AL, United Kingdom.

James S. Koopman: Department of Epidemiology, Center for the Study of Complex Systems, University of Michigan School of Public Health, 611 Church St., Ann Arbor, MI 48104, United States of America.

Kevin S. McCann: Department of Zoology, University of Guelph, Guelph, Ontario, N1G 2W1, Canada.

Jay Odenbaugh: Department of Philosophy, Lewis and Clark College, 0615 SW Palatine Rd, Portland, OR 97219, United States of America.

Robert Paine: Department of Biology, University of Washington, Seattle, WA 98195-1800, United States of America.

Lennart Persson: Department of Ecology and Environmental Science, Umeå University, SE-90187 Umeå, Sweden.

Garry D. Peterson: Department of Geography & McGill School of the Environment, McGill University, 805 Sherbrooke St. W., Montreal, Quebec, H3A 2K6, Canada.

Eric R. Pianka: Denton A. Cooley Centennial Professor of Zoology, Integrative Biology, University of Texas at Austin, Austin, TX 78712-1064, United States of America.

Kim Sterelny: Philosophy Program, Australian National University and the Victoria University of Wellington, P.O. Box 600, Wellington, New Zealand.

C.J. Wuennenberg: University of Wisconsin–Madison, Madison, WI 53706-1381, United States of America.

A.J. Zellmer: University of Wisconsin–Madison, Madison, WI 53706-1381, United States of America.

1 | WHY A HISTORY OF ECOLOGY? AN INTRODUCTION

Beatrix E. Beisner and Kim Cuddington

What is the use of knowing the history of a discipline? Why should ecologists, in particular, be concerned with how ideas have developed? At a recent Ecological Society of America meeting, we were dismayed by a graduate student's complete lack of knowledge of early ecological theory. Anything published before online journals and indexes did not exist for this and many other students we met that year. We wondered whether we were just becoming curmudgeons before our time or whether there was some scientific and philosophical value in a thorough historical grounding in our discipline. There might be value in the argument that we are better off if we ignore our history. For example, in social science many battles were fought as a result of profound historical examination. On the other hand, fields such as physics have traditionally had little focus on history outside of a morbid curiosity of outdated ideas. What is the appropriate mix of current research with older ideas in the discipline of ecology? It seems to us that no one has yet answered this fundamentally philosophical question.

When ecologists turn to the philosophy of science, one of the most commonly cited works is Thomas Kuhn's classic thesis *The Structure of Scientific Revolutions* (1962). Kuhn's ideas emerged mainly from the study of theory development in physics, which he argued occurred by revolution in which an outdated paradigm was replaced by another paradigm. Use of similar language, through which theory changes are described as *paradigm shifts*, may imply many ecologists accept that scientific development occurs in such revolutionary shifts. If this is the case, then ecologists can ignore the older history of their discipline. Such language use may also indicate that ecologists believe a paradigm shift is something other than what Kuhn described, perhaps simply a change in commonly used models, without any accompanying change in conceptual framework. In any event, it is clear that the use of Kuhnian language has increased dramatically in the last 30 years (Fig. 1.1).

The trend of using philosophical language has had a recent dénouement in a special issue of *Ecology* (Volume 83, Issue 6), where several aspects of ecological theory were painted in the colours of Kuhnian philosophy. Graham and Dayton (2002) led off by claiming that Kuhnian paradigms are a useful way to describe the change in ecological ideas. Their description of the history of ecology, however, is only Kuhnian in that they describe different schools of thought as paradigms. Although Kuhn's views may have changed in later years, invoking early Kuhn to describe conceptual change that proceeds along evolutionary lines, as these authors do, stretches this philosophy of science to a breaking point. Revolutions do not apply when one describes theory change as evolutionary, and neither is it clear to what extent incommensurability does. On the other hand, we think that Graham and Dayton, and the other authors in this special feature, do excellent work in describing the development of several aspects of theory, and there are many other models of evolutionary theory change in the philosophical literature (see Chapter 18). To this end, the use of the Kuhnian conception of paradigm provides a jumping-off point for the understanding of scientific development.

There are two major goals to this book. Both emerge from an examination of the history and development of some major theoretical ideas in ecology. The first goal is to provide an overview of the history of some traditional subdisciplines of ecology by asking eminent ecologists to review theory change in the field. The second goal is to determine whether the development of these ideas is one of paradigm shifts or one of more gradual buildup from simple to complex. For this second goal we not only have asked scientists for commentary and opinion but also have included analyses by philosophers of science.

There are some important practical reasons ecologists should pay more attention to the history of ideas and theories in our field. First, a familiarity with older ideas will help prevent "reinvention of the wheel". A rediscovery of old ideas is a largely unproductive use of research dollars and time, especially given the incredible responsibility we now have to come up with meaningful predictions and prescriptions for threatened natural ecosystems. An ethical use of public research funds would therefore require a good knowledge of what has come before in the discipline, so that useful progress can be made. Second, a good

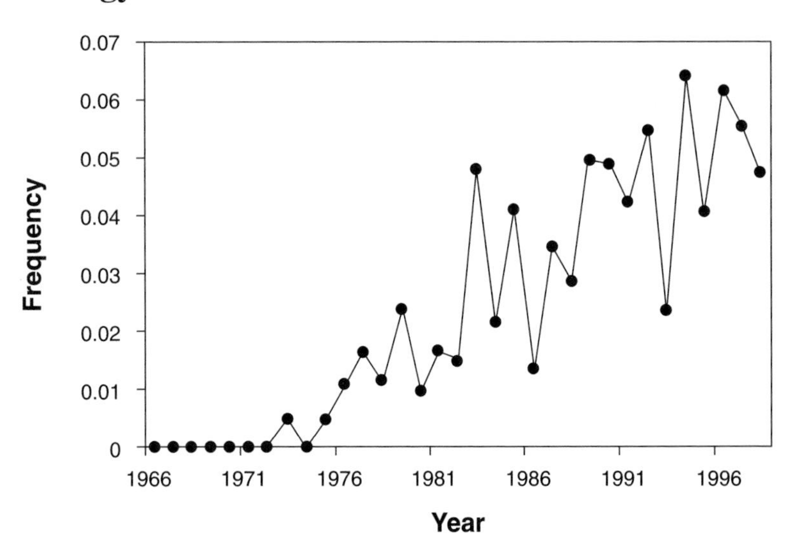

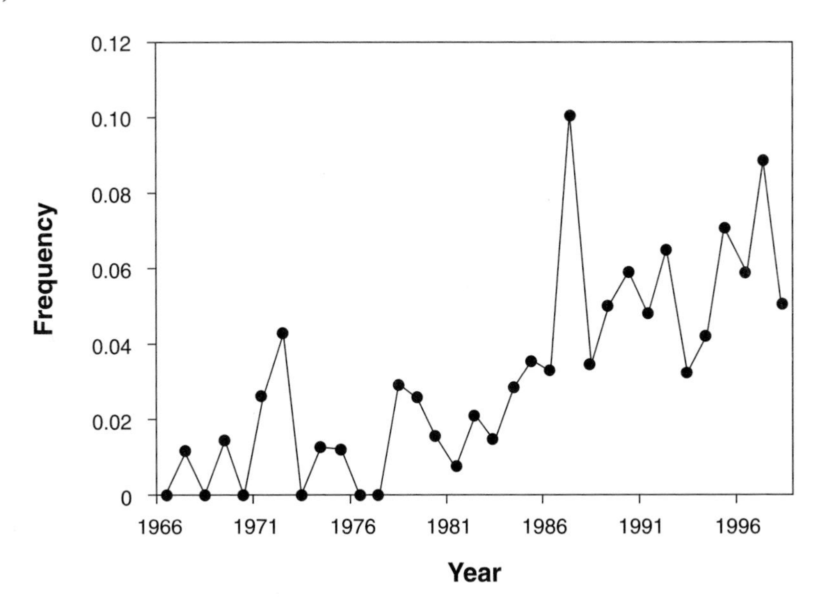

FIGURE 1.1 | An analysis of the content of articles in (A) *Ecology* and (B) *Evolution* shows that, per article, there has been an increasing tendency to use the word *paradigm* to describe areas of ecological and evolutionary research following the publication of Kuhn's thesis in 1970.

knowledge of history provides perspective in research programmes. By seeing how various researchers have dealt with different issues, we can develop a good toolbox of approaches to the discipline. Finally, knowledge of what has come before allows us to better pinpoint important key questions and defermine where missing data or knowledge may be missing. In this way, faster progress can be made.

How do we take history of our discipline into account? Luckily, the science of ecology is still young enough that most practitioners are only two to three academic generations removed from those who first laid down the major theories upon which much of ecology rests. This means there is not as much history to learn as there is in older sciences. In terms of the history in ecology, Daniel Haydon (now at the University of Glasgow) hit upon a useful analogy in a debate during the late 1990s at a retreat of the ecologists of the University of British Columbia. We (Beisner, Haydon, and Ferguson) were to argue that there had been no significant new ideas in ecology since the 1960s. Haydon's analogy was that of a chess game in which the major ideas of how ecological systems work, developed in the early days of the discipline, were represented by the chess pieces and their different ways of moving across the board. In the intervening decades, Haydon argued, ecological research has been about figuring out the different ways of playing the game using these rules. If we accept this analogy, it is easy to see that understanding the roles of the pieces and their movements is essential to understanding the discipline. Incidentally, Haydon stumbled across a different chess analogy used by Richard Feynman several years later for the laws of physics (Feynman, Leighton, & Sands 1963). Feynman argued that studying physics was like trying to learn the rules of chess just by studying chess games, which could also be applied to ecology. If these analogies hold, it is imperative that any serious ecologist become familiar with the rules of the game, which are mostly laid down in the classics of the field. There already are some excellent resources available, including *Foundations of Ecology*, edited by Les Real and James Brown (1991), and various reissues of other classics in ecology such as Charles Elton's books. Important historical perspectives on theory in ecology include Sharon Kingsland's *Modeling Nature* (1995) and Robert McIntosh's *The Background of Ecology* (1985).

Our goal with this volume is to add another layer to a historical perspective. In this work, we provide philosophical analyses rather than just historical perspectives and attempt to address whether a paradigmatic view (*sensu* Kuhn) can apply to ecology. In addition to a historical overview of major areas of ecological research, we have asked both practicing scientists and philosophers of science to provide analyses of the development of the field. Our hope is that this book will provide new perspective on the ideas in ecology. To accomplish these goals, we structured the book in an unusual way. For each category of ecological research (population, epidemiological, community, evolutionary, and ecosystem ecology) we have asked eminent ecologists of both the younger and more weathered variety to provide historical overviews of and opinion on a major issue in their field. We have then paired these "science" chapters with commentary by a practicing philosopher of science.

The chapters that follow cover the development of theory in the major traditional disciplines of ecology. In the section on population ecology, Alan Hastings provides an overview of the major types of unstructured population models, from nonspatial and deterministic to spatial and stochastic types, and describes how they emerged. The complementary chapter by André De Roos and Lennart Persson examines in detail some important theoretical developments that have arisen from simple structured models. Analysis of these contributions is provided in the chapter by Jay Odenbaugh. Hans Heesterbeek covers unstructured epidemiological models, and Matt Keeling outlines the extension of these models to allow for spatial considerations. Philosophical analysis in the epidemiological ecology section is covered by James Koopman. In the community ecology section, chapters by Anthony Ives and Kevin McCann provide two perspectives on the stability-diversity debate in ecology that has been a subject of vigorous research for several decades. A commentary is provided by philosopher of science David Castle. Part V covers evolutionary ecology and includes excellent chapters with opinion, perspective, and an overview of model types by Robert Holt and Troy Day, with commentary by Kim Sterelny. Included are reflections on the genius and influence of Robert MacArthur by Eric Pianka and Henry Horn, without which any book considering the recent history of theory change in ecology would be incomplete. Finally, in a section on ecosystem ecology, Timothy Allen and his colleagues and Garry Peterson provide overviews of major questions and approaches in this field and describe how these have been used in management questions. In their chapters, there is a strong focus on how philosophical perspectives may be limiting the implementation of ecological knowledge in management practices. Kevin de Laplante provides a philosophical perspective on these issues.

It is our hope that this book will provide interesting reading and perspective to both young ecologists and older ones. As with any viewpoint, a historical one will always be tainted by the perspective of the person telling the story. That is why we have sought several perspectives on each topic and allowed the expression of personal opinion in these chapters. In the end, we hope that this book will be a catalyst for others to pause periodically to review assumptions, goals, and methods in the busy day-to-day world of research so that we can all practice a better science.

REFERENCES

Feynman, R.P., Leighton, R.B., & Sands, M. (1963). "The Feynman Lectures on Physics." Addison-Wesley, Boston.

Graham, M., & Dayton, P. (2002). On the evolution of ecological ideas: Paradigms and scientific progress. *Ecology* **83**, 1481–1489.

Kingsland, S.E. (1995). "Modeling Nature: Episodes in the History of Population Ecology," 2nd Edition. University of Chicago Press, Chicago.

Kuhn, T. (1962). "The Structure of Scientific Revolutions." University of Chicago Press, Chicago.

MacIntosh, R.P. (1985). "The Background of Ecology." Cambridge University Press, Cambridge.

Real, L.A., & Brown, J.H., Eds. (1991). "Foundations of Ecology: Classic Papers with Commentaries." University of Chicago Press, Chicago.

I | POPULATION ECOLOGY

2 | UNSTRUCTURED MODELS IN ECOLOGY: PAST, PRESENT, AND FUTURE

Alan Hastings

2.1 | INTRODUCTION

The use of mathematical models in ecology was greatly advanced by work in the early twentieth century (e.g., by Lotka 1926, 1932; Volterra 1926, 1931; Kermack & McKendrick 1927; Gause 1934, 1935; Nicholson & Bailey 1935) that established a core set of models focusing on the numbers of individuals in a species as a function of time. In this chapter I focus on these models, in which the only dependent variables are overall species numbers, and call these *unstructured models*. Although many of the basic principles were discovered early, these models have continued to form the basis for much of the reasoning used to elucidate basic ecological principles. Therefore, it is revealing to understand the basis on which these models were developed, some of the most influential conclusions drawn from the models, the usefulness of the inferences drawn from them, and the future prospects for unstructured models in ecology.

The basic ecological goals of understanding distribution and abundance, predicting the numbers of populations, and managing populations have been well advanced by the use of unstructured models (e.g., see May 1974). In particular, questions of persistence of species have typically been addressed using unstructured models, and many ecological issues involving interacting species have been clearly addressed using unstructured models. These include predator-prey coexistence, trophic cascades, and much of food web dynamics. Many issues of complex dynamic behaviour were first elucidated in the context of unstructured models.

Any attempt to review the use of unstructured models in ecology needs to ask whether unstructured models are only of historical interest (Kingsland 1985) or whether there is a reason to continue the study and use of models that include only species numbers. In many cases, structure, such as spatial or age structure, is essential for ecological understanding. Yet, despite the great focus on the role of structured models, there has been continued interest and research using both discrete time and continuous time unstructured models (e.g., see Hassell 1978, 2000; McCann, Hastings, & Huxel 1998; and Murdoch, Briggs, & Nisbet 2003). The advances, and interest in, using statistical approaches to make quantitative and more precise connections between data and theory in ecology (Burnham & Anderson 2002, Johnson & Omland 2004) have revived the use of and interest in unstructured models. In addition, continuing interest in the dynamics of food webs has relied greatly on the use of unstructured models (e.g., see McCann, Hastings, & Huxel 1998). More generally, despite the introduction of structured models into the study of species interactions, the complexity of these problems has led to continued use of unstructured models.

As I will illustrate, unstructured models still play an important role in elucidating general principles and therefore will continue to play a role in future ecological investigations. When developing simple ecological models including only population size, researchers could start from essentially more complex models with more structure and with interactions derived from first principles. Simplifying these is one way to obtain models that are mechanistic, using functions and parameters that can be clearly mapped to ecological interactions and to measurements that can be made in the field or in experiments. As I will also illustrate, even for simple models, those that are mechanistic are often likely to be more useful and have usually had a more lasting influence than those that are heuristic, in which functions and parameters do not have clear ecological interpretations. This role of mechanistic models will be amply illustrated by epidemiological models and goes along with approaches that try to estimate the parameters. The increasing emphasis on mechanistic models, structured or unstructured, is an important way that ecological theory has continued to develop.

2.2 | THE BASIC (DETERMINISTIC) UNSTRUCTURED MODELS

Before going into the details of the models and how the different examples can help provide ecological understanding, I will first briefly describe the models that will form the core of this chapter and give some idea of their interrelationships. The basic unstructured models of population biology begin with the single-species models. The underlying assumptions for all these models are similar, so an understanding of the simplest one-species models underlies an understanding of the more complex multispecies models. The simplest one-species models of ecology are those for exponential growth, and even from the time of Malthus it was clearly recognized that such models could not be good long-term predictors of the dynamics of populations (e.g., see the information in Hutchinson 1978). Thus, I will explore various density-dependent models both in continuous and in discrete time. The discrete time models that I will describe will not only have a richer behaviour but also will, in some sense, be more mechanistic; consequently, they can be much more useful both in the context of the statistical questions raised in the introduction and in developing more general principles.

After considering the simple models of a single species, I will turn to the models for two interacting species, which, although they obviously involve more than one species, are typically viewed as being within the realm of study of population biology. These classic models were the focus of the early work of Lotka, Volterra, Gause, and Nicholson and Bailey. Within the context of interactions between two species, the classic ecological interactions of competition and predation can be studied, as well as the dynamics of mutualism. These models are of great importance because they have been used to understand the kinds of biological forces that might lead to preventing exponential growth and to the kinds of forces that might produce long-term outcomes other than an approach to a stable equilibrium. Again, both continuous and discrete time approaches have been used, and I will consider the importance of using these different approaches. The work of Nicholson and Bailey on host-parasitoid dynamics will deserve special attention because many of the themes they raised have been the subject of research that has carried on until today, although with new names given to some of the ideas they described. A special kind of exploiter-victim dynamics, namely that caused by diseases, is notable because even the earliest articles focused on short-term dynamics, used different timescales, and considered the match between data and theory. Because these early models underlie much important recent work, I will consider them carefully.

Some early work, notably that of Volterra, gave a great deal of attention to the dynamics of more than two species but in ways that may have perhaps proved less useful for later research, with an emphasis on mathematical rather than biologically important research. Some of the interesting implications of more recent work on more than two species will be described here.

2.3 | SINGLE SPECIES

Single-species models of populations were part of the early development of ecological theory and, in some sense, can be traced back to the work of Malthus (and perhaps Fibonacci) as described by Hutchinson (1978). But, the first model of a form that corresponds to modern work is the continuous time logistic equation first developed by Pierre Verhulst. I focus on this equation first.

2.3.1 | Continuous Time

As noted, the simplest model for a single population is the exponential growth equation. Letting *N* be the population size at time *t*, this model can be written in the following form:

$$\frac{dN}{dt} = (b - d)N. \tag{2.1}$$

Here, *b* is a per capita birth rate and *d* is a per capita death rate; the intrinsic rate of increase is $r = b - d$. Writing this equation using the per capita rate of increase alone is simpler and causes no problems here, but it potentially causes problems later. This equation has the unrealistic behaviour that populations grow without bound because density dependence is ignored (as well as other factors that could prevent exponential growth).

A more general way to proceed is to write the description of the rate of growth as:

$$\frac{dN}{dt} = Nf(N) \tag{2.2}$$

where $f(N)$ is the per capita rate of increase of the population. What is noteworthy about this is already twofold. First, in what follows there are few mechanistic derivations of the form of $f(N)$—perhaps because this form makes it difficult. Second, as written, the births and deaths, which are the ingredients of a potential mechanistic derivation, do not appear separately. This lack of a mechanistic basis for equation (2.2) was well recognized by Lotka (1926) who derived the logistic equation simply by expanding $f(N)$ in a Taylor series, using a mathematical, rather than a biological, approach to derive a specific form for (2.2).

Perhaps because of these potential shortcomings of the model formulation, this model is most useful in terms of qualitative rather than quantitative conclusions. One dramatic failure of the quantitative approach with the density-dependent logistic model is illustrated by attempts to predict the US population by fitting the population with a logistic form, which did a poor job of predicting future population sizes despite good fits of the US population until 1940 (Pearl, Reed, & Kish 1940).

Yet, despite the difficulties with the form of $f(N)$, several specific forms have played an important role in qualitative arguments. Take, for example, ideas embodied in the following logistic form:

$$f(N) = r\left(1 - \frac{N}{K}\right), \tag{2.3}$$

or in forms with Allee effects, with $f(0) < 0$, have played an important role in a tremendous number of ecological studies, beginning notably with some of the work by Gause (1934, 1935). This early work by Gause is also important because he sought to match the quantitative and qualitative predictions of the early models with data and results from experimental systems. This work foreshadowed the current theme of using statistical tools to make this match.

The logistic model in (2.3) can be solved explicitly—the solution $N(t)$ can be written as an explicit function of time. More usefully, you can summarize the behaviour by saying that there is an equilibrium at $N = K$ that is globally stable. What this means is that for any starting condition with N positive, the population will approach the population level K. Moreover, it is easy to show that the equilibrium is approached in a monotonic fashion through time; that is, there are no oscillations. Thus, any observed oscillations in population numbers, sustained or not, would suggest that (2.3) is not a good description of the forces controlling population dynamics. Many populations are known to oscillate; therefore, the logistic model can be shown to be a poor description on qualitative grounds in addition to the quantitative example of fitting this equation to US population growth described previously.

2.3.2 | Discrete Time

As noted previously, the discrete time models for single-species dynamics are potentially more easily interpreted in a mechanistic fashion than the continuous time ones and thus not surprisingly are playing a larger role in current research. An understanding of the discrete time models again can start with the form describing exponential growth:

$$N(t+1) = RN(t). \tag{2.4}$$

Here, R is the mean number of individuals (offspring if generations are not overlapping) next year per individual this year. If you view this model as a description of the dynamics of a population in which generations do not overlap, then it does not have the preceding problem of combined births and deaths because the deaths are separated implicitly.

To see the connection to the continuous time description, write the dynamics in terms of the change in population size from one year to the next:

$$N(t+1) - N(t) = F(N(t)). \tag{2.5}$$

If the population was censused instead at intervals of Δt, the equation for the change in population size would be written as:

$$N(t+\Delta t) - N(t) = N(t)f(N(t))\Delta t, \tag{2.6}$$

where $f(N(t))\Delta t$ is uniquely determined by $F(N(t))$ and should be viewed as the change in population size in an interval of time, Δt. What is particularly important to note is that even different choices for the discrete time dynamics of $F(N(t))$ may have the same corresponding infinitesimal per capita growth rate of $f(N(t))$. To change to a continuous time approximation, you can divide both sides of equation (2.6) by Δt to obtain:

$$\frac{[N(t+\Delta t) - N(t)]}{\Delta t} = N(t)f(N(t)), \tag{2.7}$$

which has the limit:

$$\frac{dN}{dt} = N(t)f(N(t)). \tag{2.8}$$

It is also important to note that this "derivation" shows that for a particular mechanistic discrete time model there is only one corresponding continuous time model, because the limiting procedure used to go from (2.6) to (2.8) always gives a unique answer. However, there also may be many different discrete time models that correspond to the same continuous time model because of the many-to-one relationship between F and f. This correspondence has several important implications because of the wealth of dynamic behaviour possible for the different discrete time models and because the possible dynamic behaviour is different for different models.

For the discrete time model, partly because of the implicit separation of births and deaths, there were several influential developments of mechanistic formulations for the growth rate of the population. For example, two classic discrete time models were developed to understand fisheries dynamics by Ricker (1954) on one hand and by Beverton and Holt (1957) on the other. In the fisheries context these models can be used to describe recruitment, but these models have also been successfully applied to understand single-species population dynamics. The Ricker model takes the following form (here I follow tradition and use b to designate the parameter measuring the strength for density dependence and assume that there will be no confusion with the meaning of b in the previous subsection):

$$N(t+1) = N(t)\exp(r - bN(t)), \tag{2.9}$$

and incorporates overcompensatory density dependence, where the number of individuals produced the following year drops if the number of individuals this year is large enough. Here, $N(t)$ is the population size at time t, r is a measure of the exponential density-independent growth rate, and b is a measure of density dependence. This model can be derived by assuming that births take place over a fixed period and that young are cannibalized by adults.

In contrast the Beverton–Holt model is

$$N(t+1) = R\frac{N(t)}{1 + bN(t)}, \tag{2.10}$$

only incorporates compensatory density dependence, in which the number of individuals next year per individual this year decreases as the number of individuals this year rises, but the total population size next year still increases. Here, R is a measure of the density-independent growth rate, and b is a (different) measure of density dependence.

The continuous time approximation of both the Ricker and the Beverton-Holt models would be the logistic equation. The potential dynamics of the Beverton-Holt model are simple and match those of the continuous time logistic equation. The only possible long-term outcome is a monotonic approach to a stable equilibrium point. However, the form of the density dependence in the Ricker model is more extreme, and this leads to the other striking aspect of the dynamics of discrete time models when overcompensatory density dependence is included.

That single-species discrete time models can have complex dynamics was first emphasized in work by May (1974, 1976) and others. First, unlike the corresponding continuous time model, cyclic dynamics are possible. Essentially, if the population regulation (density dependence) is too strong, the population can overshoot the equilibrium and this can lead to sustained cycles. Yet, the dynamics can be much more complex than simply cycles. For example, chaotic dynamics are possible (Hastings *et al.* 1993) with the resultant sensitive dependence on initial conditions. This is behaviour that may be of great importance for population biology, but the jury is still out. Discerning that the different ways that density dependence is incorporated into these two formulations is the key to whether chaos appears easy within the context of these simple unstructured models.

2.4 | TWO SPECIES

Although, during the classical period of the development of theoretical ecology, there were advances that included the development of structured populations, the inclusion of structure was almost completely limited to the single-species, density-independent case. However, the role of other species was not neglected, and I now turn to the models that helped explain that role and that raised fundamental issues concerning the persistence and coexistence of species, which are still the subject of research. Almost all current research on the dynamics of single-species models in ecology uses structured population models, but the dynamics of multiple species in the context of unstructured models is still an area of research. I will begin, however, with a summary of the classical results.

I will first focus on the dynamics of two-species models. Here, the focus in the first part of the twentieth century was primarily on the two classes of interactions that can be broadly characterized as exploiter-victim or competition. As you will see, the issues that arose in the context of single-species models carry over to these models in considering the differences between the discrete time and the continuous time models. In particular, the more mechanistic models are the ones that continue to have the largest influence.

2.4.1 | Continuous Time Exploiter-Victim Models

Among the first two-species models considered by Lotka (1926) and Volterra (1931) were those describing the interaction between an exploiter (predator or parasite) and a victim (prey or host). Lotka's (1926) description of this model is particularly interesting because it raises some important issues about the kind of nonlinear interaction between predator and prey that are the subject of current research. In the simplest version of the predator-prey model, the dynamics of the prey are assumed to be exponential growth in the absence of the predator, and the dynamics of the predator are assumed to be exponential decline in the absence of the prey. The interaction term that described the rate of killing of victim by exploiter was assumed to be of the form gN_1N_2, where N_1 and N_2 were the numbers of individuals in the victim and exploiter species, respectively, and k was a *function* of the numbers of the two species. Victim deaths led to an increase in the number of exploiters, assumed to be proportional to the number of deaths, where G is a measure of this proportionality. Thus the equations of dynamics were written as:

$$\frac{dN_1}{dt} = r_1N_1 - gN_1N_2 \tag{2.11}$$

$$\frac{dN_2}{dt} = GN_1N_2 - d_2N_2, \tag{2.12}$$

where r_1 and d_2 are positive constants.

Lotka's analysis of this model focused on the case in which the function g (and therefore G) was taken as a constant, appropriate for a short time. His analysis of this case found the conserved quantity that showed that under the simplifying assumptions the model (2.11–2.12) produced only cyclic behaviour that continued forever, with the amplitude of the cycle determined by the initial conditions. As Lotka noted, deviations from these simple assumptions could produce cycles that either decreased or increased in amplitude. Thus, the model has a kind of neutral stability, and any small perturbation would change the long-term behaviour, although the results of Lotka would still describe the short-term behaviour.

However, the neutral stability of (2.11–2.12) has been used by ecologists as a starting point to investigate conditions for persistence. The idea that the species combinations observed in nature correspond to the stable solutions of simple models has been powerful when combined with looking for changes in the predator-prey model away from the neutrally stable case analyzed by Lotka. Starting from even the classic work of Gause (1934), ecologists have argued, as clearly stated by Murdoch and Oaten (1975) that a way to understand the features of ecological interactions that permit coexistence is to look at deviations from the neutrally stable model that yield a stable equilibrium. Similarly, it was Gause who, through his experimental work, emphasized the importance of looking for those factors that would permit coexistence of predator and prey rather than either the predator eating too little food and starving or the predator first driving the prey extinct, then starving.

For example, as clearly explained in Murdoch and Oaten (1975), two simple modifications of the neutrally stable model (2.11–2.12) are to include density dependence in the equation for the dynamics of the prey species and nonlinear functional response (Holling 1959; also see Gause 1934) in the predation term. The former can be a stabilizing influence and may be responsible for coexistence. The latter assumption differs from the simplest Lotka-Volterra model of predation in which the rate of prey killed per predator is a linearly increasing function of the number of prey. However, if, as is reasonable, the rate of prey killed per predator saturates as the number of prey increases, this is a destabilizing influence. In this context, stability and persistence versus instability can be understood by comparing (mathematically) the strengths of these two competing influences.

The paradox of enrichment is a particular example of this research approach. Starting with the form of the predator-prey model that incorporates both density dependence and functional response, Michael Rosenzweig (1971) investigated the consequence of increasing the carrying capacity of the prey. Not surprisingly, increasing the carrying capacity of the prey reduces the effect of density dependence. This is a destabilizing influence, which struck some as counterintuitive.

The results of these simple models, if they are to have any utility for understanding natural systems, must at least describe the dynamics of laboratory systems. Gause (1934) showed that in a simple laboratory system of a predator-prey interaction between *Paramecium* and *Didinium* the ultimate outcome was always extinction of the predator and possibly extinction of the prey. Gause also demonstrated in separate experiments that the prey species alone would reach equilibrium because of density dependence. Given the simplicity of the models that would predict coexistence with density dependence in the prey, the lack of coexistence in the experiments is perhaps not surprising—something important might be missing in the models. Yet analyzing the predator-prey models with type II functional response using the kind of approach advocated in the Murdoch and Oaten (1975) article would suggest that increasing the role of density dependence (i.e., reducing the carrying capacity of the prey) and reducing the destabilizing aspect of the functional response (by reducing the predation rate) would increase the likelihood of coexistence. In a series of elegant experiments with the same species, Luckinbill (1973, 1974, 1979) showed just those stabilizing effects and obtained coexistence (at least over relatively long times).

Especially in the context of more than two species, similar research is continuing that emphasizes the role of stability as a way to understand persistence. It is also important to note that many explanations for coexistence in exploiter-victim systems (e.g., see Murdoch, Briggs, & Nisbet 2003) invoke aspects that can only be incorporated in structured models with age, stage, or space included.

2.4.2 | Nicholson-Bailey Discrete Time Models

As in the single-species case, the discrete time models of two-species exploiter-victim systems are more mechanistic than the corresponding continuous time models. The original formulation of Nicholson and Bailey (1935) for a host-parasitoid system is both insightful and innovative and is the basis of much ongoing work.

Nicholson and Bailey begin with a static model then developed their model using a more dynamic approach, but it is simplest to start with the dynamic approach. As in a single-species model, the approach starts with the concept of exponential, unchecked growth for the herbivore species. The model then is based on a description of the way that parasitoids attack hosts. Rather than duplicate their derivation, I will present the model in the modern form. Let $H(t)$ and $P(t)$ be the number of hosts and parasitoids, respectively, in generation t. For simplicity, and to match many biological systems, assume both hosts and parasitoids have one generation per year and die after laying eggs. In the absence of parasitism, assume that the host grows exponentially so that it obeys the following equation:

$$H(t+1)=\lambda H(t), \tag{2.13}$$

where λ is the mean number of hosts produced per host. Then, the key idea is to represent the dynamics using the probability that a host egg escapes from parasitism, which is expressed as $F(H(t), P(t))$, indicating that this probability may depend on the density of either or both the hosts and the parasitoids. Finally, noting that, therefore, the probability that a host egg is parasitized is then $1 - F(H(t), P(t))$ and letting c be the mean number of parasitoids emerging from a parasitized host, we obtain the following dynamic equations:

$$H(t+1)=\lambda H(t)F(H(t),P(t)) \tag{2.14}$$

$$P(t+1)=cH(t)(1-F(H(t),P(t))). \tag{2.15}$$

Several points are worth emphasizing. First, unlike the corresponding continuous time model, if there are no hosts, then the parasitoids are dead within a single time step. Second, the biology is all contained within the function F, which specifies the probability that an egg is not parasitized. This can be at least described from a mechanistic point of view, as in the original article by Nicholson and Bailey (1935).

The model (2.14–2.15) is too general to analyze without specifying the function F. The case to look at first assumes that searching is random; so with this mechanistic assumption about parasitoid behaviour, the probability of escaping parasitism is given by

$$F=\exp(-aP(t)), \tag{2.16}$$

which is the zero term from a Poisson distribution and the parameter a is called the area of search. This simplest version of the host-parasitoid model was carefully analyzed in the original article by Nicholson and Bailey (1935). The authors used two important concepts, those of equilibrium and stability. They focused first on whether there was equilibrium and, more importantly, on how the equilibrium levels depended on the parameters in the model. In particular, in the simplest case, as long as λ was greater than one, there was equilibrium with both hosts and parasitoids at positive densities (in addition to the equilibrium with both species absent, which always exists). This was taken to imply that parasitoids could control the growth of host populations.

They then turned carefully to questions of the stability of the non-trivial equilibrium and made several noteworthy conclusions. First, they demonstrated that under the assumptions they made, this equilibrium is always unstable. This is the core of a problem that continues to be an area of active research, namely, the question of what could stabilize the interaction between host and parasitoid. Since the interaction is always unstable in the simplest case, this is an important question because hosts and parasitoids do persist in nature.

They also were not afraid to draw important biological conclusions based on the unstable equilibrium of the simple model. Some of these conclusions, about the importance of the efficiency of the parasitoid and that the equilibrium host level drops as the parameter rises, are still used in thinking about determining effective biological control agents.

They also, importantly, considered a variety of other interactions, including several interactions among three or more species. In particular, they considered a system with one parasitoid and two hosts. For this system, they demonstrated that no equilibrium with all three species is typically possible because only an equilibrium with the host species that alone would produce the highest equilibrium level of parasitoids would typically be found. This result foreshadows the concept of apparent competition emphasized by Holt (1977).

It is again important to note that suggestions by Nicholson and Bailey of factors that might stabilize the interaction and permit long-term coexistence included, as one possibility, explanations that relied on spatial aspects that could only be dealt with in the context of models with structure. Other potential explanations rely on different assumptions about parasitoid search behaviour, which would change the form of the function F.

2.4.3 | *SIR* Epidemiological Models

At roughly the same time as the models just described on predator–prey and host-parasitoid systems were developed, a similar interaction was developed in a modelling context in a series of remarkable articles (beginning with Kermack & McKendrick 1927) focusing on the interactions between hosts and pathogens. One significant aspect of this study was the natural emphasis on short-term dynamics. The studies of models of interacting species emphasized the long-term dynamics and stability, especially in continuous time, except perhaps for Lotka's analysis of predator-prey dynamics. But, the studies of epidemiological models, especially the *SIR* models (defined in the next paragraph), emphasized the dynamics over short timescales. The models themselves reflected assumptions appropriate for short timescales only. This is an example of a model that straddles the line between structured and unstructured and is clearly mechanistic.

In the simplest form the models assume that the population size of the host population is constant and focus on the number (or equivalently the fraction) of the host population that is susceptible (S), infective (I), or removed (R). This is appropriate if you wish to follow the time course of a single epidemic or outbreak of a disease. Initially, you can assume that the population consists entirely of susceptible individuals. The simplest assumptions are that the rate at which susceptible individuals become infective is given by a mass action term βSI, where β is an infectivity parameter. There is then a rate γ at which infective individuals leave the infective class and are then called removed (which could be immune or dead). These assumptions lead to the equations:

$$\frac{dS}{dt} = -\beta SI \tag{2.17}$$

$$\frac{dS}{dt} = -\beta SI - \gamma I \tag{2.18}$$

$$\frac{dR}{dt} = \gamma I. \tag{2.19}$$

In this case, there is no long-term equilibrium, and the biological questions raised initially focused on several issues. One primary question was, will there be an epidemic? That is, if a small number of infectives are introduced into a population that is initially all susceptibles, will the number of infective individuals initially increase? A second question is, what stops an epidemic, if there is an epidemic—will the epidemic stop if there are still susceptibles? This leads to the third question: what is the state of the population after an epidemic? Surprisingly, it was not until recently that the role of pathogens in controlling the dynamics of populations was emphasized.

The initial questions can be answered relatively simply through the use of phase plane techniques. Because the population size is assumed constant, the dynamics are determined by equations (2.17) and (2.18). Divide equation (2.18) by (2.17) to obtain:

$$\frac{dI}{dt} = -1 + \frac{\gamma}{\beta S}. \tag{2.20}$$

From this equation one can essentially read the first fundamental result, namely, that an epidemic will occur if the initial population size (assumed to be essentially S) and the parameters satisfy the relationship:

$$\frac{\beta S}{\gamma} > 1. \tag{2.21}$$

The interpretation of this is that $\frac{\beta S}{\gamma}$, typically denoted R_0, is the mean number of infections caused by a single infective individual (before entering the removed class); therefore, it needs to be larger than one for an epidemic to occur. This is the basis for the famous threshold result of work by Kermack and McKendrick (1927). The basic *SIR* model is an example of a model with just the right level of complexity to be truly influential.

Still more recently, the *SIR* model and its derivatives have played a paramount role in understanding human diseases and control measures (e.g., see Anderson & May 1991). Once again, however, structured versions are needed to get at some of the important questions, such as using the mean age at which an individual gets a disease to estimate parameters. Also, models for sexually transmitted diseases have required the structuring of the population into groups based on the number of contacts.

2.4.4 | Competition

Competition is a different kind of interaction than predation with different dynamic consequences. Here, in the two-species systems, as carefully studied by Lotka (1926) and Volterra (1931), the only possible outcome is a stable equilibrium. The basic questions are again of coexistence—what permits two species to coexist, and why does one species not outcompete the other? Simple analyses of these basic models are possible and show that the asymptotic behaviour is simple, as explained in virtually every elementary ecology textbook.

The analyses of the simple models, in particular Gause's (1934) exposition of the principle that competing species had to differ to coexist, set the stage for much ecological research over the next three or four decades. The search for features that permitted coexistence and the role played by competition was a prominent feature of field research (e.g., see Connell 1983 and Schoener 1983) and laboratory research in ecology (e.g., see Park 1948, 1954). Although the initial models were not mechanistic, attempts were made to provide a mechanistic basis for the models (see the review in May 1974). However, perhaps because of the simplicity of the interaction, there is much less recent research on the dynamics of two-species competition models than on the corresponding exploiter-victim models.

2.5 | MORE THAN TWO SPECIES

Unstructured models continue to play a large role when more than two species are considered. Here, I focus on two important avenues of research, both of current interest and beginning to converge. First, ecologists began to recognize several decades ago that issues of coexistence of species needed to be studied in the context of more species. As exemplified by May's classic book (1974), the study of coexistence both of large numbers of competitors and of species in food webs could usefully be framed as a question of stability of models with large numbers of species. These questions are the subject of continuing study, and I outline some issues in this chapter.

Other specific and important ecological issues have been well studied using unstructured models, applying the approach of focusing on modules with a small number (often three or four) tightly interacting species. As in the study of two-species exploiter-victim interactions, much insight has been gained by looking for equilibria and determining conditions for stability. In particular, issues such as the role of trophic cascades as a way of understanding the effects of human removal of top predators in lakes (Carpenter & Kitchell 1993) have been effectively studied using models of this kind. Similarly, the role of predators in mediating competition among their prey (Holt 1977), denoted apparent competition, has been looked at this way.

However, there are other aspects of the dynamics of unstructured models with more than two species that are important to consider, namely, the possibility that complex dynamics may emerge. This is related to the issues raised previously concerning the dynamics of single-species models in discrete time, but the importance of complex dynamics emerges only with structure.

Another exciting development has been to emphasize the role and importance of complex dynamics arising in models describing small sets

However, it is important to recognize that there are clearly many times when structured models will be needed and can play a valuable role. In particular, the importance of space as a structuring variable has become almost universally recognized (Okubo & Levin 2002), and it is becoming clearer that there are few circumstances in which space can be ignored. However, even in this context, much recent research has been devoted to simplifying the representation of space as much as possible. Thus, the use of mean field models, or perhaps even some of the approximation methods used to simplify fully spatial models, can be seen as extensions of the spirit of the work by Lotka and Volterra, who recognized the importance of making models as simple as possible.

ACKNOWLEDGEMENTS

This research has been supported by grants from the National Science Foundation. I thank Kim Cuddington and an anonymous referee for helpful comments.

REFERENCES

Anderson, R.M., & May, R.M. (1991). "Infectious Diseases of Humans." Oxford University Press, Oxford.

Bererton, R.S.H., & Holt, S.J. (1957). "On the Dynamics of Exploited Fish Populations." Chapmen and Hall, London.

Burnham, K.P., & Anderson, D.R. (2002). "Model Selection and Multimodel Inference," 2nd Edition. Springer-Verlag, New York.

Carpenter, S.R., & Kitchell, J.F., Eds. (1993). "The Trophic Cascade in Lakes." Cambridge University Press, Cambridge.

Connell, J.H. (1983). On the prevalence and relative importance of interspecific competition. *Am. Nat.* **122**, 661–696.

Gause, G.F. (1934). "The Struggle for Existence." Williams & Wilkins, Baltimore.

Gause, G.F. (1935). "Vérifications Expérimentales de la Théorie Mathématique de la Lutte pour la Vie." Hermann, Paris.

Hassell, M.P. (1978). "The Dynamics of Arthropod Predator–Prey Systems." Princeton University Press, Princeton, NJ.

Hassell, M.P. (2000). "The Spatial and Temporal Dynamics of Host–Parasitoid Interactions." Oxford University Press, Oxford.

Hastings, A. (2004). Transients: the key to long-term ecological understanding? *Trends Ecol. Evol.* **19**, 39–45.

Hastings, A., Hom, C.L., Ellner, S., Turchin, P., & Godfray, H.C.J. (1993). Chaos in ecology: Is mother nature a strange attractor? *Annu. Rev. Ecol. Syst.* **24**, 1–33.

Hastings, A., & Powell, T. (1991). Chaos in a three-species food chain. *Ecology* **72**, 896–903.

Holling, C.S. (1959). Some characteristics of simple types of predation and parasitism. *Can. Entomol.* **91**, 385–398.

Holt, R.D. (1977). Predation, apparent competition, and the structure of prey communities. *Theor. Popul. Biol.* **12**, 197–229.

Huisman, J., & Weissing, F.J. (2001). Biological conditions for oscillations and chaos generated by multispecies competition. *Ecology* **82**, 2682–2695.

Hutchinson, G.E. (1978). "An Introduction to Population Ecology." Yale University Press, New Haven, CT.

Johnson, J.B., & Omland, K.S. (2004). Model selection in ecology and evolution. *TREE* **19**, 101–108.

Kermack, W.O., & McKendrick, A.G. (1927). A contribution to the mathematical theory of epidemics. *Proc. Roy. Soc. Lond. A* **115**, 700–721.

Kingsland, S. (1985). "Modeling Nature." University of Chicago Press, Chicago.

Levins, R. (1966). The strategy of model building in population biology. *Am. Sci.* **54**, 421–431.

Lotka, A.J. (1926). "Elements of Physical Biology." Williams & Wilkins, Baltimore.

Lotka, A.J. (1932). The growth of mixed populations: Two species competing for a common food supply. *J. Wash. Acad. Sci.* **22**, 461–469.

Luckinbill, L.S. (1973). Coexistence in laboratory populations of *Paramecium aurelia* and *Didinium nasutum*. *Ecology* **54**, 1320–1327.

Luckinbill, L.S. (1974). The effects of space and enrichment on a predator–prey system. *Ecology* **55**, 1142–1147.

Luckinbill, L.S. (1979). Regulation, stability, and diversity in a model experimental microcosm. *Ecology* **60**, 1098–1102.

May, R.M. (1974). Biological populations with nonoverlapping generations: Stable points, stable cycles, and chaos. *Science* **186**, 645–647.

May, R.M. (1974). "Stability and Complexity in Model Ecosystems," 2nd Edition. Princeton University Press, Princeton, NJ.

May, R.M. (1976). Simple mathematical models with very complicated dynamics. *Nature* **261**, 459–467.

McCann, K., Hastings, A., & Huxel, G.R. (1998). Weak trophic interactions and the balance of nature. *Nature* **395**, 794–798.

Murdoch, W.W., Briggs, C.J., & Nisbet, R.M. (2003). "Consumer–Resource Dynamics." Princeton University Press, Princeton, NJ.

Murdoch, W.W., & Oaten, A. (1975). Predation and population stability. *Adv. Ecol. Res.* **9**, 1–131.

Nicholson, A.J., & Bailey, V.A. (1935). The balance of animal populations. *Proc. Zool. Soc. Lond.* **1**, 551–598.

Okubo, A., & Levin, S.A. (2002). "Diffusion and Ecological Problems: A Modern Perspective." Springer-Verlag, New York.

Park, T. (1948). Experimental studies of interspecific competition: I. Competition between populations of the flour beetle *Tribolium confusum* Duval and *Tribolium castaneum* Herbst. *Ecol. Monogr.* **18**, 267–307.

Park, T. (1954). Experimental studies of interspecific competition: II. Temperature, humidity, and competition in two species of *Tribolium*. *Physiol. Zool.* **27**, 177–238.

Pearl, R., Reed, L.J., & Kish, J.F. (1940). The logistic curve and the census count of 1940. *Science* **92**, 486–488.

Ricker, W. (1954). Stock and recruitment. *J. Fish. Res. Bd. Can.* **11**, 559–623.

Rosenzweig, M.L. (1971). Paradox of enrichment: Destabilization of exploitation ecosystems in ecological time. *Science* **171**, 385–387.

Schoener, T.W. (1983). Field experiments on interspecific competition. *Am. Nat.* **122**, 240–285.

Volterra, V. (1926). Variazioni e fluttuazioni del numero d'individui in specie animali conviventi. *Mem. del Acad. Lincei* **2**, 31–113.

Volterra, V. (1931). "Leçons sur la Mathématique de la Lutte pour la Vie." Marcel Brelot, Paris.

3 | UNSTRUCTURED POPULATION MODELS: DO POPULATION-LEVEL ASSUMPTIONS YIELD GENERAL THEORY?

André M. De Roos and Lennart Persson

3.1 | INTRODUCTION

Over the last century mathematical models and their results have played an influential role in our way of thinking about ecological systems. A short anecdote about the work of the Italian mathematician Vito Volterra may serve as an excellent example. In 1926 Volterra wrote an article entitled "Fluctuations in the abundance of a species considered mathematically," inspired by a question from his son-in-law, U. d'Ancona. The question posed by d'Ancona was, why had the predatory fish species in the Adriatic Sea increased in abundance after fisheries ceased during World War I, but the opposite was true for the prey species these predators fed upon? Volterra (1926) developed a model to describe the interaction between a predator and a prey based on a few simple assumptions. Without stating any equations, Volterra concluded on the basis of his analysis that prey and predator would fluctuate periodically in abundance, where the period of the fluctuations depended only on the coefficients of increase and of destruction of the two species and on the initial numbers of individuals of each species. Volterra's conclusion laid the basis for one of the cornerstones of current ecological theory:

the idea that two species engaged in a consumer-resource interaction have an intrinsic propensity to fluctuate in abundance. This idea significantly influenced the ecological studies, carried out over the last 75 years, that aimed to understand what mechanisms drive the dynamics of populations and ecological communities.[1]

The core models of our current body of ecological theory include the Lotka-Volterra competition model, the Lotka-Volterra predator-prey model, and the Fretwell-Oksanen food chain model. They represent three fundamental types of species interaction: the antagonistic interaction between two competitors, the consumer-resource interaction between a predator and its prey, and the indirect mutualism in chains of species. The analyses of these models have brought forward important theoretical concepts (e.g., competitive exclusion principle, predator-prey cycles, and trophic cascades), inspired new experimental and empirical studies, and advanced our understanding about ecological systems. The status of these models is comparable with the Navier-Stokes equations in physics, which form the basis for the study of fluid mechanics, aerodynamics, and hydrodynamics, and with the differential equations developed by van't Hoff (1884), which describe the kinetics of chemical reactions as a function of the concentration of the reactants and products and of the ambient temperature. The core ecological models describe dynamics of populations and communities much like the kinetics of chemical reactions, as if predator and prey are particles (the "reactants") that on encounter would form newborn predators as a product. The rate law of these interactions is described with the *law of mass-action*, according to which the reaction "forces" are proportional to the product of the concentrations of the reactants. This law of mass-action was first developed in chemistry (Guldberg & Waage 1864), which was the original field of science of one of the founders of ecological modelling, Alfred Lotka.

The ecological systems that the Lotka-Volterra type models are supposed to describe, however, are far more replete with complex and subtle mechanisms than physical and chemical systems. In addition, these mechanisms generally are poorly understood and are not constrained by first principles. Hence, no ecologist will claim that the representation "particle encounter leads to reaction" is an appropriate description of

---

[1] Volterra closed his short article with the following: "Seeing that a great number of biological phenomena are characteristic of associations of species, it is to be hoped that this theory may receive further verification and may be of some use to biologists." This was a very modest statement given the role his work has played.

any ecological process. The conceptualization contrasts with reality in two important aspects:

- *Ecological interactions are more than random encounters.* Many, often subtle, mechanisms are playing a role in the encounter, either because predators have developed intelligent ways to detect, locate, ambush, or even attract the prey or because prey have developed skills to evade getting caught.
- *Individuals are unique.* Biological organisms are not uniform, standardized particles but rather individuals with their own set of traits and their own specific physiology. Moreover, these traits are changing in time (i.e., individuals grow and develop during life), and these changes often depend on the ecological interactions, as exemplified by the (feeding) interaction between a consumer and its resources.

A considerable part of theoretical ecological research over the last 75 years has addressed the fact that interactions are more than random encounters. For example, the original Lotka-Volterra model was modified to include logistic growth for the prey individuals and a functional response of the predators by MacArthur and Rosenzweig, which led to the discovery of the paradox of enrichment (Rosenzweig 1971). Other examples include the extension of the competition models to account explicitly for resources (MacArthur & Levins 1967, MacArthur 1972, Tilman 1982), the consideration that interactions may be of a mixed type, such as intraguild predation (Pimm & Lawton 1978, Holt & Polis 1997), and more recently the incorporation of behaviour into population dynamic models (see Bolker *et al.* 2003 for a review). The recognition that distributions of populations are never homogeneous over space and that ecological interactions are modified by spatial heterogeneity can be seen as another example: it has given rise to the broad field of metapopulation theory (Hanski 1999) and more recently to spatially explicit (simulation) models (e.g., see Hassell, Comins, & May 1991; De Roos, McCauley, & Wilson 1991; and Rand & Wilson 1995). Without providing a complete catalogue, these selected examples illustrate that, in general, theoretical ecologists have been well aware that there is more to an ecological interaction than the random encounter of two (or more) individuals. Many ecological studies have hence investigated the implications of more complicated interaction mechanisms.

In comparison, the consequences of the *identical-individuals assumption* for the predictions and insights that result from the core ecological models have been much less debated and investigated. The question of whether the core ecological models yield an accurate or plausible perspective on the working of ecological systems, given that they ignore variation among individual organisms, has rarely been posed or investigated (e.g., see Lomnicki 1988). In this respect, even to the present day, ecological theory has followed the seminal ideas first presented by Elton (1927) and later elaborated by Lindeman (1942). In Elton's view, ecological communities are conceptualized as networks of interacting trophic units in which the unit is a population of a particular species or even a collection of populations of different species.

This view of populations and communities contradicts the ideas of Darwin, who, despite focusing in his analysis on an evolutionary timescale, emphasized the uniqueness of the individual organism; discarded the idea that species are unchanging, uniform entities; and pointed out the far-reaching consequences that variation among individuals might have on the dynamics of ecological communities. The recognition of individual variation has been suggested as the major advancement of the Darwinian revolution, replacing the Aristotelian view of *ideal types* to which observed objects were *imperfect approximations* (Lewontin 1964). By neglecting the variation among individuals, the uniqueness of individual characteristics, and the fact that individuals change in these characteristics during life with concomitant changes in ecological performance, the core ecological models neglect the unique feature that makes an ecological community different from a chemical reaction network! We will refer to the population and community models based on the identical-individuals assumption as top-down, unstructured, or population-level models; models that explicitly account for individuals and individual-level processes are referred to as bottom-up or individual-based models (Bolker *et al.* 2003).

In this chapter, we reflect on the extent to which unstructured population models, which currently make up the core of ecological theory but ignore variation among individuals, yield valuable insights into the working of ecological systems. In line with the central theme of this book, we propose a major shift in paradigm. We challenge the established approach that starts with formulating and analyzing strategic, population-level models for broad insights and subsequently adds details to the emerging picture by studying more mechanistic, individual-based models. We argue that starting a modelling process from a set of general, rather unspecific, population-level assumptions—that is, aiming at "capturing the essential dynamics" (May 1979) and ignoring the detailed, individual-level mechanisms that ecological interactions are based on—is no guarantee that the resulting theory is applicable to a range of systems. Which aspects of a system are essential and which are details cannot be determined *a priori.* Rather, they can only be assessed by an in-depth examination of the system dynamics.

We strongly argue for a paradigm that starts with the bottom-up modelling of an ecological system and carefully accounts for the individual-level mechanisms that make the system tick. The level of the individual organism is the natural, basic scale at which this modelling process should be started because at this scale the elementary processes of population dynamics take place. Whether the results of these bottom-up, individual-based models are system specific or extend to other systems can then be assessed by stripping them down to more aggregated population-level representations, by identifying crucial mechanisms, and by discarding assumptions that leave the essential outcomes of the model unchanged. In our view, bottom-up and top-down models should hence be the start and the end point, respectively, of a theoretical development, which should progress from complex to simple as opposed to the other way around. Only such a bottom-up to top-down approach will ensure that the required generality of models will be in their results and not in their assumptions. In the following sections we elaborate on, provide support for, and give examples of this new paradigm we envision.

3.2 | CORE THEORY OR LIMITING CASE?

Theory in general is only relevant if it has some level of generality, such that it is not system specific but applies to a range of systems and situations. An additional requirement of theory is testability, especially in a Popperian scientific milieu. Testability of model predictions is a topic of a later section. Here we focus on the required generality of ecological theory.

When requiring ecological theory to be general, we aim to have the insights stemming from verbal, logical, or mathematical models apply to as broad a range of systems and situations as possible. In practice, this has implicitly been translated into requiring that the assumptions on which ecological models are based should be *general*. General in this context means that they should not be too detailed, such as an assumption that lions eat zebras, and should only involve a small number of parameters. Models that make more explicit assumptions about mechanisms at the level of individual organisms are interpreted as being specialized and system specific. Consider the following quote from Bolker *et al.* (2003):

> A final set of open theoretical questions, which we have neglected in this review, has to do with the connections between the general theoretical top-down models (p-state), which are defined in terms of populations and average states, and bottom-up (i-state) models that forgo aggregation in favour of a detailed description of individual states and behaviours. . . .

> The shortcomings of bottom-up models include a tendency towards over-parameterization, and the difficulty of drawing broad conclusions that extend beyond a particular system, but they will form an essential ingredient in connecting observations of natural and manipulated empirical systems with top-down analytical models.

The quote expresses the widespread trust that population-level (top-down or p-state) models will yield broad conclusions because of their supposed generality and that predictions of more mechanistic, individual-based (bottom-up or i-state) models will apply only to the particular system for which they are parameterized. In this view, population-level models form the core of ecological theory, and mechanistic, individual-based models provide only system-specific variations and modifications of the core ideas. Whether model assumptions are general and unspecific or detailed but mechanistic is, however, somewhat irrelevant. What counts is whether the results and insights gleaned from the model analysis are general and broadly applicable.

We argue that there is no *a priori* reason to expect population-level assumptions to yield general theory. Top-down models are phrased in terms of populations, their abundances, and their average state, and they purposely ignore the individual-level mechanisms of ecological interactions. These models specify the interactions between populations as a function of N replicates of some *average individual*, where N denotes population abundance. Jensen's inequality teaches us, however, that the expected value of a function is not equal to the function of the expected value in case of ecological interactions because these interactions are typically nonlinear. Murdoch, Briggs, and Nisbet (2003, p. 355) showed how this may influence dynamics in a spatially heterogeneous population model and lead to stabilization of predator-prey cycles (see Chesson 1990 for another example). The insights gained from the population-level models are therefore likely to apply only to those populations in which either within-population variability is low (individuals are identical) or ecological interactions and their consequences are relatively independent of the characteristics and traits of the individuals involved. Both these conditions are generally not the rule but the exception in ecological systems. These considerations would suggest an alternative view on the relevance of unstructured population-level models. They do not constitute a core body of theory to which bottom-up models add details as variations on a theme; rather, they are limiting cases only applicable under the exceptional conditions mentioned previously.

The uncertain status of ecological theory based on unstructured, population-level models is exemplified by the occurrence of predator-prey or paradox-of-enrichment cycles (Rosenzweig 1971). Models that incorporate a logistic growth process of the prey population and a saturating functional response of the predators generally predict that cycles of this particular type should be ubiquitous in consumer-resource systems, especially in more productive environments. As a consequence, ecology as a science has been dominated by questions about their occurrence and concomitant questions about the stability of equilibria in consumer-resource systems. Luckinbill (1974) indeed showed predator-prey cycles to occur in simple protist systems of *Paramecium* and *Didinium*. In contrast, systems of *Daphnia* and algae only exhibited true predator-prey cycles under special experimental conditions (McCauley & Murdoch 1990, McCauley *et al.* 1999). Several empirical studies have indicated that most populations do not cycle (Kendall, Prendergast, & Björnstad 1998). In addition, Murdoch *et al.* (2002) found that among populations of 40 different species, only 15 exhibited dynamics that, judging from the cycle period, could possibly be classified as predator-prey cycles.

Moreover, theoretical studies have shown that even a little bit of variability among individuals, especially when pertaining to the susceptibility of the prey, will prevent predator–prey cycles from occurring. Murdoch *et al.* (1987) showed that an invulnerable life stage in a host-parasitoid system represented a stabilizing mechanism, suppressing the occurrence of predator-prey cycles. Analogously, we have shown that classical predator-prey cycles do occur as expected in a three-link food chain, in which the complete size structure of the middle trophic level (the consumer) is accounted for, but only if all consumers are vulnerable to predation (De Roos, Persson, & McCauley 2003). As soon as the topmost 10% (in terms of biomass) of all consumers are protected from being preyed upon, all cyclic dynamics disappear. In short, despite the prominence of predator-prey cycles in predictions of unstructured models, they do not seem to abound in natural systems, and their ubiquitous occurrence in models might be strongly related to the special set of assumptions incorporated in the population-level, predator-prey model (Rosenzweig & MacArthur 1963) for which they were first identified.

3.3 | DERIVING GENERAL POPULATION MODELS: STARTING WITH THE INDIVIDUAL

The question of how to derive useful and appropriate population dynamic models has been addressed many times (e.g., see

Levins 1966; Holling 1966; May 1974, 1979; Nisbet & Gurney 1982; Murdoch *et al.* 1992; Murdoch & Nisbet 1996; and Murdoch, Briggs, & Nisbet 2003). All these contributions wrestle with the fundamental dilemma of model building: how do we develop models that are general and testable given that these two goals are typically incompatible? Stated another way, the fundamental dilemma relates to building models that have all three components: generality, realism, and precision (Levins 1966). The solution to this dilemma has generally been found in the distinction between *strategic* and *tactical* models (Holling 1966, May 1974, Nisbet & Gurney 1982). Strategic models aim to capture the "essence" of a broad class of ecological systems by leaving out the large amount of detail that makes every member of this class so distinctly unique (May 1979). In contrast, tactical models are usually highly system specific, incorporate lots of detailed mechanisms, and mainly aim at predicting the future dynamics of the system in detail, as is required, for example, in many applied management situations.

Murdoch *et al.* (1992) describe the role of strategic models as models "that try to reach past the tangle of unique detail associated with each particular ecological system to grasp and abstract the features that account for those aspects of its dynamics that are generic . . . Generality can be attained because, one hopes, many systems have a similar essence." This is a true and fundamental statement that expresses that the patterns, which population ecology should aim to explain, are the generic features in the dynamics of the system—those that are observable in a range of systems. The essence of a system, on the other hand, is what brings these generic features about and hence relates to the mechanisms responsible for the dynamics.

Logically, to identify aspects of dynamics that are generic and the essence of a system responsible for these aspects, be it a model or a natural system, the following steps are required:

1. get a detailed overview of the dynamics that the system could possibly display,
2. determine which features of these dynamics are generic (This might be a subjective choice.), and
3. identify which mechanisms are responsible for these features (the essence).

We suggest that this sequence of steps necessitates that the modelling of the population or community of interest starts at the level at which the dynamics originate: the level of individual organisms. Individuals are the fundamental entities engaged in all population dynamics processes (such as reproduction and death) and in the ecological interactions that form the basis of every ecological community. The analysis should thus start with a bottom-up model that explicitly accounts for individuals, their characteristics, and the mechanisms of the ecological processes of interests. Only after identification of the prominent features of the dynamics of such a model will it be possible to address the question of which individual-level mechanisms and assumptions are responsible and essential for their occurrence and should hence be considered the essence of the system.

This sequence of steps contrasts strongly with the approach, in vogue, in which the essence of a particular system is decided upon *a priori* and invariably includes only population-level aspects. The hope, and even the implicit belief, is that the resulting "general" population-level assumptions (in the sense of lacking mechanistic detail and involving few parameters) lead to "generic" dynamics. These assumptions could, however, also be considered unspecific, simplistic, or restrictive. For example, the assumption that lions eat zebra may seem general, but in a modelling context it more accurately translates into an assumption that *all* lions eat *all* zebra *with the same probability.* In the latter interpretation, the assumption is clearly restrictive and specific and moreover in conflict with reality (Sinclair, Mduma, & Brashares 2003).

As another example, population-dynamic modelling studies of flexible behaviour (see Bolker *et al.* 2003 for an overview) have invariably assumed that all individuals of a consumer population make the same choices with respect to where and on which resource to forage, because the flexible behaviour process is incorporated into a population-level model and the individuals exhibiting the flexible behaviour are not explicitly accounted for. The choices in these models usually follow optimal foraging rules, leading to predictions of ideal-free distributions of all (identical) consumers. However, with the slightest variation among consumers, ideal-free distributions are not the default expectation. Rather, consumers with different properties will segregate over different resource patches or specialize on different resources (Adler, Richards, & De Roos 2001). The consequences of such a segregation process substantially change the predictions about the influence of flexible behaviour on population dynamics (see De Roos *et al.* 2002). When incorporated into an unstructured, population-level model, the "general" assumption that consumers exhibit flexible behaviour hence only applies to a specific, limiting case and it is questionable whether the dynamics that result are in any sense generic.

In our view, general theory can only be derived if we understand what is lost or neglected by not including specific aspects or mechanisms into a model. In other words, generality can only be achieved by selectively deleting details from more complex, mechanistic bottom-up models as opposed to by deciding *a priori* what the essence of a system is and refraining from including any mechanism for the sake of being too specific. The road towards general theory is then a stepwise process, which most importantly starts with formulating a mechanistic bottom-up model, analyzing its dynamics, and deriving a broad picture of its properties. The outcome of the model-stripping analysis should be the identification of the essence of the system. The exact form of the resulting simplified population model may, however, depend on which features of the system dynamics we consider crucial and want to preserve, and which we deem to be details.

3.4 | THREE CASE STUDIES

In the following section, we illustrate the paradigm we put forward in this chapter, which starts with detailed bottom-up models explicitly representing individuals and their interactions and progresses to more aggregated population-level models, but preserves generic dynamic features. We compare the results obtained from fully structured models belonging to the class of physiologically structured population models (PSPMs) (Metz & Diekmann 1986; Metz, De Roos, & Van Den Bosch 1988; De Roos 1997; see the next section), with those obtained from stripped-down versions for three scenarios: a size-structured consumer feeding on a non–size-structured prey, a tritrophic food chain with size structure in the consumer, and cannibalism with size structure in cannibals and victims. This is done with the idea that stage-based models, compared with PSPMs, represent strategic vehicles that may abstract basic mechanisms of the latter (Murdoch *et al.* 1992). We address the issues of the extent to which stripped-down, stage-based models successfully preserve the essential mechanisms of more complex models and whether stripped-down models generate more generality in terms of insights about how a particular process (e.g., cannibalism) affects system dynamics. The latter question turns out to be less straightforward to answer than suggested by Murdoch *et al.* (1992).

3.4.1 | Consumer-Resource Interactions

The dynamics of fully size-structured consumer-resource models with a size-structured consumer feeding on a prey of fixed size have been studied for both *Daphnia*-algae and planktivorous fish-zooplankton interactions

(Metz, De Roos, & Van Den Bosch 1988; De Roos *et al.* 1990; Persson *et al.* 1998). In these models every individual can have a unique body size, but individuals with identical sizes at a particular time, for example, when they are born, will never diverge in size. The focus of the model that we studied (Persson *et al.* 1998) was on how the body size scaling of the foraging (attack) rate, as parameterized by the allometric scaling exponent α, affected population dynamics. In addition to the size dependency of the attack rate, this model included a size dependency in metabolic demands and maximum consumption capacity.

Essentially, three dynamic regimes were found by changing the size scaling of the attack rate. With an attack rate that increased relatively slowly with body size, large amplitude cohort cycles were observed in which recruiting individuals outcompeted older or larger individuals (Fig. 3.1, upper left). The characteristic aspect of these cohort cycles, which are comparable to the single-generation cycles identified by Gurney, Nisbet, and Lawton (1983), is that at any particular time the population is dominated by a single cohort of individuals born at roughly the same time. Cohort cycles were also observed when the attack rate of the consumer increased rapidly with body size. However, in this case, the cohort dynamics were driven by a strong cohort of older or larger individuals that prevented newly born individuals from entering the system (Fig. 3.1, bottom left). Finally, for intermediate values of the size-scaling exponent, stable fixed-point dynamics prevailed.

The analysis of the model further showed that the dynamics could be *a priori* predicted by the form of the curve relating the critical resource demands (the resource level at which energy gains equal metabolic costs) to body size (Persson *et al.* 1998). Recruit-driven cohort cycles occur when smaller individuals are competitively superior such that critical resource demands increase monotonically with body size (Fig. 3.1, upper left). In contrast, cycles driven by larger juveniles occur when larger individuals are competitively superior and critical resource demands first decrease with body size (Fig. 3.1, bottom right).

The same types of cohort cycles were found for comparable relationships of intraspecific competitiveness in a stripped-down version of the consumer-resource model in which the consumer population consisted of two size classes, juveniles and adults (De Roos & Persson 2003). In this model, the development rate of juveniles into adults was food dependent as was adult reproduction. In addition, the mortality of both size classes was inversely related to energy assimilated. This simplified size-structured model allowed a more complete analysis of the dynamic properties compared with the fully structured model, including stability analysis of the equilibrium.

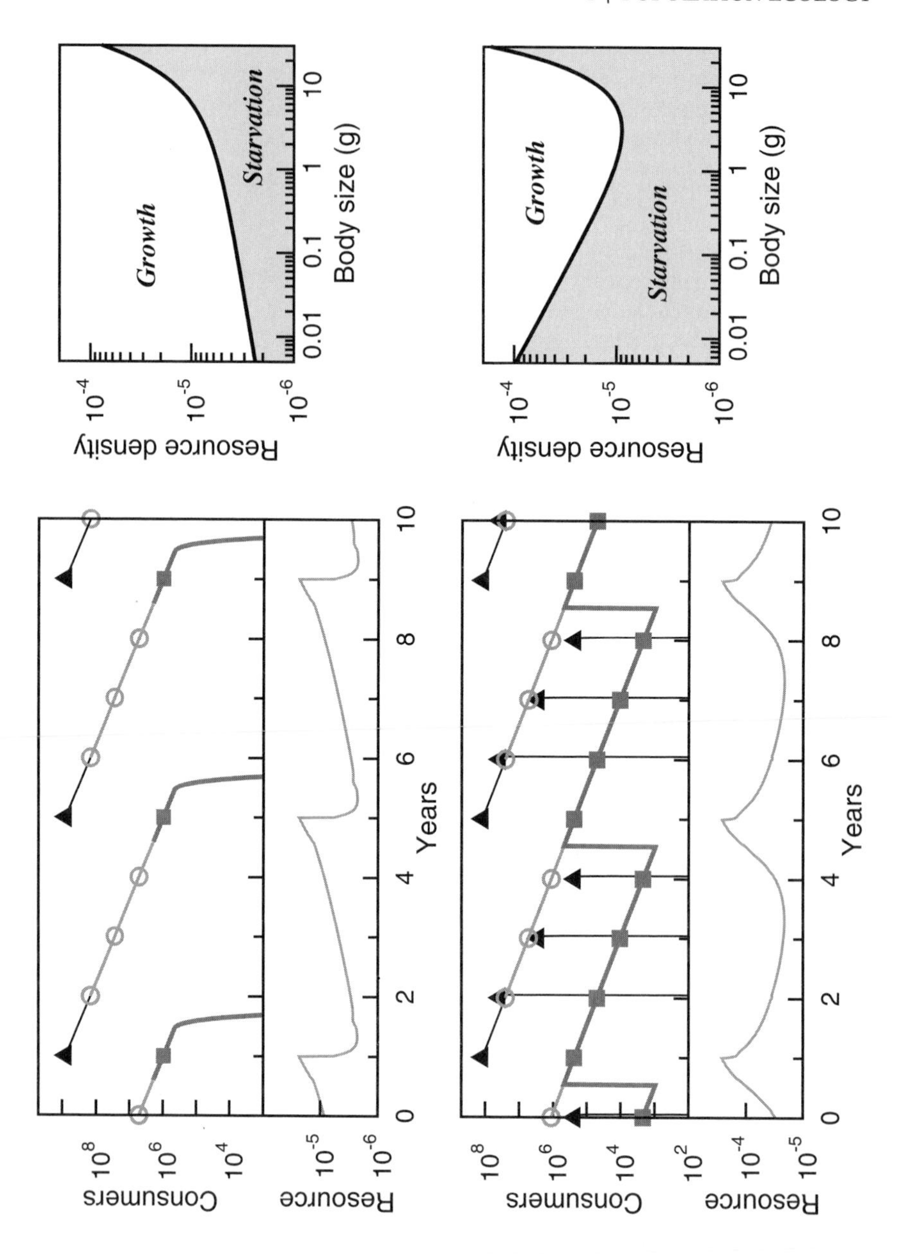
Consumers
Resource
Years
Resource density
Body size (g)
Growth
Starvation
Consumers
Resource
Years
Resource density
Body size (g)
Growth
Starvation

FIGURE 3.1 | Representative examples of the juvenile- (*top*) and adult-driven (*bottom*) cycles, as predicted by our model (Persson *et al.* 1998). The model represents a fully size-structured consumer population feeding on an unstructured resource. Reproduction occurs as a pulsed event at the beginning of summer, giving rise to new cohorts of individuals. Feeding, growth, mortality, and resource production are continuous-time processes. Left panels show the dynamics of newborn (age between 0 and 1; *black solid triangles and thin lines*), juvenile (*light-grey open circles*), and adult (*dark-grey solid squares and thick lines*) consumer individuals and resource density. Right panels show zero-growth isocline (*solid lines*), defined as the food density at which assimilation rate equals metabolic requirements for individual consumers as a function of their body size.

To investigate the effects of size-dependent competition on population dynamics analogously to the fully structured model, a parameter q, which described the ratio between adult and juvenile feeding rates, was introduced. Overall, changing q had the same dynamic consequences as had been found when changing the size scaling of the attack rate in the fully structured model (Fig. 3.2). This also concerned the presence of two alternative types of population fluctuations when juveniles were competitively superior. This means that the larger complexity and the higher number of parameters of the fully structured model (20) had not prevented us from identifying all dynamic regimes also found in the simplified model, containing only 7 parameters (De Roos & Persson 2003). Further analysis of the conditions for the different cycles to occur showed that cohort cycles driven by recruiters also occurred when development rate was independent of food density. In contrast, cycles driven by larger/adult individuals were primarily induced when the juvenile period changed with food density and disappeared when the juvenile delay was fixed.

Compared with the fully structured model, the stripped-down model preserved essentially the same dynamic patterns in relation to the ratio of juvenile/adult competitiveness and provided new general insights into the essential mechanisms necessary for the different cycles to occur. At the same time, the fully structured model yielded a more general understanding of the implications of different individual body size scalings (attack rate, consumption capacity, and metabolic rate) for population dynamics.

3.4.2 | Tritrophic Food Chain

We presented a tritrophic food chain model that included a size-structured consumer population feeding on a non-size-structured resource and being preyed upon by a non–size-structured top predator

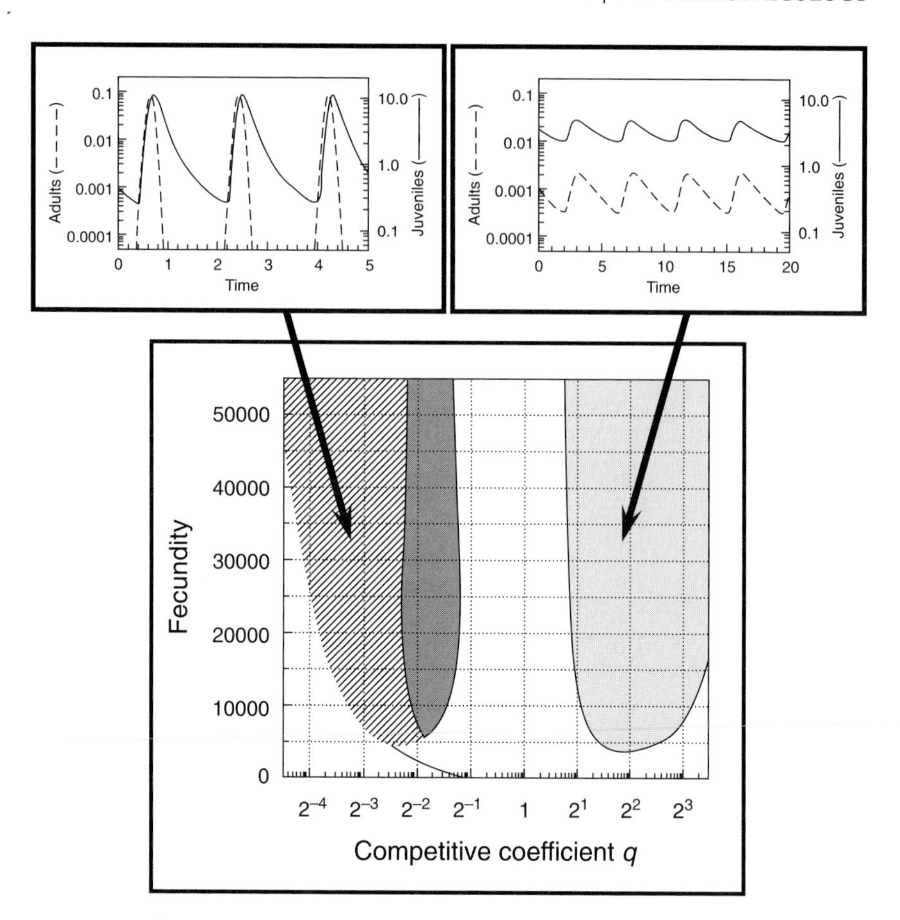

FIGURE 3.2 | Stability diagram for the two-stage consumer–resource model that we analyzed (De Roos & Persson 2003) as a function of the ratio between adult to juvenile competitiveness (q on the ordinate) and the adult fecundity coefficient (β on the abscissa). Population cycles occur for parameter values in coloured regions. When juveniles are competitively superior (low values of q), large amplitude cycles occur (*left inset*) in which a single juvenile cohort (*solid line*) dominates the dynamics throughout its life and adult consumers (*dashed line*) are only temporarily present, despite reproduction being a continuous process. When adults are competitively superior (high values of q), low-amplitude cycles occur in which both juveniles and adults are present continuously and fluctuate rather synchronously. Stable equilibria only occur when individuals are competitively similar. In the hatched region of parameter space with low values of q, two alternative types of populations cycles occur (see De Roos & Persson 2003 for details).

(De Roos & Persson 2002). We made two basic assumptions. First, we assumed that the top predator was size selective in its feeding on the consumer by only feeding on consumer individuals up to a certain size threshold. Second, we assumed that individual consumer growth rate was food dependent.

This tritrophic food chain model, in contrast to non-structured tritrophic food chains, exhibited alternative states and, more importantly, an emergent Allee effect that resulted from pure exploitation of the size-structured consumer by the top predator. The mechanism behind this emerging Allee effect was essentially that increased predation rate caused an increased maturation rate of juveniles, and this increase more than compensated for the mortality imposed by the predator on juvenile consumers. As a result, the population fecundity of adult consumers increased, leading to an increase in juveniles upon which the top predator fed. A key requirement for this Allee effect was a food-dependent development rate, because the effect disappeared when this food dependency was removed (De Roos & Persson 2002). Changing the assumption of negative size selectivity in the predator to assuming positive size selectivity showed that an Allee effect was also present in this case although for a narrower parameter range (De Roos & Persson, in press).

A stripped-down version of the tritrophic level models with three stages (two juvenile stages and one adult stage) showed, as in the fully structured model, that both negative and positive predator size selectivity can lead to the presence of an emergent Allee effect (De Roos, Persson, & Thieme 2003). Furthermore, this stripped-down model clarified that an Allee effect was only present when the prey was regulated by density dependence in development rate and when for the regulating stage this rate exhibited overcompensation. The Allee effect does not occur when the prey is regulated by adult fecundity. Van Kooten *et al.* (2005) further showed that bistability was also possible between two states in which both predator and consumer are present. As for the consumer-resource interaction, the stripped-down model thus provided new general insights into the mechanisms necessary for an Allee effect to occur. Also similar to the consumer-resource system, the fully structured model has the potential of generating more general insights into how the overall individual size scalings of predator-prey interactions will affect the likelihood for an Allee effect to occur compared with the stripped-down model.

3.4.3 | Cannibalism

To analyze the dynamics of cannibal-victim interactions, functions need to be derived that specify both the size limits within which cannibal-victim interactions take place and the intensity of this interaction within these limits. Claessen, De Roos, & Persson (2000), analyzed a fully structured cannibal model in which these functions were developed. Analyses of this model showed that two main parameters affected population dynamics. One parameter (β) described how fast the cannibalistic attack rate increased with cannibal size, and the other parameter (δ) defined the smallest victim a cannibal of a specific size could perceive and therefore attack.

For high δ values, newborn victims were able to strongly depress the shared resource before they reached a size vulnerable to cannibals, thereby causing the starvation death of cannibals before they could start to consume victims. As a result, cohort cycles, as described earlier, prevailed (Fig. 3.3, *left*). In contrast, for low δ values, cannibals encountered victims at small victim sizes and thereby through heavy cannibalism prevented victims from depressing the shared resource. This led to a situation with cannibal control and low-amplitude fluctuations or fixed-point dynamics (Fig. 3.3, *right*). For intermediate values of δ, a mixture of periods with low-amplitude dynamics and cannibal control and periods with cohort cycles in which recruits outcompeted most of the older individuals was present (Fig. 3.3, *middle*) (Claessen *et al.* 2002; Persson, De Roos, & Bertolo 2004). The few larger individuals that survived intercohort competition profited substantially from the strong, newly recruited cohort and became giants, but because of their low numbers these giant individuals only inflicted a negligible mortality on this cohort.

Increasing β resulted in major shifts in population dynamics. For δ values for which high-amplitude dynamics prevailed, an increase in β first collapsed cohort cycles to fixed-point dynamics in a narrow parameter regime. Further increases in β dramatically increased population fluctuations, with cohort cycles and cannibal-driven dynamics interchanging (Fig. 3.3, *middle*) (Claessen, De Roos, & Persson 2000).

Essentially, a similar shift in dynamics from competition control to cannibal control as observed in the fully structured model with increas-

FIGURE 3.3 | Representative dynamics of the cannibalistic model (Claessen, De Roos, & Persson 2000; Claessen *et al.* 2002) for different values of the minimum ratio between victim-to-cannibal length, referred to as δ (*left:* $\delta = 0.08$; *middle:* $\delta = 0.056$; *right:* $\delta = 0.0$), the minimum ratio between victim-to-cannibal length at which cannibalism can occur. The top row shows the dynamics of newborn (age between 0 and 1; *black solid triangles and thin lines*), juvenile (*light-grey open circles*), and adult (*dark-grey solid squares and thick lines*) cannibals; The bottom row shows length-age relations for cannibals born at different times. Note that for $\delta = 0.0$ cannibals mature within 1 year; hence, juvenile cannibals do not show up in the top-right panel.

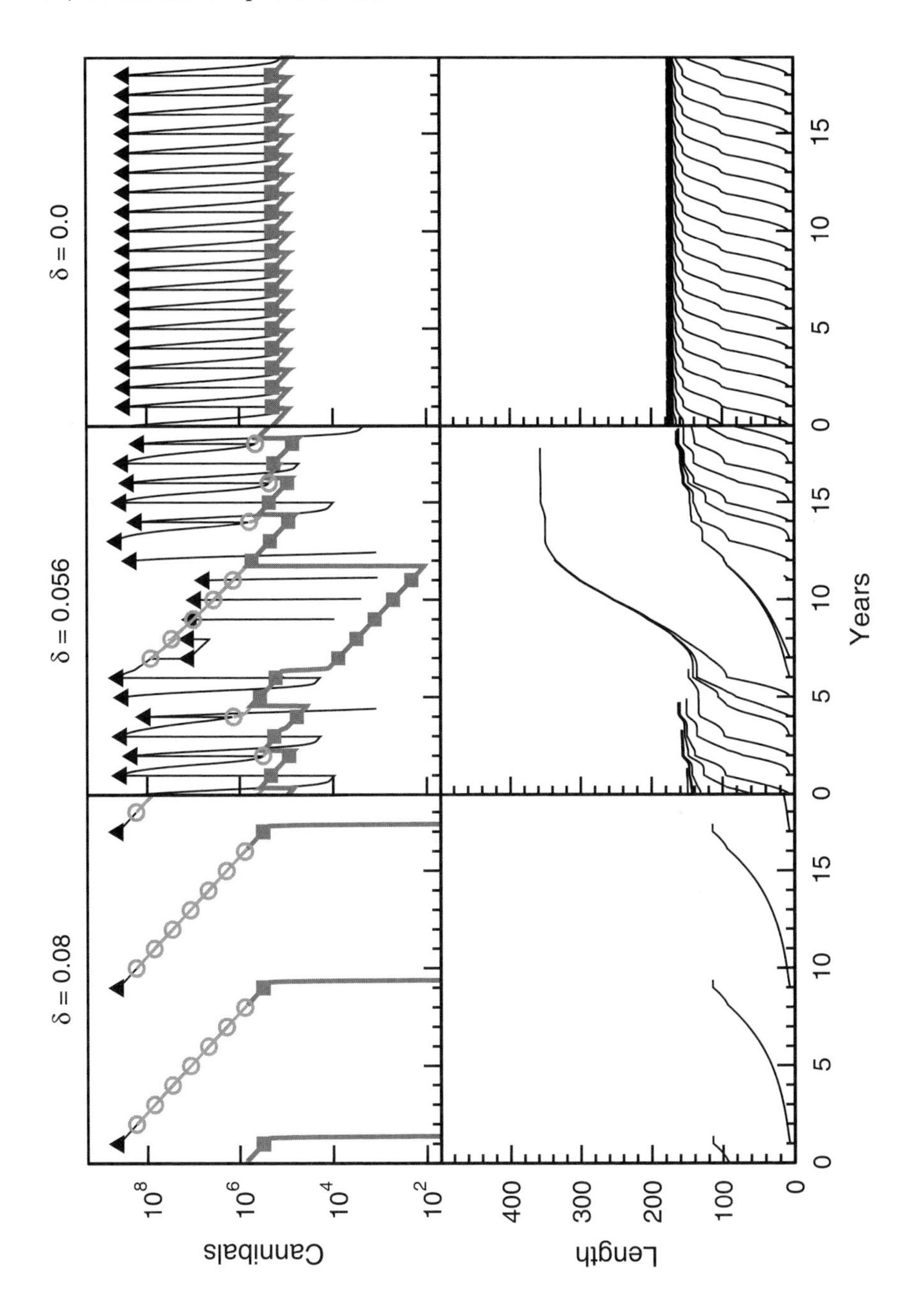
δ = 0.08
δ = 0.056
δ = 0.0
Cannibals
Length
Years
10^8
10^6
10^4
10^2
400
300
200
100
0
5
10
15

ing cannibal efficiency (decreasing δ as explained previously) can be found in a stripped-down, stage-based cannibalistic model with juveniles (J), subadults (S), and adults (A) (Fig. 3.4). In the latter model, the regime with mixed dynamics is expressed as alternative stable equilibria. The stripped-down stage-based model thus preserves the essential dynamics of the fully structured models. At the same time, the stage-based model does not provide further general insights. Cannibal efficiency is reduced to one parameter (c) in the stage-based model, whereas two parameters, β and δ, were used to model cannibal efficiency in the fully structured model. This reduction in parameter numbers in the stage-based model took place at the expense of general understanding regarding the effect of increased cannibal efficiency on population dynamics. An increase in cannibal efficiency in terms of increased ability to see small victims (reduced δ) indeed led to increased cannibal control overall. However, increasing β ultimately led to high-amplitude dynamics in which cannibals lost control over victims and were outcompeted (Claessen, De Roos, & Persson 2000). Thus, besides being readily measured empirically, handling cannibal efficiency by two parameters describing different aspects of cannibalism resulted in more general insights into the effects of cannibalism on population dynamics than did a one parameter description of cannibal efficiency. Compared with the cases of consumer-resource interactions and the tritrophic food chain, it is therefore unclear whether the stripped-down version has any advantages over the fully structured models in terms of providing more general insights.

3.4.4 | Overall Conclusions

In line with the strategy advocated by Murdoch *et al.* (1992), our results suggest that stripping down more complex models to simpler models, where the latter preserve the essential mechanisms of the former, is both a possible and useful route to take for the purpose of deriving generality in insights. This reduction in model complexity, for at least the consumer-resource and the tritrophic food chain models, generated new insights into the necessary conditions for a specific dynamics or phenomenon to occur. The question of whether the stripped-down versions resulted in more general insights is still difficult to answer. At the population level, this was usually the case, although it is noteworthy that all the different dynamics found in the stripped-down models had already been shown in the fully structured models. However, because the fully structured models also provided individual-level information, they gave general insights into how different individual life-history characteristics

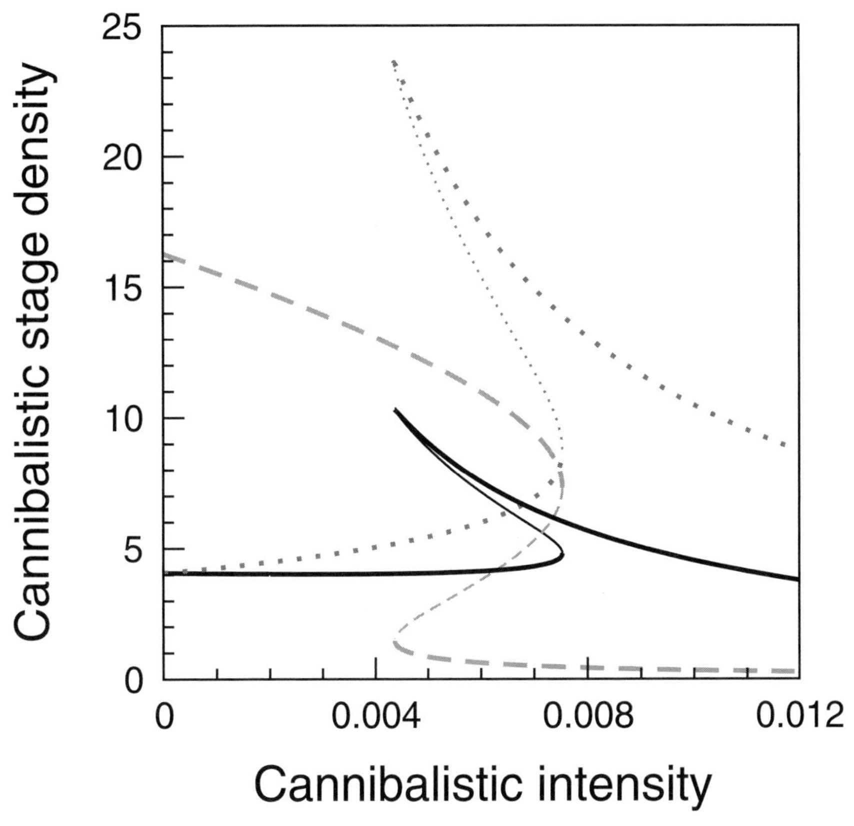

FIGURE 3.4 | Bifurcation plot for a three-stage cannibalism model showing juvenile (*J; black solid line*), subadult (*S; light-grey dashed line*), and adult (*A; dark-grey dotted line*) numbers as a function of the intensity, *c*, with which adults cannibalize on juveniles. Stage dynamics are governed by the equations $dJ/dt = \beta A - \mu_J J - \rho J - cJA$, $dS/dt = \rho J - \mu_S S - \phi S/(1 + dS^2)$, and $dA/dt = \phi S/(1 + dS^2) - \mu_A A$, respectively. In these equations, μ_i is the per capita death rate of juveniles, subadults, and adults, β is the per capita fecundity, ρ is the transfer rate of juveniles to subadults, and ϕ is a rate constant affecting the maturation rate of subadults. The recruitment from the subadult to the adult stage is assumed to depend on subadult density following a hump-shaped relationship, reflecting competition for resources (not explicitly modelled; De Roos, Persson, & Thieme 2003). The parameter *d* determines the degree of overcompensation in this relationship. The plot shows a shift from competition-controlled equilibrium states with high densities of subadults at low cannibalistic intensity to cannibal-driven equilibrium states with low densities of subadults at high cannibalistic intensity. These two domains are separated by a mixed regime with alternative states. Thick and thin line segments represent stable and unstable equilibria, respectively. Parameter values: $\beta = 0.08$, $\mu_j = \mu_s = 0.01$, $\mu_a = 0.03$, $\rho = 0.07$, $\phi = 1.0$, and $d = 0.5$.

affect dynamics, which the simpler stage-based models, by their nature, are unable to provide. Murdoch *et al.* (1992) described ways of simplifying complex models into models of more generality by deriving development indices of different complexity. Our analyses suggest that the parameters, which determine the ecological, intraspecific, and interspecific interactions that an individual is engaged in (α, β, and δ in our examples), are the crucial ones upon which to focus. This circumstance also means that, although the fully structured models are rich in parameters, the number of parameters that have a significant effect on population dynamics may be relatively restricted.

Overall, our three examples suggest that the fully structured models we used to study the dynamics of consumer-resource, cannibal-victim, and the tritrophic food chain systems in themselves do allow strategic and general questions to be addressed to a substantially larger extent than has been previously anticipated. Our results suggest that two reasons, and because only a few parameters are key, can be advanced for why insights of substantial generality may be gained from fully size-structured models. First, the modelling approach that we have used, PSPMs, does not follow each separate individual in time and space but rather lumps individuals into size distributions; thus, they represent intermediate models in the range from purely individual-based models to unstructured models (Metz & De Roos 1992, De Roos 1997). Second, and connected to the previous point, several mathematical tools have been developed during the last decade, including tools for equilibrium and bifurcation analyses, that substantially increase our ability to analyze the dynamics of more complex models beyond mere simulations. These issues are covered in the next section.

3.5 | AN APPROPRIATE MODELLING FRAMEWORK: PHYSIOLOGICALLY STRUCTURED POPULATION MODELS

Several attempts have been made to handle variation among individuals in population dynamic models. In most cases, the representation of individual variation has been restricted to population-level formulations without explicit consideration of processes at the individual level (cf. Lomnicki 1988). The approach we put forward in this chapter, however, requires the use of population dynamic models that explicitly account for individual organisms and that incorporate mechanistic representations of the life history processes and the ecological interactions they are engaged in. At least two modelling frameworks, known as PSPMs and individual-by-individual simulation models (IBSMs), allow for these features. PSPMs (Metz & Diekmann 1986, De Roos 1997), which have also been referred to as i-state distribution models (Caswell & Meridith-John 1992), are individual based in the sense that all modelling and hence all assumptions pertain to the individual life history. To this end, PSPMs are based on two different state concepts, the individual or i-state and the population or p-state. The i-state represents the state of the individual in terms of a collection of characteristic physiological traits such as size, age, and energy reserves, and the p-state is the frequency distribution over the space of all possible i-states. The model formulation process consists of deriving a mathematical description of how individual performance (growth, survival, and reproduction) depends on the physiological characteristics of the individual and the condition of the environment. Handling the population-level dynamics is subsequently just a matter of bookkeeping for all individuals in different states, and, importantly, no further model assumptions are made at this level. The core of PSPMs is thus the individual state and the model of the individual life history. Although bookkeeping provides the link from the individual to the population level, the reverse link is through population feedback on individual life history, behaviour, or both. For example, in a consumer-resource system the population influence on consumer life history operates through an increased or decreased density of resource, which affects individual growth, mortality, and reproduction.

The model of the individual life history, which forms the core of the PSPM, is deterministic and assumes that individuals born at the same time, with the same state at birth, will remain identical throughout their lives. This assumption implies that PSPMs incorporate less flexibility to describe the fate of an individual compared with IBSMs that account for each individual organism as a distinct entity and describe its dynamics by means of rules. The latter have also been referred to as "i-state configuration models" (Caswell & Meridith-John 1992). Despite their often daunting mathematical representation, PSPMs are hence intermediate in conceptual complexity between the unstructured, ordinary differential equation models that ignore individual life history and population structure altogether and the IBSMs. Moreover, compared with the often complex set of rules implemented in IBSMs, most PSPMs are based on a highly idealized representation of the individual life history, only accounting for basic processes such as feeding, growth, reproduction, and mortality.

An important advantage of the rigorous, although often intimidating, mathematical formulation of PSPMs is that it allows for the development of mathematical tools to analyze them. The dynamics predicted by

PSPMs can be studied using reliable numerical integration methods (De Roos, Metz, & Diekmann 1992; De Roos 1997). IBSMs require many replicates for the same parameter set to assess the ultimate dynamics, whereas the (deterministic) PSPMs only require a single numerical integration. In addition, there is a general problem of resolving whether alternative types of dynamics exist for the same parameter set. Theoretically, these could be found by starting numerical simulations or integrations from different initial conditions. In practice, however, the high degree of freedom in specifying the initial state prevents the detection of alternative attractors with this approach.

Fortunately, for PSPMs it is now possible to compute equilibria numerically without analytical manipulation and to carry out a bifurcation analysis (Kirkilionis *et al.* 2001; Diekmann, Gyllenberg, & Metz 2003; Claessen & De Roos 2003). For systems of differential and difference equations, methods to numerically locate an entire branch of equilibria have been available longer. These techniques allow ecologists to trace the value of the equilibrium when a single parameter is varied (Kuznetsov 1995; Kuznetsov, Levitin, & Skovoroda 1996). Both stable and unstable equilibria can be detected, as can the ranges of parameters in which the model predicts multiple equilibria to occur. Such numerical equilibrium and bifurcation analysis techniques more readily detect the occurrence of alternative dynamic attractors. With the same methods developed for PSPMs (Kirkilionis *et al.* 2001; Diekmann, Gyllenberg, & Metz 2003; Claessen & De Roos 2003), it is now possible to efficiently develop a full picture of the possible model outcomes, which otherwise would have required huge numbers of numerical simulations starting from many initial conditions. In addition, these techniques can also be used to study evolutionary dynamics, invasions of mutants in resident populations, and species divergence through evolutionary branching (Diekmann, Gyllenberg, & Metz 2003; Claessen & Dieckmann 2002). Their availability and the expectation that they will be developed further to investigate, for example, stability properties of the equilibrium (Kirkilionis *et al.* 2001) offer PSPMs a distinct advantage over the rule-based IBSMs. As a consequence, PSPMs may be more suited to yield general ecological and evolutionary theory about the interplay of individual life history and population or community dynamics.

3.6 | ON TESTABILITY

PSPMs also allow more extensive testing, both qualitatively and quantitatively, of model predictions against empirical data and make it possible to progress towards a closer interaction between modelling and empirical work. Two properties of PSPMs are responsible for the increased testing potential. First, comparisons between model predictions and empirical data have a higher discriminatory power because most parameter estimates are derived independently from the empirical data set and because assumptions and parameter values required as input exclusively pertain to individual-level processes. Second, model predictions include both population-level predictions (overall dynamics, cycle length, and amplitude) and individual-level predictions (growth, fecundity, and mortality). Specifically, individual-level processes are predicted in a population dynamic context, which, because of the population feedback, are somewhat independent of the assumptions upon which the models are based. Moreover, these predictions generally allow more stringent tests against empirical data than predictions about the population dynamics alone. The appearance of giants and dwarfs in the cannibalism model (see the section on cannibalism and Claessen, De Roos, & Persson 2000) is an example of an individual-level pattern emerging in a population-level context, which cannot be predicted from knowledge of individual growth capacities. The presence of giants in both model predictions and empirical perch data represents a particularly nice illustration of how physiologically structured models can be used to critically test model predictions (Persson *et al.* 2003). Physiologically structured models thus allow a much more critical testing of model predictions than is possible with unstructured, population-level models. This increased testing power is a direct result of the clear distinction between the individual and the population level, with their associated state concepts and the formalization of their interrelation in terms of bookkeeping and feedback. The distinction between individual and population also means that the generality of the predictions of physiologically structured models, in contrast to unstructured and stage-structured models, can be evaluated at two levels: the extent to which generality is present in terms of population dynamic insights per se and the extent to which generality is present in insights about individual size scaling effects.

3.7 | DISCUSSION AND CONCLUDING REMARKS

Without exception, ecologists will agree that population and community dynamics originate at the level of the individual organisms because it is their birth, growth, development, and death that change the population or community state. The level of the individual organism is hence the natural, basic scale to start modelling these dynamics. Developing theory and understanding about population dynamics relates to what Levin (1992) has described as the central problem in ecology: "the problem of pattern and scale". Explaining the population-level patterns requires the study of how these patterns originate from mechanisms at the scale of individuals, the study of how they change with the scale of description, and the development of laws for simplification, aggregation, and scaling (Levin 1992). In this chapter we have essentially argued that for too long population ecology has neglected these fundamental questions about the link between population-level patterns and individual-level mechanisms as the scientific discipline followed Elton in his abstraction of the population as the trophic unit. We have advocated an approach that allows addressing these questions systematically.

Elton's view has given rise to an approach of top-down modelling in which the dynamics of populations are defined in terms of populations and average states. This starting point incorporates the implicit assumption that variability among individuals in a population is not important because the average state suffices for an appropriate model description. We have challenged this *population-level approach* on several grounds. First, we have shown in all three examples that the simplification, aggregation, and scaling of a fully individual-based population model does not lead to the type of top-down models that ecologists would formulate when starting from a population-level perspective, primarily because the top-down models fail to account for individual growth and development and they do not capture the differences in ecological interactions that individuals in different stages of their life history are involved in. Judging from the examples described, we hypothesize that especially the plasticity of the individual life history through its dependence on environmental conditions (food) plays a crucial role in the dynamics of populations and communities. Second, the predictions of the top-down models may not apply to many of the natural systems we wish to understand, an argument we illustrated with the difficulty of identifying predator-prey cycles in natural populations. We have suggested that this discrepancy might occur because the supposedly general assumptions of top-down models, for example, predators eat prey, are in essence restrictive, representing *limiting cases*, because they mathematically translate into *all* predators eat *all* prey.

Top-down, population-level models are based on population-level, supposedly general assumptions, which are *a priori* formulated and not the outcome of some simplification or aggregation process. Our central thesis is that studying the dynamics and predictions of such models yields theoretical insights, but their status and relevance is unclear.

There is, in our opinion, no guarantee that the insights broadly apply to many systems. Just as likely, they might represent special, limiting cases with a small domain of application. These issues can only be resolved by starting with a bottom-up model of an ecological system and slowly progressing to a more aggregated, top-down representation by eliminating nonessential assumptions that do not relate to the generic features of the dynamics.

The same philosophy was already proposed by Murdoch *et al.* (1992; Murdoch & Nisbet 1996; Murdoch, Briggs, & Nisbet 2003), although these authors argued that simple, top-down models play an important role. Indeed, Murdoch, Briggs, and Nisbet (2003) elegantly show how a hierarchy of stage-based extensions of a Lotka-Volterra-type consumer-resource model generates insight about the influence of differences between individuals on population dynamics. The crucial role of the simple, top-down models here might be related, however, to that the authors largely focus on dynamic phenomena archetypical for such models: predator-prey cycles and the stabilization thereof. Murdoch, Briggs, and Nisbet (2003) even argue, “Simple models are not only useful, they are essential in understanding complex models” (p. 422). We doubt whether this latter point holds generally, a position supported by the case studies we have described.

For the consumer-resource interaction, the bottom-up model that we analyzed revealed a clear relationship between the scaling of individual competitiveness with their developmental state and the characteristics of the single-generation cycles, characterizing the population dynamics (Persson *et al.* 1998, Fig. 3.1). We subsequently verified these findings using a two-stage, juvenile-adult model in which the individual life history was highly abstracted (De Roos & Persson 2003). Similarly, the emergent Allee effect that we reported in a model with a continuous size structure of consumers (De Roos & Persson 2002) also occurs in a model that only accounts for three size classes (De Roos, Persson, & Thieme 2003) but includes overcompensation in the maturation process as an important mechanism. Finally, a comparable stage-based model, which included a juvenile-adult cannibalistic interaction, captured the biphasic dynamics observed in a more detailed, individual-based cannibalism model and clearly identified in a field population of perch. We expect these findings to apply in more general settings than the one in which they were first identified. In all three examples, we were able to derive simplifications of the more detailed, bottom-up models that preserved the dynamic or structural patterns of interest. These more aggregated models are, however, not easily derived by a modelling process that starts with population-level assumptions. In addition, in none of the examples did the stripped-down model play a crucial role in understanding and unravelling the mechanisms behind the observed population dynamics, although they provided additional insight about the necessary conditions for its occurrence in the consumer-resource and tritrophic case.

Even though we argue in favour of detailed bottom-up models, we do not argue for including everything known about a particular system in a model. Such praxis will lead to highly complicated models and will prevent us from gaining insight into the essential mechanisms. Modelling an ecological system will always involve making judicious choices about which aspects of a system are important and which are not. What we emphasize, however, is that the entities involved in the processes, the individual organisms, should be *explicitly* accounted for and that the mathematical representation of the ecological processes in the model should be mechanistic, even when these representations are greatly simplified. Necessarily, the resulting model will be more complicated than the top-down models, which are used most often to investigate ecological scenarios.

In practice, a modelling analysis of a particular ecological question may start with lumped stage-based models as a first step to gain some preliminary insight. However, bottom-up models with an explicit handling of individual processes are essential ingredients in the modelling process for evaluating the generality and relevance of the insights thus derived. We are fully aware that we propose a difficult approach and that the analysis of PSPMs is harder and requires more technical skills than the stability analysis of a model that is formulated in terms of differential or difference equations. However, it is perhaps too much to expect that systems as complex as ecological communities can be disentangled and understood with simple or even simplistic tools. Some minimum level of technical sophistication might be necessary given the complexity of the study objects.

In this chapter we focused on size structure because it exemplifies how the characteristics of individuals determine the type and strength of their ecological interactions yet change as a result of these interactions. Behavioural structure of a population could provide another such example, but research in this direction is virtually non-existent. Recent studies on spatial population structure have largely addressed questions about the link between individual-level mechanisms and population-level models, for example, by using pair-approximation and moment-closure methods (see Dieckmann, Law, & Metz 2000 for a review) to derive low-dimensional approximations to spatially explicit, individual-based simulation models. In a study comparing several of such approximation methods, Pascual and Levin (1999) showed that an approximation based on average values poorly captured the dynamics of the full, individual-based model but higher order methods performed significantly better. These findings stress the importance of accounting for variability among individuals. In addition, Durrett and Levin (1994) derived an exact population-level model from its individual-level mechanisms. The result, however, differed significantly from a nonspatial population-level model with additional diffusion terms, which would be the obvious choice when the modelling process started at the population level. Theoretical developments on spatial population dynamics have thus followed an approach that closely resembles the one we advocate in this chapter for dealing with population life-history structure. These developments have moreover proved some of the points (e.g., the importance of individual variability and of rigorously deriving population-level models on the basis of individual-level mechanisms) we made in the foregoing section.

Lastly, from a more philosophical point we have criticized the population-level approach *a priori* for denying the importance of variability among individuals. Growth and development, which leads to continuous changes in the state of individuals during their life history, is in our opinion a uniquely biological and ubiquitous process, which sets ecological systems apart from other physical systems. Mainly for this reason, we advocate starting with a complex, bottom-up model of the ecological system and simplifying this to a more aggregated representation, preserving the important dynamic patterns. This approach contrasts with the approach commonly applied in physics and chemistry, which first considers the simplest representation of a system. Compared with physical or chemical systems, the interactions among the elements (individual organisms) making up an ecological system are likely to be more complex and multifaceted.

More importantly, however, physical and chemical systems may consist of many types of entities or particles that react with one another, but these entities never exhibit the gradual changes that individual organisms go through during the course of their existence (life) as they grow and develop. The interactions that individuals are engaged in and that eventually are the origin of population and community dynamics strongly depend on the characteristics and properties of the individuals. At the same time, it is through these interactions (e.g., through foraging) that the individual characteristics change themselves. Hence, there is a two-way interplay between the individual state and the interaction mechanisms, which on the one hand are influenced by the individual state but on the other induce gradual changes in this state. We know of no physical or chemical system that embodies a similar type of intertwined dependence between the state of fundamental units and the interactions they are engaged in.

Because of the differences among individuals, a population is more than a one-dimensional object with its magnitude (abundance) as the only degree of freedom. Populations also have a shape (composition), for example, a size distribution, and are therefore infinite-dimensional with unlimited degrees of freedom. Changes in population composition are central to the findings described in the three case studies. For example, changes in population abundance and population composition may amplify each other for certain subgroups of individuals in the population and may dampen or even more than compensate one another for other subgroups. This is exemplified by our model (De Roos & Persson 2002), in which the net effect of a predator feeding on the smallest prey individuals is an eventual increase in the density of these smallest prey as overall prey abundance decreases, but the change in size distribution compensates for this decline in both the low and the high end of the prey size distribution. The extent to which the population and community patterns that ecology seeks to explain result from processes in which population abundances play the dominating role or from processes that crucially involve the population composition is the challenging question for the future.

ACKNOWLEDGEMENTS

We thank Bill Murdoch, Simon Levin, and the editors for their thoughtful and constructive comments, which helped improve this chapter considerably. This research was funded by the Netherlands Organization for Scientific Research, the Swedish Research Council, and the Swedish Research Council for Environment, Agricultural Sciences, and Spatial Planning.

REFERENCES

Adler, F.R., Richards, S.A., & De Roos, A.M. (2001). Patterns of patch rejection in size-structured populations: Beyond the ideal free distribution and size segregation. *Evol. Ecol. Res.* **3**, 805–827.

Bolker, B., Holyoak, M., Krivan, V., Rowe, L., & Schmitz, O. (2003). Connecting theoretical and empirical studies of trait-mediated interactions. *Ecology* **84**, 1101–1114.

Caswell, H., & Meridith-John, A. (1992). From the individual to the population in demographic models. *In* "Individual-based Models and Approaches in Ecology:

Populations, Communities, and Ecosystems" (D.L. DeAngelis & L.J. Gross, Eds.). Chapman & Hall, New York, pp. 36–61.

Chesson, P.L. (1990). Geometry, heterogeneity, and competition in variable environments. *Phil. Trans. Roy. Soc. Lond. B* **330**, 165–173.

Claessen, D., & De Roos, A.M. (2003). Bistability in a size-structured population model of cannibalistic fish: A continuation study. *Theor. Popul. Biol.* **64**, 49–65.

Claessen, D., De Roos, A.M., & Persson, L. (2000). Dwarfs and giants: Cannibalism and competition in size-structured populations. *Am. Nat.* **155**, 219–237.

Claessen, D., & Dieckmann, U. (2002). Ontogenetic niche shifts and evolutionary branching in size-structured populations. *Evol. Ecol. Res.* **4**, 189–217.

Claessen, D., van Oss, C., De Roos, A.M., & Persson, L. (2002). The impact of size-dependent predation on population dynamics and individual life history. *Ecology* **83**, 1660–1675.

De Roos, A.M. (1997). A gentle introduction to physiologically structured population models. *In* "Structured Population Models in Marine, Terrestrial, and Freshwater Systems," (S. Tuljapurkar & H. Caswell, Eds.). Chapman & Hall, New York, pp. 119–204.

De Roos, A.M., Leonardsson, K., Persson, L., & Mittelbach, G.G. (2002). Ontogenetic niche shifts and flexible habitat use in size-structured populations. *Ecol. Monogr.* **72**, 271–292.

De Roos, A.M., McCauley, E., & Wilson, W.G. (1991). Mobility versus density-limited predator–prey dynamics on different spatial scales. *Proc. Roy. Soc. Lond. B* **246**, 117–122.

De Roos, A.M., Metz, J.A.J., & Diekmann, O. (1992). Studying the dynamics of structured population models: A versatile technique and its application to *Daphnia*. *Am. Nat.* **139**, 123–147.

De Roos, A.M., Metz, J.A.J., Evers, E., & Leipoldt, A. (1990). A size-dependent predator–prey interaction: Who pursues whom? *J. Math. Biol.* **28**, 609–643.

De Roos, A.M., & Persson, L. (2002). Size-dependent processes promote the catastrophic collapse of top predators. *Proc. Natl. Acad. Sci. USA* **99**, 12,907–12,912.

De Roos, A.M., & Persson, L. (2003). Competition in size-structured populations: Mechanisms inducing cohort formation and population cyclos. *Theor. Popul. Biol.* **63**, 1–16.

De Roos, A.M., & Persson, L. (2005). The influence of individual growth and development on the structure of ecological communities. In "Dynamic Food Webs-multispecies assemblages, ecosystem development and environmental change" (P.C. de Ruiter, V. Wolters & J.C. Moore, Eds.). Elsevier, Amsterdam (in press).

De Roos, A.M., Persson, L., & McCauley, E. (2003). The influence of size-dependent life history traits on the structure and dynamics of populations and communities. *Ecol. Lett.* **6**, 473–487.

De Roos, A.M., Persson, L., & Thieme, H. (2003). Emergent Allee effects in top predators feeding on structured prey populations. *Proc. Roy. Soc. Lond. B* **270**, 611–618.

Diekmann, O., Gyllenberg, M., & Metz, J.A.J. (2003). Steady state analysis of structured population models. *Theor. Popul. Biol.* **63**, 309–338.

Dieckmann, U., Law, R., & Metz, J.A.J. (2000). "The Geometry of Ecological Interactions: Simplifying Spatial Complexity." Cambridge University Press, Cambridge.

Durrett, R., & Levin, S.A. (1994). The importance of being discrete (and spatial). *Theor. Popul. Biol.* **46**, 363–394.

Elton, C. (1927). "Animal Ecology." Methuen & Co., London.

Guldberg, C.M., & Waage, P. (1864). Forhandl. i Videnskabs-Selskabet Christiania, 35–45.

Gurney, W.S.C., Nisbet, R.M., & Lawton, J.H. (1983). The systematic formulation of tractable single-species population models incorporating age structure. *J. Anim. Ecol.* **52**, 479–495.

Hanski, I. (1999). "Metapopulation Ecology." Oxford University Press, Oxford.

Hassell, M.P., Comins, H.N., & May, R.M. (1991). Spatial structure and chaos in insect population dynamics. *Nature* **353**, 255–258.

Holling, C.S. (1966). The strategy of building models of complex systems. *In* "System Analysis in Ecology" (K.E.D. Watt, Ed.). Academic Press, New York, pp. 195–214.

Holt, R.D., & Polis, G.A. (1997). A theoretical framework for intraguild predation. *Am. Nat.* **149**, 745–764.

Kendall, B.E., Prendergast, J., & Björnstad, O.N. (1998). The macroecology of population dynamics: Taxonomic and biogeographic patterns in population cycles. *Ecol. Lett.* **1**, 160–164.

Kirkilionis, M.A., Diekmann, O., Lisser, B., Nool, M., Sommeijer, B.P., & De Roos, A.M. (2001). Numerical continuation of equilibria of physiologically structured population models: I. Theory. *Math. Mod. Meth. Appl. Sci.* **11**, 1101–1127.

Kuznetsov, Y.A. (1995). "Elements of Applied Bifurcation Theory." Springer-Verlag, New York.

Kuznetsov, Y.A., Levitin, V.V., & Skovoroda, A.R. (1996). "Continuation of Stationary Solutions to Evolution Problems in Content." Report AM-R9611, Centre for Mathematics and Computer Science, Amsterdam, The Netherlands.

Levin, S.A. (1992). The problem of pattern and scale in ecology. *Ecology* **73**, 1943–1967.

Levins, R. (1966). The strategy of model building in population biology. *Am. Sci.* **54**, 421–431.

Lewontin, R.C. (1964). "The Genetic Basis of Evolutionary Change." Columbia University Press, New York.

Lindeman, R.L. (1942). The trophic–dynamic aspect of ecology. *Ecology* **23**, 399–417.

Lomnicki, A. (1988). "Population Ecology of Individuals." Princeton University Press, Princeton, NJ.

Luckinbill, L.S. (1974). The effect of space and enrichment on a predator–prey system. *Ecology* **55**, 1142–1147.

MacArthur, R.H. (1972). "Geographical Ecology." Harper & Row, New York.

MacArthur, R.H., & Levins, R. (1967). The limiting similarity, convergence, and divergence of coexisting species. *Am. Nat.* **101**, 377–385.

May, R.M. (1974). "Stability and Complexity in Model Ecosystems," 2nd Edition. Princeton University Press, Princeton, NJ.

May, R.M. (1979). "Theoretical Ecology: Principles and Applications." Blackwell Science, Oxford.

McCauley, E., & Murdoch, W.W. (1990). Predator–prey dynamics in environments rich and poor in nutrients. *Nature* **343**, 455–457.

McCauley, E., Nisbet, R.M., Murdoch, W.W., De Roos, A.M., & Gurney, W.S.C. (1999). Large-amplitude cycles of *Daphnia* and its algal prey in enriched environments. *Nature* **402**, 653–656.

Metz, J.A.J., & De Roos, A.M. (1992). The role of physiologically structured population models within a general individual-based modelling perspective. *In* "Individual-based models and approaches in ecology: Populations, communities and ecosystems." (D.L. DeAngelis, L.A. Gross & T.G. Hallam, Eds.). Chapman & Hall, New York, pp. 88–111.

Metz, J.A.J., De Roos, A.M., & Van Den Bosch, F. (1988). Population models incorporating physiological structure: A quick survey of the basic concepts and an application to size-structured population dynamics in waterfleas. *In* "Size-structured Populations: Ecology and Evolution" (B. Ebenman & L. Persson, Eds.). Springer-Verlag, Berlin, pp. 106–126.

Metz, J.A.J., & Diekmann, O. (1986). "The Dynamics of Physiologically Structured Populations," Springer Lecture Notes in Biomathematics, Vol. 68. Springer-Verlag, Heidelberg.

Murdoch, W.W., Briggs, C.J., & Nisbet, R.M. (2003). "Consumer–Resource Dynamics." Princeton University Press, Princeton, NJ.

Murdoch, W.W., Kendall, B.E., Nisbet, R.M., Briggs, C.J., McCauley, E., & Bolser, R. (2002). Single-species models for many-species food webs. *Nature* **417**, 541–543.

Murdoch, W.W., McCauley, E., Nisbet, R.M., Gurney, W.S.C., & De Roos, A.M. (1992). Individual-based models: Combining testability and generality. *In* "Individual-based Models and Approaches in Ecology: Populations, Communities, and Ecosystems" (D.L. DeAngelis & L.J. Gross, Eds.). Chapman & Hall, New York, pp. 18–35.

Murdoch, W.W., & Nisbet, R.M. (1996). Frontiers of population ecology. *In* "Frontiers of Population Ecology: Essays to Celebrate the Centenary of the Birth of A.J. Nicholson." (R.B. Floyd, A.W. Sheppard, & P.J. De Barra, Eds.). CSIRO Publishing, Melbourne, pp. 31–43.

Murdoch, W.W., Nisbet, R.M., Gurney, W.S.C., & Reeve, J.D. (1987). An invulnerable age class and stability in delay-differential parasitoid–host models. *Am. Nat.* **129**, 263–282.

Nisbet, R.M., & Gurney, W.S.C. (1982). "Modelling Fluctuating Populations." John Wiley & Sons, New York.

Pascual, M., & Levin, S.A. (1999). Spatial scaling in a benthic population model with density-dependent disturbance. *Theor. Popul. Biol.* **56**, 106–122.

Persson, L., De Roos, A.M., & Bertolo, A. (2004). Predicting shifts in dynamical regimes of cannibalistic field populations using individual-based models. *Proc. Roy. Soc. Lond. B* **271**, 2489–2493.

Persson, L., De Roos, A.M., Claessen, D., Byström, P., Lövgren, J., Svanbäck, R., Wahlström, E., & Westman, E. (2003). Gigantic cannibals driving a whole lake trophic cascade. *Proc. Natl. Acad. Sci. USA* **100**, 4035–4039.

Persson, L., Leonardsson, K., De Roos, A.M., Gyllenberg, M., & Christensen, B. (1998). Ontogenetic scaling of foraging rates and the dynamics of a size-structured consumer–resource model. *Theor. Popul. Biol.* **54**, 270–293.

Pimm, S.L., & Lawton, J. (1978). On feeding on more than one trophic level. *Nature* **275**, 542–544.

Rand, D.A., & Wilson, H.B. (1995). Using spatiotemporal chaos and intermediate-scale determinism to quantify spatially extended ecosystems. *Proc. Roy. Soc. Lond. B* **259**, 111–117.

Rosenzweig, M.L. (1971). Paradox of enrichment: Destabilization of exploitation ecosystems in ecological time. *Science* **171**, 385–387.

Rosenzweig, M.L., & MacArthur, R.H. (1963). Graphic representation and stability conditions of predator–prey interaction. *Am. Nat.* **97**, 209–223.

Sinclair, A.R.E., Mduma, S., & Brashares, J.S. (2003). Patterns of predation in a diverse predator–prey system. *Nature* **425**, 288–290.

Tilman, D. (1982). "Resource Competition and Community Structure." Princeton University Press, Princeton, NJ.

Van Kooten, T., De Roos, A.M., & Persson, L. (2005). Bistability and an allee effect as emergent consequences of stage-specific predation. *J. Theor. Biol.* (in press).

van't Hoff, J.H. (1884). "Études de Dynamique Chimique." Frederik Muller, Amsterdam, The Netherlands.

Volterra, V. (1926). Variations and fluctuations of the number of individuals in animal species living together. *J. Cons. Per. Int. Ent. Mer.* **3**, 3–51.

4 | THE "STRUCTURE" OF POPULATION ECOLOGY: PHILOSOPHICAL REFLECTIONS ON UNSTRUCTURED AND STRUCTURED MODELS

Jay Odenbaugh

4.1 | INTRODUCTION

In 1974, John Maynard Smith wrote the following in his little book, *Models in Ecology*:

> A theory of ecology must make statements about ecosystems as a whole, as well as about particular species at particular times, and it must make statements that are true for many species and not just for one . . . For the discovery of general ideas in ecology, therefore, different kinds of mathematical description, which may be called models, are called for. Whereas a good simulation should include as much detail as possible, a good model should include as little as possible.

The aspiration of many population ecologists has been to devise minimalist models that describe general ecological patterns. They have worked hard at finding such minimalist models but not without difficulty. (There is difficulty in finding general ecological patterns as well.) Thus, some recent ecologists argue that population ecology theory should consist less of Maynard Smith's models and more of his simulations. De Roos and Persson (Chapter 3) argue that we must begin again. Since the 1980s, a "paradigm shift" has been occurring. In this chapter, I want to consider the rationale of one part of such a revolution and the search of general theory. I consider the case for building structured population ecological models to the near exclusion of unstructured models as presented by De Roos and Persson. First, I sketch what unstructured and structured models are in population ecology. Second, by way of clarifying and strengthening their argument, I contend that the traditional argument for trade-offs between realism, generality, and precision of models is problematic and I sketch what generality means in population ecology. Third, I argue that De Roos and Persson assume an unjustified form of mechanistic individualism and ignore pluralistic model building strategies. Thus, I suggest that they have not made their case for individual-based modelling over the traditional approach.

4.2 | MODELS, MODELS, AND MORE MODELS

Models in population ecology consist in two types—unstructured and structured. Unstructured models, or what are sometimes called p-state models, describe the population with stated variables such as population abundance or density N. In choosing such variables, we make an *identical-individuals assumption*—we assume that individuals can be treated as *nearly* identical. That is, we suppose that when they differ we lose little by aggregating or averaging those differences. Most of traditional theoretical population ecology consists of p-state models.

Structured models, or i-state models, come in one of two types. In *individual distribution models*, we relax the assumption of identical individuals. We represent differences among individuals by placing them in different classes with respect to age, sex, or size, for example. By grouping individuals by age, sex, or size, we are assuming that the individuals are identical within each class. This is an idealization, although one that is less capricious than the identical-individuals assumption. In *individual configuration models*, we represent each individual of the population and track their changes by simulating their behaviour computationally. Hence, in individual configuration models we find no reminisce of the identical-individuals assumption.

As noted previously, traditional population ecology consisted in largely unstructured population models as Hastings (Chapter 2) masterfully describes. This classic theory includes density-independent and density-dependent population growth, interspecific competition, predatory-prey, epidemic, and host-parasitoid models (see Hastings 1997). As one example of such models, consider the Lotka-Volterra predator-prey model. To derive the model, make the following assumptions:

- growth of prey population is exponential in absence of predators;
- predator declines exponentially in absence of prey;
- individual predators can consume an infinite number of prey;
- predator and prey encounter each other randomly in a homogenous environment;
- individuals in the predator and prey populations are ecologically and genetically identical, respectively.

So, if I let r represent the intrinsic growth rate of the prey, a represent the capture efficiency of the predator, b represent the conversion efficiency of the predator, and q represent the mortality rate of the predator, then I have the following model, where V is the prey population and P is the prey population:

$$
\begin{aligned}
\frac{dV}{dt} &= rV - aVP \\
\frac{dP}{dt} &= baVP - qP.
\end{aligned}
\tag{4.1}
$$

In effect, I have used a *law of mass-action* in deriving the model. The interactions between predator and prey are proportional to their respective abundance. However, ecologists know that encounters between predator and prey are not random and individuals are not identical. Much ink has been spilled in attempts to deal with the first assumption. The preceding Lotka-Volterra model was modified so that the prey grows logistically and the predator is satiated. Now there are lots of models that incorporate spatial heterogeneity, the most recent being metapopulation models and spatially explicit models involving diffusion equations (Tilman & Karieva 1997). However, the assumption of identical individuals has largely been ignored.

There are two important anomalies for models including this assumption. First, ecologists know that individuals are not identical in their ecological and genetic properties. Moreover, those properties change through time. As Darwin noted, and proponents of structured models remind us, variation is the rule not the exception. Thus, the identical-individuals assumption is patently and obviously false. However, we are accustomed to idealizations in models, so the question is whether this falsehood makes a difference and this takes us to the next problem.

Second, the bulk of predator-prey theory predicts predator-prey cycles. These cycles are, according to De Roos and Persson, rarely found in nature or at least outside the lab. This is just a special case showing that unstructured models have not been terribly successful empirically according to the critics. So why not relax this assumption of identical individuals? De Roos and Persson suggest that we do just that. Given that the identical-individuals assumption is false and that the models based on it are empirically deficient, they argue that we need to reorient model building in population ecology.

As an example of an individual distribution model, consider the following simple density-independent population growth model. Let $n_i(t)$ represent the number of individuals at time t in age class i. Suppose there are k age classes in the population. Thus, the age structure at time t consists of the following vector of abundances:

$$\mathbf{n}(t) = \begin{bmatrix} n_1(t) \\ n_2(t) \\ \vdots \\ n_k(t) \end{bmatrix}. \tag{4.2}$$

Using information from fertility and survivorship schedules from traditional life table analysis, we can predict how the age structure changes from $\mathbf{n}(t)$ to $\mathbf{n}(t+1)$. Let P_i be the probability that an individual in age class i survives to age $i+1$. Similarly, let F_i represent the average number of offspring produced by an individual of age class i. For example, if the population has k age classes, then we have:

$$\begin{aligned} n_1(t+1) &= F_1 n_1(t) + F_2 n_2(t) + \ldots + F_k n_k(t) \\ n_2(t+1) &= P_1 n_1(t) \\ n_3(t+1) &= P_2 n_2(t) \\ &\vdots \\ n_k(t+1) &= P_{k-1} n_{k-1}(t) \end{aligned} \tag{4.3}$$

Thus, for k age classes, then we have a $k \times k$ Leslie matrix:

$$\mathbf{A} = \begin{bmatrix} F_1 & F_2 & F_3 & \cdots & F_k \\ P_1 & 0 & 0 & 0 & 0 \\ 0 & P_2 & 0 & 0 & 0 \\ 0 & 0 & \ddots & 0 & 0 \\ 0 & 0 & 0 & P_{k-1} & 0 \end{bmatrix}. \tag{4.4}$$

Finally, we have the following population growth equation:

$$\mathbf{n}(t+1) = \mathbf{A}\mathbf{n}(t). \tag{4.5}$$

It is difficult to provide an example of an individual configuration model given that they are often written in code in a programming language of choice, so I will not attempt to present such a model. However, individual distribution models consist of matrix models, delay–differential equations in which the classes are discrete and time is continuous, or partial differential equations in which both are continuous (Caswell *et al.* 1997).

De Roos and Persson suggest that modelling in ecology should consist of several stages:

- determine the dynamic patterns of the system of interest,
- determine which features of these dynamics are generic—which form general patterns,
- identify the mechanisms responsible for those patterns.

Supposing that such general patterns exist and can be identified, they suggest that the mechanisms generating such patterns reside at the level of individual interaction and behaviour. For De Roos and Persson, this provides an account of the structure of modelling strategies. Instead of beginning with *top-down* unstructured models, we should begin with *bottom-up* structured models. These unstructured models are not then *complexified* by adding more realistic assumptions but are limiting cases of structured models in which individual variation is low or interactions among individuals are approximately random. They write:

> In our view, general theory can only be derived if we understand what is lost or neglected by not including specific aspects or mechanisms into a model. In other words, generality can only be achieved by selectively deleting details from more complex, mechanistic bottom-up models as opposed to by deciding *a priori* on what the essence of a system is and refraining from including any mechanism for the sake of being too specific.

Put a different way, we begin with a "richly mechanistic" structured model "and successfully simplifying it until it fails to produce realistic dynamics" (Judson 1994).

In the remainder of this chapter, I want to examine the case for structured population models. I will first try to clear some of the brush in favour of their defence. This requires returning to the issue of trade-offs in model building because many ecologists still believe there are necessary trade-offs among realism, generality, and precision (see Hastings in

Chapter 2 as an example). Put differently, we must choose between *strategic* and *tactical* models. This potentially begets problems for structured population models because they purport to be realistic, general, and precise. I argue that things look brighter than they did when Richard Levins first argued for this claim in 1966. Likewise, ecologists need to be clearer about what generality is and what makes a model general. There is a prejudice against structured models because it is believed that they are less general than unstructured models; once we are clearer as to what *generality* is, we will find that this is not so. However, there are two points that I believe count against De Roos and Persson's case. In effect, they make the assumption that general population-level patterns are generated by individual mechanisms and not by population-level mechanisms. This may be true, but they need an argument for this claim. Second, in assuming that we start with structured models and only move to unstructured models later, they ignore a pluralistic model building strategy that investigates unstructured, individual distribution, and individual configuration models. I claim it retains the strength of their own proposal and avoids its weaknesses (see Hastings in Chapter 2 as well).

4.3 | REVISITING MODELLING TRADE-OFFS

In 1966, Levins argued that there are necessary trade-offs among the generality, the precision, and the realism of ecological and evolutionary models. Ultimately, only two of these model properties can be maximized per model. These trade-offs are necessary because of the psychological and computational constraints of modellers and the complexity of the systems of interest. An optimally general, precise, and realistic model would require an enormous number of parameters in an enormous number of coupled partial differential equations. A model of this form would be analytically insoluble, and the terms of the equations would be unmeasurable and uninterpretable. The model would be of no use to scientists. So there are inescapable trade-offs among the generality, the precision, and the realism of the mathematical models if they are to be of any use to ecologists and evolutionists (see Orzack & Sober 1993, Levins 1993, and Odenbaugh 2004).

Levins own solution to this problem was to suggest modellers maximize two model dimensions at a time, giving his famous trichotomy of type I, type II, and type III models. We can devise models that are general and realistic but sacrifice precision, we can devise models that are general and precise but are not realistic, or we can devise models that are precise and realistic but not general. His predilections leaned towards type III models. However, he was a pluralist, recognizing that different strategies yielded different fruit. Historian of population ecology Sharon Kingsland (1985) writes:

> Levins was arguing here not only for the place of modelling in population biology as the only way to cope with complexity, but for the need to foster a pluralistic style of ecology to piece together a general theory of community structure from many sides. Of course he realized that the choice of different models or strategies would reflect conflicting goals and even conflicting aesthetic standards on the parts of biologists. For this reason he regarded disagreements about methods as basically irreconcilable. But he saw the alternative approaches, even of opposing schools, as partaking of a larger "mixed strategy" which would fit different pieces of the puzzle of community structure into a coherent whole.

In his view, theories in population biology consist of "families of models". Models should be articulated on the basis of each of the strategies removing the deficiencies of the others. Although I will explore this notion of a mixed strategy in the section on structural pluralism, I think there is room for resisting his conclusion that there is a necessary trade-off. The limitations of 1966 are not the limitations of 2004.

First, it does not follow that because ecological models have no closed-form solutions the models will be useless. It is well known in physics, chemistry, *and* biology that many models do not have analytically tractable solutions. However, this does not mean that we cannot use these differential or difference equations. Rather, it is our computational abilities that must be augmented. The most successful way of doing this is through computer-based numerical techniques (see Humphreys 2004 for information about computational science). In effect, we are *inductively* exploring the behaviour of models for a variety of variable and parameter values. We can then find regularities or patterns in the model's behaviour. This is the sort of analysis that Caswell and his colleagues (1997) recommend. In simulating structured models computationally, we must determine the asymptotic behaviour, determine the transient behaviour, and perform a perturbation analysis.

It would be unfair to blame Levins for overlooking these methods given that they were not available to biologists in 1966. These computational methods are all the more important because many of the obstacles facing structured models and individual-based models are computational limitations.

Second, I would agree with Levins that if a model is so complex as to be uninterpretable then it will be useless. However, I would deny that the equations in population biology need to be this complex. Levins would have us imagine a model with an enormous number of variables and parameters. Notice, however, that the individual distribution models under consideration do not have numerous distinct variables and parameters *types*. Even if these models consist of many equations, we would know what the equations *mean*. This is even more apparent because the number of equations crucially depends on the mathematical formalism chosen. In matrix algebra, we reduced k equations to a single equation in the preceding age-structure model. This objection has more force when we consider individual configuration models. There, parameters represent each individual and their properties. However, if we can summarize this information informally or formally, then it is possible to at least describe the assumptions of the models. Nonetheless, I am sure that the *non-transparency* of these models is one reason ecologists have been sceptical of individual-based models. Hence, I would claim that the problem of interpretation is manageable, contrary to Levins' argument.

Hastings writes:

> As emphasized in a variety of classic articles describing models in ecology, models are useful because they are simplifications. In any use of models in ecology or population biology, there is an important trade-off between simplicity and biological realism (Levins 1966). The advantages of biological realism incorporated into structured models may be offset by difficulties in matching to data and in proliferation of parameters, as well as by complications precluding complete analysis . . . An important conclusion is that continuing in the tradition started by Lotka and Volterra is not a dead end.

If there is a problem that makes mechanistically rich mathematical models extremely difficult to use in ecology, it is that the sheer number of variables and parameters are difficult to measure and estimate in natural systems. As we increase the realism of our models, the number of variables and parameters increases. Our models must be relatively system specific or phenomenological if our laboratory and field studies are as limited as Levins and Hastings would have us believe. This is what I believe to be the chief difficulty for structured and individual-based models.

4.4 | GENERALITY?

Ecologists love a general, testable theory. There are those who doubt the possibility of such a thing (Dunham *et al.* 1999). However, many important ecologists have claimed that science requires it, so we had better find it. Most famously, Robert MacArthur wrote in *Geographical Ecology* (1972), "To do science is to search for general patterns. Not all naturalists want to do science; many take refuge in nature's complexity in a justification to oppose any search for patterns. This book is addressed to those who do wish to do science." However, it is crucial to "suss out" exactly what a *general* theory or model consists of.

There are several ways in which a model can be general. Better yet, *generality* is a comparative term; hence, there are several ways in which a model can be more general than another model. First, here is a common definition implicitly used by ecologists: a model is more general than another model if the former applies to more systems than the latter (see Orzack & Sober 1993 for a similar proposal). Clearly, from a philosopher's point of view, this will not do because we must more carefully determine what *applies* means here. There are two ways to flesh out this notion of application:

> A model is more general than another model if the former *represents* a larger number of systems than the latter, or
> a model is more general than another model if the former *successfully represents* a larger number of systems than the latter.

There are two ways of construing "larger number of systems". If it simply means that one model represents a larger number of specific populations than another model, then we are tallying individual populations. However, this is not the only interpretation available. Consider consumer-resource models and host-parasitoid systems. Every host-parasitoid system is a consumer-system but not vice versa. Thus, "larger number of systems" can also be construed as "larger number of system *types*." Ideally, general models would accomplish both, but it is important to note the difference.

To determine what a model represents, we scout for the dependent variables. If the dependent variables are host population H and parasitoid population P, then the model represents host and parasitoid populations given the usual conventions. If the dependent variables are consumer population C and resource population R, then the model represents consumer and resource populations given the conventions. However, it is one thing to represent some system and it is another to successfully represent the population. To successfully represent a system I mean nothing more than *either* the assumptions of the model are approximately true *or* the model's dynamics bear a sufficient goodness of fit to the population through a time series or some other data set. Assumptions of models include the functional form of the equations, the number and type of variables, and the number and type of parameters. Statistical model selection criteria such as the Akaike and Bayesian Information Criteria allow us to evaluate models based on their assumptions *and* fit to data (see Mikkelson, unpublished, for an interesting description of Akaike Information Criteria in ecology). I believe that ecologists are most interested in successful representation and thus are most interested in generality as characterized in the last form.

What is important to recognize is that choosing variables or parameters has little to do with generating general models in the aforementioned sense. That the abundance N includes individual organisms does not guarantee an unstructured model will be more general than a structured model. It all depends on how successfully the model represents the phenomena of interest. To suppose that a model with more general variables (dependent variables represent broader system types) will be a more successful representation than a model with less general variables (dependent variables represent narrower system types) is just to assume *a priori* what surely must be determined *a posteriori*. Moreover, De Roos and Persson argue that the evidence suggests just the opposite, although I have not argued that here. By being careful about the sense of generality we use, we can at least clear the way for a defence of structured population models.

4.5 | REDUCTIONISM REDUX

Now I want to turn to worries I have about De Roos and Persson's proposal. They explicitly assume that the mechanisms that generate population patterns occur at the level of individuals. They write, "This sequence of steps [their modelling strategy] necessitates that the modelling of the population or community of interest starts at the level at which the dynamics originate: the level of individual organisms" (Chapter 3). From their point of view, individuals are born, die, migrate, develop, feed, forage, and reproduce after all, and *these* are the processes that determine population rates. However, De Roos and Persson simply ignore the possibility that there are population-level mechanisms that drive population change. There is a popular metaphysical account of levels of organization that agrees with De Roos and Persson that individual mechanisms give rise to population rates but also countenances population-level mechanisms. It is that account that I want to sketch here.

Philosophers are familiar with the idea of *multiple realization*. A higher-level property is multiply realizable if it can be realized by more than one lower-level property. An ecological example of such a functional property is that of a Simpson's species diversity, or $\Sigma_i \log p_i n_i$. It is a property of communities and is determined by the species richness and the relative abundance of those species. However—and this is the salient point—two communities can have the same Simpson diversity index yet have different relative abundances or different species richness (see Levins 1968 for a similar notion of a *sufficient parameter*).

I should also note that this does not mean that multiply realizable properties are *emergent* in some dubious sense. Given an appropriate set of lower-level properties, the higher-level property is determined. Philosophers often couple the notion of multiple realization with the concept of *supervenience*. One way characterizing supervenience is the following: a family of properties Q: $\{Q_1, Q_2, \ldots, Q_n\}$ supervenes on a family of properties P: $\{P_1, P_2, \ldots, P_n\}$ if, and only if, any two objects that have the same P properties also have the same Q properties; however, the converse does not hold. It is because of these multiply realizable properties that there can be generalizations about populations that cannot be explained simply by considering the individuals of the population. If there are population-level multiply realizable properties and those properties are causally related to other such properties, then there can be general patterns generated at the population level and explained by that level's dynamics. Thus, it would be a mistake to assume that all of the mechanistic work is occurring at the individual level.

Whether there are such broad, multiply realizable properties is an empirical issue, and I have not defended that possibility. To put this in a less metaphysical way, whether there are general population patterns that can be causally explained by population-level generalizations depends on whether such population-level generalizations exist. It may turn out that the contingencies of history and complexity (Sterelny 2001, Odenbaugh 2004) render any such generalizations false. However, there has been a resurgence of the view of ecological laws (Ginzberg & Colyvan 2004, Mikkelson 2004). MacArthur, an admirer of general patterns if there ever was one, was ironically sympathetic to such a position. He argued (MacArthur 1971) that there may be no population-level general patterns or mechanisms; rather, the general patterns and mechanisms occur at the level of the community:

> The question is not whether such communities exist but whether they exhibit interesting patterns about which we can make generalizations. This need not imply that communities are superorganisms or have properties not contained in the component parts and their interactions. Rather it implies simply that we see patterns of communities and that, at this stage

> of ecology, the patterns may be more easily related than the complex dynamics of the component species. There is nothing mysterious about this: the relations between the temperature, pressure, and volume of a gas were made into "laws" long before their molecular interpretation was known, and Mendel's laws were clear before his inferred genes attached to the visible chromosome.

Let me close this section with two implications of this view. First, if there are no multiply realizable properties at the population level, there will be no *general* population-level patterns. Thus, there will be no bottom-up explanations of such patterns. Even if there are population-level multiply realizable properties, it does not follow that there are causal relations among such properties. After all, the general population-level patterns could be epiphenomena of the mechanisms governing the individuals. Therefore, it does not follow that if there are general population-level patterns, there are population-level mechanisms. This leads to my final criticism.

4.6 | STRUCTURAL PLURALISM

De Roos and Persson argue that we should begin with bottom-up models and only try to devise top-down models when we have evidence that the identical-individuals assumption is roughly correct when encounters are approximately random. They claim that the top-down approach has been assumed *a priori*. We can understand this criticism in one of two ways. First, we can see those early theoreticians as believing that unstructured models *must* succeed independent of those models having been tested. One might have believed this on the basis of analogies with the success of physics and chemistry in making similar assumptions with the theory of gases and chemical reactions. Interestingly, Huston, DeAngelis, and Post (1988) argue that laws of mass-action do not always work in physics either. They claim that phenomena such as the "spontaneous onset of ferromagnetism near critical temperature, the condensation of water vapour into liquid droplets, and the development of galactic structure" are caused by strong local interactions and are not susceptible to identical-individuals assumptions. On the other hand, one might believe that unstructured models *might* succeed and no other available types of models can be offered that can fulfil the goals at hand. Imagine you are Vito Volterra or Alfred Lotka. You devise the simplest models you can and hope that nature cooperates. This was not only a *plausible* place to start but, given the computational limitations that existed, it was the *only* place to start. Thus, De Roos and Persson write,

"This sequence of steps contrasts strongly with the approach, in vogue, in which the essence of a particular system is decided upon *a priori* and invariably includes only population-level aspects" (Chapter 3). I think this would not be an apt criticism of Lotka and Volterra.

I suggest that the latter is the most charitable way of reading the history of these early efforts; thus, I think this was imminently reasonable for Volterra and Lotka to have done. Nonetheless, given that those limitations have been pushed back at least to some degree, we can now pursue bottom-up models.

However, should we not pursue *both* strategies at once? That is, given some phenomenon we want to understand or anticipate, why can we not allocate theoretical resources among unstructured, structured, and individual-based models? Why should we choose one of these model types as the 'null strategy'? Hastings agrees when he writes, "Despite the obvious need for and the importance of structured models in ecology, unstructured models focusing only on population sizes and ignoring space not only have played an important role in ecology but also will clearly continue to do so" (Chapter 2).

This mixed strategy in its simplest form would be to apportion theoretical resources among each model type equally. I suspect we should not do so because the unstructured models have been tried repeatedly. We are less likely to garnish radically new insights with unstructured models than with structured models. However, as Hastings argues, given the new tools of model selection, we can test unstructured models in powerful ways.

Consider a literal science fiction (see Kitcher 1993 and Strevens 2003). Suppose that the probability of success in explaining general population patterns is a function of effort with diminishing returns. You can think of effort consisting of the number of hours devoted to a model type or the number of researchers working on a model type per unit of time. In effect, the reward per unit of effort in exploring unstructured models has flattened, and we are at the steepest part of the reward curve with respect to the latter two types of models. Thus, we maximize change of success by apportioning more effort with respect to the latter model types than to the former. This I find to be a plausible argument; however, we should never put all of our eggs in one basket. Hence, the most efficient strategy seems to be to pursue the structured models but to continue work on the unstructured models.

In effect, I am recommending a mixed strategy that I believe is precisely what Levins had in mind. What is crucial is that we integrate these model types in an attempt to understand what ecological patterns we can discern.

4.7 | CONCLUSION

In this chapter, I have attempted to defuse the charge that structured models are problematic because they are realistic, general, and precise. I have also attempted to clarify the nature of generality that population ecologists pursue. Nonetheless, it appears that De Roos and Persson have not yet made their case for a mechanistic, individual-based modelling strategy. Their preferred strategy contains an unwarranted (or at least undefended) mechanistic individualism, and they have ignored a plausible pluralistic, mixed strategy that combines unstructured and structured models. Thus, I contend that the best "structure" for theoretical population ecology is still left open.

ACKNOWLEDGEMENTS

Bill Wimsatt and Ed McCauley both encouraged me to explore individual-based models and their ramifications philosophically as a graduate student. This investigation has been long overdue, and I thank them for their helpful advice. I also thank Kim Cuddington for helpful comments and patience as an editor.

REFERENCES

Caswell, H., De Roos, A., Nisbet, R., & Tuljapurkar, S. (1997). Structured population models: Many methods, a few general principles. *In* "Structured Population Models in Marine, Terrestrial, and Freshwater Systems," (S. Tuljapurkar & H. Caswell, Eds.). Chapman & Hall, New York, pp. 3–17.

Durham, A.E., & Beaupre, S.J. (1997). Ecological experiments: Scale, phenomenology, mechanism, and the illusion of generality. *In* "Issues and Perspectives in Experimental Ecology" (W. Regitarits & J. Bernerdo, Eds.). Oxford University Press, Oxford.

Ginzburg, L., & Colyvan, M. (2004). "Ecological Orbits: How Planets Move and Populations Grow." Oxford University Press, New York.

Hastings, A. (1997). "Population Biology: Concepts and Models." Springer-Verlag, New York.

Humphreys, P. (2004). "Extending Ourselves: Computational Science, Empiricism, and Scientific Method." Oxford University Press, New York.

Huston, M., DeAngelis, D., & Post, W. (1988). New computer models unify ecological theory. *BioScience* **38**: 682–691.

Judson, O. (1994). The rise of the individual-based model in ecology. *TREE* **9**, 9–14.

Kingsland, S. (1985). "Modeling Nature." University of Chicago Press, Chicago.

Kitcher, P. (1993). "The Advancement of Science." Oxford University Press, New York.

Levins, R. (1966). The strategy of model building in population biology. *Am. Sci.* **54**, 421–431.

Levins, R. (1968). "Evolution in Changing Environments." Princeton University Press, Princeton, NJ.

Levins, R. (1983). A Response to Orzack and Sober: Formal Analysis and the Fluidity of Science. *Q. Rev. Biol.* **68**, 547–555.

MacArthur, R.H. (1971). Patterns of terrestrial bird communities. *In* "Avian Biology," Vol. 1, (D.S. Farner & J.R. King, Eds.). Academic Press, New York, pp. 189–221.

MacArthur, R.H. (1972). "Geographical Ecology." Princeton University Press, Princeton, NJ.

Maynard Smith, J. (1974). "Models in Ecology." Cambridge University Press, New York.

Mikkelson, G. (2004). Biological diversity, ecological stability, and downward causation. *In* "Philosophy and Biodiversity," (M. Oksanen & J. Pietarinen, Eds.). Cambridge University Press, New York, pp. 119–132.

Mikkelson, G. (unpublished). Realism vs. instrumentalism in a new statistical framework.

Odenbaugh, J. (2003). Complex systems, trade-offs, and mathematical modeling: A response to Sober and Orzack. *Phil. Sci.* **70**, 1496–1507.

Odenbaugh, J. (2004). Ecology. *In* "Encyclopedia of the Philosophy of Science," (S. Sarkar, Ed.). Routledge, Oxford.

Orzack, S., & Sober, E. (1993). A critical assessment of Levins' "The Strategy of Model Building," (1966). *Q. Rev. Biol.* **68**, 534–546.

Sterelny, K. (2001). "The Evolution of Agency and Other Essays." Cambridge University Press, New York.

Strevens, M. (2003). The role of the priority rule in science. *J. Phil.* **100**, 55–79.

Tilman, D., & Karieva, P. (1997). "Spatial Ecology: The Role of Space in Population Dynamics and Interspecific Interactions." Princeton University Press, Princeton, NJ.

II | EPIDEMIOLOGICAL ECOLOGY

5 | THE LAW OF MASS-ACTION IN EPIDEMIOLOGY: A HISTORICAL PERSPECTIVE

Hans Heesterbeek

5.1 | INTRODUCTION

The law of mass-action as a paradigm for describing the contact rate of individuals in a population has been in widespread and heavy use in ecology and epidemiology for almost 100 years. The law roughly states that the rate at which individuals of two types, X and Y, meet is proportional to the product of the (spatial) densities of the respective subpopulations: $\propto xy$.

The law originates in the theory and practice of chemical reaction kinetics. The analogy between individuals meeting when moving around in space and molecules meeting when moving around in a gas or solution is intuitively pleasing. Moreover, the simplicity of the interaction term widens the possibilities to analytically study the behaviour of the systems of differential equations incorporating this description of the contact process. Both of these factors have undoubtedly been major determinants in creating the success of mass action as a modelling concept.

As with many paradigms, its relation to the process it supposedly describes is strained at best. In contrast to the situation in chemical reaction kinetics, there is little evidence that in ecology any form of contact among individuals abides closely to this law. If ecologists scrutinize what they are assuming about the behaviour of individuals and the population they constitute, then it is clear that no contact process follows these rules. It is, however, a forceful paradigm if it, despite all its flaws, has had such an effect on the development of ecological and epidemiological theory over a broad range of applications. It is surely one of the most powerful and most useful old metaphors of ecology. The best of these metaphors are easy to grasp, and through their use, profound and robust biological insights can be gained. Picasso once remarked: "Art is the lie that helps us to discover the truth." Ecologists can replace "art" with "modelling" and see the value of this statement when they realize that the law of mass action is the most basic of lies about the contact process in a population.

In this chapter I concentrate mainly on epidemiological mass action. These developments precede the introduction to ecology but not necessarily because one influenced the other. I will show how mass-action was introduced into the study of interactions between susceptible and infected individuals, who was responsible for this, and which of the various "discoveries" of this description of interaction was crucial for the success of the metaphor. I will show that, contrary to what most researchers in epidemic theory today have been led to believe—by chains of papers copying one another's references—the celebrated paper by Hamer from 1906 was not instrumental in the success of mass-action. Mass-action in epidemiology did not originate with Hamer but with Ronald Ross and Anderson McKendrick, who both arrived at the concept simultaneously but by different routes. Only one of these consisted of translating the original chemical definition into interaction among individuals.

Because of its simplicity, and because it made analytical results possible, mass-action has, from the start, made deeper forays into epidemic phenomena possible. Indeed, it can be stated that the metaphor made it possible for the theory to take off. A few particular strands of the mass-action story—the developments started by Ross and, mainly, McKendrick—set in motion a long chain of researchers writing papers and reacting to one another's ideas and led to a long series of developments and insights. In short, mass-action turned epidemiology into a science.

5.2 | CATO MAXIMILIAN GULDBERG AND PETER WAAGE

It is insightful to first go into the origins in chemical reaction kinetics. The law of mass-action was discovered by two Norwegians, Cato

Maximilian Guldberg (1836–1902) and Peter Waage (1833–1900): two brothers-in-law. A wealth of historical information on the law can be found in the book published in 1964 by the Norwegian Academy of Sciences to celebrate the centenary of the original paper (Bastiansen 1964, which includes a facsimile of the 1864 paper). Guldberg was a mathematician who also studied physics and chemistry, and he was, among other things, professor of applied mathematics at the University of Christiania (Oslo) from 1869. Waage studied mineralogy and chemistry, after becoming disappointed in medicine, and was professor of chemistry at Christiania from 1866 (for details about their lives see their entries in the *Dictionary of Scientific Biography*, Gillispie 1972, and the chapter by Haraldsen in Bastiansen 1964). The technical details that follow are taken from Lund and Hassel (1964).

Guldberg and Waage's work on chemical affinities led them to formulate a principle concerning the role of the amounts of reactants in chemical equilibrium systems. They first distilled the "law" from experiments (using barium sulphate and potassium carbonate and substituting to barium carbonate and potassium sulphate) and later gave it a mathematically exact formulation. For a chemical equilibrium of substitution reaction $AB + CD = AC + BD$, they found that, for a given temperature, the product of reactant concentrations is proportional to the product of reaction-product concentrations. The proportionality constant gives the force of the formation, the magnitude of which depends on the force of attraction between the reacting substances. They call this the *affinity coefficient*. Similar statements of such a law had been made before, but this was the first time that the law was expressed clearly. They presented their paper, "Studier over Affiniteten," in 1864 to the Norwegian Academy of Sciences; it was printed in 1865 in the *Forhandlingen i Videnskabsselkabet i Christiania*. Apparently (Bastiansen 1964) the presentation provoked no comment or question from the audience. Perhaps not surprisingly, the paper remained almost unknown for a long time. An extended version of their theory published three years later, although written in French, helped matters little because it appeared in a journal of Oslo University, which was not widely read. Only when Wilhelm Ostwald reconfirmed the law experimentally and presented it in a paper in 1877 did their discovery become widely known. Incidentally, Jacobus Henricus van't Hoff independently discovered the law in 1878, something which happened several times in ecology and epidemiology as we will see later in this chapter.

It is interesting that Guldberg and Waage originally formulated their law in a different way. If the concentrations ("active masses") of the reacting molecules are given by p and q and the affinity coefficient by k, then they concluded that the interaction should be described by:

$$kp^m q^n. \tag{5.1}$$

They had noted, experimentally, that doubling the masses of the different molecules did not always have a similar result in magnitude. Their reasoning was that in the neighbourhood of two molecules taking part in a reaction, there are other molecules exerting their attracting force. The theory works from the principle that the force between the two reactants is dominating in this arena. Later Guldberg and Waage concluded that these secondary forces could be of the same order of magnitude and they ascribe the difference between reactants to these secondary forces (Lund & Hassel 1964). They conclude that it makes more sense to describe the primary affinity by the simpler relation:

$$kpq. \tag{5.2}$$

It also seemed best to use additional terms for the secondary forces when necessary. This has an interesting parallel in the modern epidemic modelling literature, where mass-action has sometimes been modified to allow exponents different from unity for the densities of infected individuals and susceptibles. One problem with the approach is that a mechanistic basis seems to be lacking. Perhaps a detailed study of the chemical literature might generate a basis for the translation to interaction in populations.

5.3 | WILLIAM HEATON HAMER

Hamer (1862–1936) studied medicine in Cambridge and London. He became Medical Officer of Health of the London County Council in 1912, one of the most important positions in British public health, a position he held until his retirement in 1925 (Greenwood 1936).

Hamer had an interest in explaining epidemic curves (i.e., curves of disease prevalence/incidence against time) and was, like several authors before him at the end of the nineteenth century (e.g., see Ransome 1880), focused on periodic behaviour. To understand the developments, discussions, and papers in that period and the first two decades of the twentieth century, we must understand the context in which the study of epidemiology arose. Briefly, the period starts with the definite proof (in 1877 by Robert Koch and Louis Pasteur) that infectious diseases were caused by living organisms. When researchers first started collecting data in earnest (notably British statistician William Farr), they could clearly document epidemics and started constructing epidemic curves. These curves turned out to be surprisingly regular in shape. Trying to explain mechanistically the shape of these curves gave rise to many scientific studies in that period. For details see Dietz & Heesterbeek (2005), where we also explain in detail that data collection on infectious diseases, and its analysis, started well before the time of Farr: by Graunt in the seventeenth century and, in the eighteenth century, in relation to, for example, smallpox.

Basically researchers can distinguish two competing approaches to explaining the epidemic curve's shape; I called these *Farr's hypothesis* and *Snow's hypothesis*. They arose because of the limited knowledge about the living organisms causing disease and the confusion that early experiments caused because of this limited knowledge. Farr's hypothesis has its roots in the work of Justus von Liebig, who advocated a purely chemical basis for infection, and asserts that epidemics end because the potency of the causal "organism" decreases with every individual it passes through (Farr 1866). John Brownlee was its main mathematical exponent. Snow's hypothesis asserts that epidemics end because the epidemic runs out of "fuel," that is, the decreasing availability of susceptibles causes the outbreak to end (Snow 1853). Ross was the main mathematical exponent of Snow's hypothesis, and ultimately Brownlee gave up his efforts to counter this theory (see Fine 1979 and Dietz & Heesterbeek 2005).

Hamer was also convinced of the validity of Snow's hypothesis. Hamer published his first paper on the epidemiology of infectious diseases in 1896: a review of the most important contributions to the study of the epidemic curve and cyclic behaviour. Until then, the papers addressing periodicity in infectious disease did so systematically—directed by data—and speculatively but not in a theoretical mathematical setting. There was some mechanistic—more descriptive—reasoning, but this reasoning was not taken to the extreme of testing its consequences either quantitatively or qualitatively by analyzing mathematical models based on these mechanisms. In this respect, Hamer's work is different from that of the authors before him; Hamer applied mathematical reasoning in his 1906 periodicity paper but not, as you will see, from a mechanistic description of underlying processes. Hamer did not have a favourable view of the disciplines of mathematics or biometry, to put it mildly, nor do his other papers contain mathematical modelling.

In 1906, Hamer delivered the so-called Milroy Lectures of the Royal College of Physicians, London, which Whitelegge, among others, had given before him. This was an honour bestowed on Greenwood later. Hamer's three lectures were called "Epidemic Diseases in England: The

Evidence of Variability and Persistency of Type" (Hamer 1906). The papers predominantly described disease *type*, by which he meant a combination of virulence, ability to spread, and overall behaviour of the disease's epidemics; the precise definition, however, remains vague. In the third lecture on March 8, 1906, in what is basically a single page (p. 734) of a long paper, the paragraphs of interest appear. The argumentation and theoretical style is in disharmony with the flowing, talkative style of the rest of the papers, which are more like the nineteenth century writing on epidemic theory in England. To strengthen this disharmony, the vital parts of page 734 are printed in a smaller typeface (footnote style), as if it needed to be accentuated that here a rather long side-remark was being made that could be skipped if so desired. Little did he realize that this part of the manuscript would become the basis of his lasting claim to fame.

Hamer starts by stating: "The persistency of type displayed by measles and small-pox is quite remarkable. For that reason they afford specially promising material for study of short-term period waves." The simplest case, he continues, is that of the short-period waves of measles. He lists several observations that might be relevant:

1. "Explosions in towns occur commonly at about biennial intervals when the accumulation of susceptible persons is sufficient and the climatic and other internal conditions offer sufficiently small resistance . . . The mean seasonal wave shows two maxima."
2. "The problem (in the case of measles in a large community) is simplified [because] we are dealing with an obligatory parasite. Hence questions of saprophytic growth, of food outbreaks, &c, do not arise."
3. "Furthermore, one attack confers almost complete protection."
4. "Again, infection spreads from person to person, population being densely aggregated and new susceptible material added in sufficient quantity and with sufficient frequency to favour stable epidemic movement."

The aim of Hamer is to construct an epidemic curve based on several assumptions about transmission and to compare the resulting curve with measles data from London. He is interested in whether such a curve could at least be qualitatively similar to observed patterns. Unfortunately, neither the data nor a curve based on them is given in the paper. Presumably it concerns data from 1880 to 1884 as reported by Whitelegge in 1892. Continuing in the main text, he states: "I have taken the London figures and assumed a case-mortality of 2.5 per cent." He proceeds to explain how he arrived at an archetypal curve for the London measles data, averaging over the curves of an unnamed number of

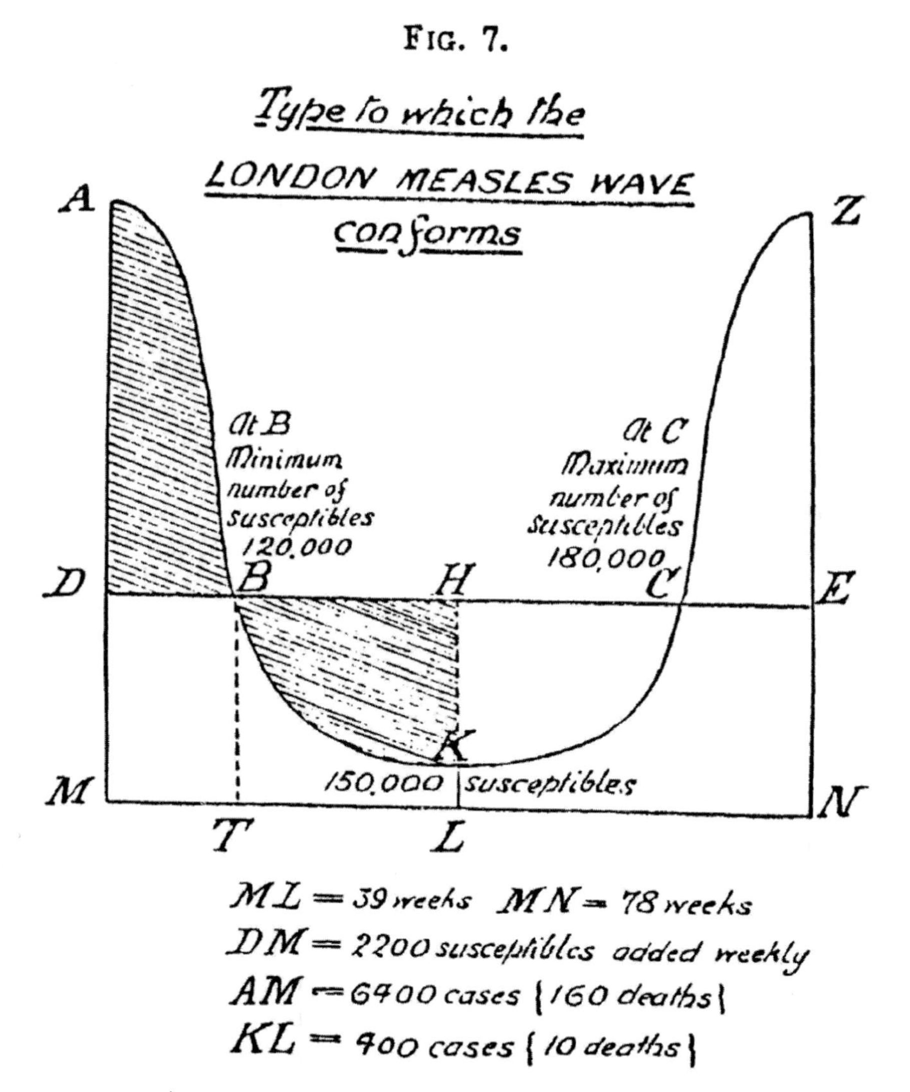

FIGURE 5.1 | Idealization of London measles wave by Hamer; see the main text for an explanation. Figure taken from Hamer 1906, Lecture III, page 734.

epidemic cycles. At this point in Hamer's text the small print starts. The curve is reproduced in Figure 5.1.

I will highlight the main ideas and refer to Figure 5.1 for the explanation of the various special points on the curve. The time-axis *MN* of the typical measles wave in Hamer's view equals 78 weeks. He takes the incubation period of measles to be 2 weeks and considers this the length of the time steps in his curve. He is interested in calculating the number of new cases arising in the next such interval from the cases in the present interval, i.e., discrete time incidence. In doing this he makes several important remarks. First, concentrate on the left half of Figure 5.1. Hamer notes that at points A and K the increase (decrease) in the number of cases changes into a decrease (increase):

> It will be apparent, therefore, that at A each case may be regarded as infecting one other case; this will also hold good at K.
>
> If the virulence of the measles organism and other factors be assumed to be the same at A and K, it will follow, inasmuch as each case is then capable of infecting one other case, that the number of susceptible persons in the population at those points of time will be identical.

Hamer notes that at points B and C it must hold that the number of new cases in the week-intervals containing these points equals the total number of new susceptibles added to the population: "Further, area ADB = area BHK, for the former represents the excess of persons attacked over susceptibles newly added in the time MT, and this must be equal to the excess of these newly added over those attacked in the time TL." He does a quick estimation using the area of the triangle ADB—knowing that at B the incidence is about 2200 and at A about 6400 with 14 weeks from point M to T—to arrive at an estimate of 30,000. Upon the assumption that the availability of susceptibles is the only mechanism at play—Snow's hypothesis—he comes to the heart of the matter:

> In the neighbourhood of B the cases fall in a fortnight from about 2500 to 2000. Assume the number of susceptibles at A to be x, the number at B will be $x - 30.000$. If the lessened ability to infect at B be solely due to diminution in number of susceptibles we may write
>
> $$\frac{x - 30.000}{x} = \frac{2000}{2500}$$
>
> i.e., $x = 150.000$ approximately.

Examine the crucial equation more closely, which Hamer neglects to do. In words, the relation states the following:

$$\begin{aligned}&[\text{cases in next interval}]/[\text{current cases}]\\&\quad = [\text{current susceptibles}]/[\text{susceptibles in which}\\&\qquad \text{one infects one}].\end{aligned} \tag{5.3}$$

This can be slightly rewritten as follows:

$$[\text{cases in next interval}] \propto [\text{current cases}][\text{current susceptibles}]. \quad (5.4)$$

Here, a proportionality constant is characteristic for the infection and the community and is taken to be the inverse of the steady-state susceptible population size when one infective gives rise to exactly one new case; in Hamer's words, the inverse of "the susceptible population when one infects one." Note that, in modern terms, Hamer shows great understanding here. If the infection is endemic in the community, then each case generates exactly one new case during its infectious period, that is, the reproduction ratio $R=1$. In a homogeneously mixed population with a constant force of infection, the basic reproduction ratio R_0 can be estimated as the inverse of the endemic susceptibles (e.g., see Diekmann & Heesterbeek 2000). The proportionality constant therefore equals the basic reproduction ratio. Because Hamer takes the infectious period to be of length 1 (i.e., takes the time step equal to the average infectious period), the proportionality constant equals the transmission rate constant of the basic SIR-epidemic model. Even though Hamer did not think in these terms, the words he uses clearly indicate that he should be credited with substantial insight into the matter.

Hamer also indicates full understanding of the problems in combination with the distinction between Snow's and Farr's hypothesis of epidemic decline. He writes, still in small print, the following:

> This examination shows the absurdity of assuming that an epidemic comes to an end because all the susceptibles have been attacked; or, again, of expecting to find explanation of the decline in loss of virulence of the organism or of its infecting power . . . The measles curve just defined sufficiently indicates that an epidemic may come to an end despite the existence of large numbers of susceptible persons in the population, merely on a "mechanical theory of numbers and density," and that the assumption of loss of virulence or infecting power on the part of the organism is quite unnecessary.

Because of its importance in chemistry, the law of mass-action had become a standard part of the education of many students in natural sciences and medicine at the end of the nineteenth century. Two key initiators in ecology and epidemiology, Lotka and McKendrick are known to have had access to such textbooks. Hamer, considering his fine education, could easily have come into contact with the law during chemistry courses. It is certain that he did so before 1906, because in his measles paper the words "mass-action" appear frequently (see a later section). However, contrary to what is commonly believed, nothing in Hamer's paper suggests that he made the link from chemical mass action to epidemic theory. Indeed, the words mass-action in his paper might have given the impression that he had made the link, but this is a misreading of Hamer's text. There are several arguments to support the view that Hamer failed to "put two and two together". First, he does not refer to mass-action or even chemical kinetics, however vaguely, in or near the small print in which his theory is unfolded, whereas "mass action" occurs frequently later in the same paper in a different context. Second, note that Hamer formulates his relation in vaguer terms than I have given previously and without the verbal interpretation that Soper would give to it 23 years later. Hamer introduced his relation in the direct neighbourhood of two particular points on his curve, *B* and *C*, solely to calculate the minimum and maximum number of susceptibles present during the passing of the epidemic wave. He does not, contrary to Ross, McKendrick, and Soper (see the next sections), introduce his relation based on assumptions that describe how individuals make contact and how frequent that contact is.

A third and important argument follows from the context in which Hamer *does* use the words "mass action." This is a context far removed from the study of measles periodicity. Following his brief theoretical small-print excursion, the remaining two-thirds of Hamer's third lecture is devoted to a purely epidemiological discourse on persistence and variability of type. On page 737 he embarks upon a description of the emerging ideas concerning "immune bodies, in number untold, each specific to its particular toxin." He predicts that the questions related to "bacilli, immune bodies, &c [will receive] a good deal of further attention in the near future." He then describes the "conception of the bacillus as a kind of host with attached 'enzymes.'" These particular "enzymes" (Hamer's quotes) would be capable of attacking the variety of sugars that could be fermented by bacteria. He writes: "We may regard the enzyme . . . as in a condition of equilibrium, influenced equally but in opposed directions by two 'mass actions.'" Later he uses the phrase "mass-action-equilibrium point" several times and he draws analogies between the enzyme example and immune reactions. Of the previous object of his study, however—where the mass-action concept would also be appropriate—no trace is left. Towards the end of the paper Hamer becomes increasingly involved in his argumentation against bacteriological epidemiology, maybe even up to a point where he would be unable to make the connection that Soper in 1929 and others later supplied. For more information on Hamer's life and work I refer to the obituary by Greenwood (1936).

5.4 | RONALD ROSS AND ANDERSON McKENDRICK

Even though Hamer's paper is almost invariably credited with introducing chemical mass action into epidemiology, we have seen that Hamer did not realize he had done so. More remarkably, the credit does not hold in a more fundamental sense: The researchers that, with hindsight, turned out to be the main players in shaping epidemic theory in the early phase of the nineteenth century show no knowledge of Hamer's paper. Ross and McKendrick never refer to Hamer in their work. Although it is true that in those days precise references, in the way we are accustomed to today, were scarce, both Ross and McKendrick are positive exceptions in that they consistently at least give the names and in most cases the journals and page numbers of the literature they use or describe. Both Ross and McKendrick were medical doctors, and it is therefore not unlikely that they had regular access to *The Lancet*, the journal that published Hamer's three papers. Ross and McKendrick performed much of their duties abroad (India, Mauritius, Africa, Sierra Leone), however, and their main interest was in diseases and infections from the tropics. Of McKendrick, for example, it is known that in 1905 and 1906 he was at the Pasteur Institute in Kasauli, India. Even if Ross and McKendrick had seen the papers, it is possible that they did not read the crucial part three because the first two instalments, dealing as they did with measles, are unlikely to have caught their attention. Hamer's paper did not start modern epidemic theory.

If it was not Hamer, who then introduced mass-action into modern epidemic theory? Ross (1857–1932) introduced his "theory of happenings" in the appendix to his famous book *The Prevention of Malaria*, but only in the second edition. Later he referred to the theory as the field of "*a priori* pathometry" (Ross 1916). For a description of the life and work of Ross see, for example, the most recent of several biographies devoted to him (Nye & Gibson 1997). None of the many sources on Ross, however, celebrate in detail his accomplishments as one of the two founding fathers of modern epidemic theory (the other being McKendrick, as explained later). A positive exception is the paper by Paul Fine (Fine 1975). For some details of his later work and its role in the genesis of the basic reproduction ratio, see my article (Heesterbeek 2002). Ross was the first to present a mechanistic theory of epidemics without having a specific infectious agent in mind. He used the term *a priori* to signify that he first makes assumptions about the way infections spread among individuals (through various "happenings"). This contrasts with a widely used approach at that time, for example, in the elaborate work of

Brownlee (see Fine 1979), of trying to gain insight into the phenomenon of epidemics by studying incidence curves of past outbreaks. The theory of Ross from 1916 is based on the 1911 appendix, but there the modelling uses discrete time evolution, whereas in 1916 he switches to ordinary differential equations (and in later papers with Hilda Hudson also integral equations, laying the groundwork for Kermack and Anderson McKendrick). In the 1916 version, Ross writes (p. 208; cf. p. 656 of Ross 1911): "The problem before us is as follows. Suppose that we have a population of living things numbering P individuals, of whom a number Z are affected by something (such as a disease), and the remainder A are not so affected; suppose that a proportion $h.dt$ [sic] of the nonaffected become affected in every element of time dt."

Transmission is thus described as $-hdt.A$ [Ross's notation]. Ross analyses various choices in the resulting systems of equations for P, A, and Z (which also include terms for recovery, birth, death, and migration), and he first devotes ample space to the cases where h is constant. He considers his theory widely applicable because for the constant case he has in mind "such happenings as many kinds of accidents and noninfectious diseases due to causes which operate, so to speak, from outside." Then on page 220 (p. 666 in 1911) he starts the analysis of the case of *dependent happenings* and first considers a *proportional happening*. To the class of dependent happenings, he considers to belong "infectious diseases, membership of societies and sects with propagandas, trade-unions, political parties, etc., due to propagation from within, that is, from individual to individual." For proportional happenings he then writes $h = cZ$. Because he has $Z = P - A$, transmission is then described as $-cAZ = -cA\ (P - A)$. We see that Ross indeed introduces mass-action in this paper, as one of the possibilities of describing transmission, but like Hamer he shows no awareness of this. Ross's mass-action comes as a natural step in the theory he is developing and not as a translation of chemical reaction kinetics.

Incidentally, in the 1917 paper by Ross and Hudson, the follow-up to Ross's 1916 paper, the authors extend the theory by allowing infectious individuals to recover into an immune state and present the model that is nowadays commonly referred to as the *SIR model*. In modern literature, the seminal paper of Kermack and McKendrick from 1927 is usually credited as the first description of the SIR model (even though they only used it as a special case of a more general model in terms of integral equations). Ross and Hudson, however, were also not the first. The credit for the SIR model, that is, the mass-action transmission term and an exponentially distributed infectious period after which the individual becomes immune to renewed infection, should go to McKendrick alone in a paper published in 1914. That paper is important for another reason: its main subject is the first clear description of an age-structured model. In a paper from 1926, McKendrick also introduces the first description of the stochastic SIR model (reprinted in Kotz & Johnson 1997 with a commentary by Dietz). It is time to look at McKendrick in more detail.

McKendrick (1876–1943) was, with Ross, by far the most prolific and influential of the pioneers in epidemic theory. For a description of his life and work (and an almost complete list of 50+ publications) see the obituary by W.F. Harvey (1943) or the chapter by Aitchison and Watson (1988) in a book about the influence of Scottish medicine. McKendrick was, like Ross, a medical doctor and a self-taught mathematician. It is without doubt that McKendrick was heavily influenced by the older scientist Ross in his interest in applying mathematical reasoning to medical problems. McKendrick served under Ross during an antimalaria campaign in Sierra Leone and they returned home together by boat in the summer of 1901. Correspondence between the two men exists in the Ross Archives at the London School of Hygiene and Tropical Medicine. From this it is evident that Ross valued highly his own efforts to establish a general mechanistic mathematical theory of epidemics and that he saw in McKendrick the perfect student to carry further what he was starting. In 1911 he wrote in a letter to McKendrick: "We shall end by establishing a new science. But first let you and me unlock the door and then anybody can go in who likes."

During their stay in Sierra Leone and their trip home, Ross must have advised several books to study, because McKendrick thanks him for this in a letter in late 1901. It is not very likely that the book, *Higher Mathematics for Students of Chemistry and Physics, with Special Reference to Practical Work* by Joseph William Mellor, was also on Ross' list (the first edition appeared in 1902). This book was to become McKendrick's "bible" for learning mathematical techniques. It is known that he studied it in great detail while he was stationed in India. The book, which went through many editions in several decades, contains a detailed exposition of the theory of chemical reaction kinetics as it had unfolded by then and included the original chemical law of mass-action. In the introduction, where Mellor lists several uses of mathematics in physical chemistry, he writes: "Wilhemy's law of mass action prepares us for a detailed study of processes of integration" (Mellor 1905). Wilhelmy studied monomolecular reactions. Mellor presents chapters on differential and integral calculus, probability theory, Fourier's theorem, calculus of variations, infinite series, and differential equations. In bold face he remarks in the latter chapter: "A differential equation, freed from constants, is the most general way of expressing a natural law." The author proceeds to treat chemical reactions of the second order in detail and mentions Guldberg and Waage. He writes: "When the system contains $a - x$ gram molecules of acetic acid it must also contain $b - x$ gram molecules of alcohol. Hence $dx/dt = k(a-x)(b-x)$."

It is therefore not surprising that McKendrick put "two and two together" and considered it natural to use the chemical description as a metaphor for interaction between susceptible and infectious individuals. In 1910 McKendrick worked on a mathematical theory of "serum dynamics", by which he meant the "multitude of phenomena of widely varying character" such as agglutination, bacteriolysis, and haemolysis. He then wrote (McKendrick 1911): "It is the object of this paper to show that the above reactions are subject to the law of mass action." The content of the paper is not of importance to epidemic theory directly, but it shows how McKendrick thought about modelling interaction phenomena. That he considered the law of mass-action of universal use becomes clear in a paper read before a conference on malaria. Where Ross initially used discrete-time systems for his theory of happenings, McKendrick, under the influence of Mellor, used differential equations from the start. Ross turned to differential equations in a paper on malaria in *Nature* in 1911 (this paper contains both the discrete time and the continuous time formulation), and McKendrick remarks in the conference proceedings in *Paludism* in 1912 that "he was glad to see . . . that his [Ross's] final results, as there expressed in the form of differential equations, were the same as his own." He does not follow in Ross's footsteps and ignores the approach through "happenings". He writes:

> I propose to develop the subject on my own lines as the argument used is easier to understand and I think more convincing. The mathematical classification of epidemic diseases is a very different classification to that arrived at from the medical standpoint. I will not deal with this in detail here, but I will in the first instance consider a type of epidemic which is spread by simple contact from human being to human being and in which there is no recovery rate. An example of this would be an epidemic of itch in a fixed population of, say, guinea pigs.
>
> The rate at which this epidemic will spread depends obviously in the number of infected animals, and also on the number of animals which remain to be infected—in other words the occurrence of a new infection depends on a *collision* [my italics, H] between an infected and an uninfected animal. If y be the number infected, and a be the total popula-

tion, then $a - y$ = the number uninfected. If we denote rate of increase of infected by the symbol dy/dt then we have *at once* [my italics] the equation

$$\frac{dy}{dt} = ky(a - y).$$

Where k is a factor which measures the chance of infection, it includes degree of dispersion of individuals, degree of intercourse, and the chance of transmission, etc.

The use of the word *collision* clearly shows that McKendrick considers this interaction to be similar to chemical mass-action (later in the paper he refers to it as "the argument of collisions"), and the words *at once* indicate that he considers this analogy to be a natural one. This seems to be the first clear statement of chemical mass action as applied to the interaction among individuals in the epidemiological-ecological context. Given that, in addition, it was McKendrick's work, rather than Hamer's, that influenced the evolution of epidemic theory as we know it today, I feel future credit for introducing mass action into epidemiology and ecology should go to McKendrick.

In addition to Ross and Hudson's and McKendrick's work, there was another important paper in which mass-action was used in an epidemic, or rather endemic, model. In 1921 Martini published a paper in which he used the mass-action formulation for an infection leading to immunity, where births and deaths of hosts are taken into account. The resulting equations were analyzed by Lotka (1923) and treated in his influential book (Lotka 1925). Lotka's initial exposure to epidenic theory came from reading the work of Ross.

5.5 | HERBERT EDWARD SOPER

There was at least one scientist with a mathematical education who did familiarize himself with Hamer's paper. Soper (1865–1930) was originally an engineer who became interested in statistics while studying under Pearson in London. In 1922 he published a book, *Frequency Arrays*, on generating functions (for details about this book and a few details about the life of Soper, see the obituary by Greenwood in 1931), and shortly after he started work as a mathematician in Brownlee's department at the National Institute of Medical Research. The aim was that he should bring some structure to biometry and "give mathematical form to

Brownlee's own doctrines" (Greenwood 1931). Brownlee had little interest in others' work; as a consequence, Soper did not receive much encouragement to develop his own views on epidemic theory. Given the content of his 1929 paper, it is clear that there would have been ample material for discussion because Soper's view of epidemics was, like Hamer's, based on Snow's hypothesis rather than Farr's hypothesis, of which Brownlee was the main advocate. It was only after the death of Brownlee in 1927, when Soper nominally passed to Greenwood's department in the same institute, that Soper received the freedom and credit he deserved. In Greenwood's department he wrote his important paper on measles periodicity, to be published in 1929. Unfortunately, there was little time to further develop his theory, because Soper died shortly after that publication, on September 10, 1930.

Although Soper's paper appeared more than 10 years after the Ross-Hudson papers, it should be treated independently because Soper does not show any knowledge of the earlier abstract work. Ross is not mentioned, either by Soper or by any of the discussants to the paper (described later in this chapter). In addition, Soper does not appear to have had any knowledge of the first Kermack-McKendrick paper, published 2 years earlier, or of the elaborate preceding work by McKendrick. It might, like in the case of Ross and McKendrick remaining oblivious of Hamer's work, be simply a case of working in two fields perceived as different: tropical infections that mainly affected the colonies and childhood infections that still affected the United Kingdom. Had Soper read any of these papers he could have formulated a large part of his theory less awkwardly.

Soper's aim is to "adopt the simplest mathematical postulate that would describe in a first measure the generally accepted mechanism of epidemic measles, if the accumulation of susceptibles were really the prime factor" and then to "compare the deduced results with the observed facts" and, if necessary, "to modify the primary hypothesis". In a nutshell, Soper describes a philosophy for research in epidemic theory in that he combines inferences from dynamic modelling with data to scrutinize the assumptions underlying the original, simple model, concluding that they need adjustment and then improving predictions by trying a slightly less simple version of the model. Soper was probably the first with this approach in the English literature; Ross and Hudson do not apply their theory to any data and Kermack and McKendrick merely use their single example as an illustration; they do connect to data in their later papers. Effectively, the honour for the first to devise an abstract mathematical model, to confront it with data and then to modify the model, should go to Piotr Dmitirievich En'ko (1844–1916), but unfortunately he published in Russian (in 1889). See Dietz (1988) for a review of his work and En'ko (1989) for a translation by Dietz of (the main part of) the paper from 1889.

Soper is humble in that he states he is "merely following up the trail blazed by Sir William Hamer . . . only in detail departing from his methods." The hypotheses on which Soper starts his investigation are clearly stated. He assumes that in a certain population there is "a perpetual flow of susceptibles possessing three characteristics, *viz.* (1) an equal susceptibility to a disease prevalent in the community, (2) an equal capacity to transmit the disease according to a law, when infected, and (3) the property of passing out of observation when the transmitting period is over." The chosen "law" is the key element that determines the course of the epidemic. Soper clearly wants to separate the law in two distinct processes. The first concerns the infectious period and infectivity, and the second concerns the opportunities for transmission. For the infectivity part, he presents a complicated-looking argument, which comes close to Kermack and McKendrick's idea of a general infectivity function. In essence he states that the "power of transmitting infection is some function of the lapse of time from a definite infection instant" and he illustrates this by drawing the typical unimodal function of measles infectivity. Soper initially chooses the extreme view that "all infecting power is concentrated" at the "definite end" of the incubation period, constituting "an instantaneous power of reinfecting". Later in the paper he assumes a geometric distribution for the infectivity function, expressing the long-popular alternative of a constant infectiousness for an exponentially distributed period.

Having concretized the infectivity, Soper turns to the contact opportunities: "The instant or point infection law being accepted, it is next assumed that a process analogous to 'mass action' governs the operations of transmission and that, other things equal, the number of cases infected by one case is proportional to the number of susceptibles in the community at the instant." The time unit chosen is 2 weeks. If zdt are the cases arising per unit of time, Soper expresses Hamer's "formula" as:

$$\frac{z_{1/2}}{z_{-1/2}} = \frac{x}{m}, \tag{5.5}$$

where $z_{1/2}$ is the number of cases in the unit interval succeeding the present instant. Unlike Hamer, Soper tries to give more meaning to the parameter m. Like Hamer, he refers to it as the "steady-state" number of susceptibles in which "one infects one". However, he then states: "Since the synthetic epidemics to be made do not depend on absolute sizes of communities, we relate m and a by the quotient s defined by $m = sa$ and think of a community as characterized by a time element s . . . The space of time is a measure of the 'seclusion' of the community, a large arguing few interminglings of the sort that conduce the infection." Here we can see that both Hamer and Soper had in mind that the constant should be a measure of the contact opportunities between susceptibles and infectives in a community.

By assuming that $a = 1000$ per 2-week interval, and taking $s = 20$, 30, 40, and 50, Soper generates a "series of epidemic curves showing all the features found by [Hamer] among them the asymmetry. I do not find any damping, and the curves appear to repeat themselves precisely." He proceeds to calculate the period of the oscillations and finds that it is $2\pi\sqrt{s\tau}$ "under the laws taken", in which τ is the length of the incubation period. For the London data that Hamer used, Soper is reasonably satisfied with the results. However, "the Glasgow curve of measles cases 1888–1927 . . . [does] not show anywhere the simple form of single repeated wave, and Glasgow appears to be rather different from London . . . in the course taken by its measles epidemics."

The only way in which he could create something similar to Hamer's composite curve was to average the epidemic over six consecutive 2-year periods, but the resulting curve still showed "a large winter peak and a small summer forepeak". Although retaining his assumption that the infectious period is a point event, Soper investigated two additions to his model to see whether the Glasgow curves could be recreated. First, he introduced a factor k_θ "representing the influence of the season θ . . . such as might be brought about by school break-up and reassembling." In addition, he took the inflow A (equal to a, which is part of the constant m) of new susceptibles in each time interval (incubation period) into account. He obtained:

$$\frac{z_{1/2}}{z_{-1/2}} = k_\theta\left(\frac{x+A}{m}\right), \tag{5.6}$$

where the incubation period was chosen as 1 month for convenience (Soper raised the left-hand side of the equation to the power i, which is the incubation period in months; he referred to the expression with $i = 1/2$ as the "infectivity"). He was not happy with the "makeshift" procedure he devised for estimating A, i, and k_θ from data and was—after detailed calculations of periods and resulting curves—not happy with the final predictive power of his equation: "The law of propagation of the disease of measles . . . is therefore not quite so simple that we can get good forecasts merely by premising a uniform inflow of susceptibles, who will all take the disease, and an infectivity depending on the accumulation of such susceptibles and on a season factor." Several factors are responsible for the poor quality of the predictions, according to Soper. Among them is, for example, the assumption of "an even inflow of susceptible persons who will become registered cases" and that "susceptibility cannot be thought of as a unitary character, but certainly varies with age". The most important point, however, "is, perhaps, a false analogy between infection in disease and the mechanism understood under the name of chemical mass action." Soper continues:

> Apart from the great differences in the statistical numbers dealt with in the two fields, in a liquid the intimate uniformity of the mixture and the conditions of intermingling and collision are likely to be more law-abiding than are similar traits in a community of persons. After all, the contacts are comparatively few and are subject to volition as well as to chance and, in addition, a single community may be quite differently constituted in respect to its seclusion or minglings in one part or another.

There may be different degrees of mingling in different sections of the community or depending on the season. Even if "something approaching the mass-action law of infection" is appropriate in each of these sections and seasons, "it is still questionable how far the different curves with their different phases and, perhaps, different periods will unite into a single curve having the same characteristics." He added that it is perhaps "the imperfect mixture and the imperfect tuning of the parts [that is] responsible for the apparent discord in the whole." It is perhaps ironic that the man who would finally provide a mathematical foundation of Hamer's ideas and would link these ideas to chemical mass-action came to the conclusion that the mass-action assumption was not realistic for epidemiology. It is a pity that Soper was not trained more as a modeller.

As is usual with papers "read" to the Royal Statistical Society, Soper's paper is followed by an elaborate commentary (12 pages). The paper is met with great acclaim from all commentators. Some of the criticism could have been valuable to Soper in steering his research. However, because Soper died soon after publication, he did not have the opportunity to follow any of the suggestions made. Hamer, who was in audience, called the analysis "extraordinarily interesting" and likened the transformation of his original idea to Soper's theory to the transformation of Eliza Doolittle into "one of the most peerless Galateas that ever stepped off a pedestal" in Bernard Shaw's *Pygmalion*. Neither Ross nor McKendrick, who never published in statistical journals, are discussants of Soper's paper and, more remarkably, are not mentioned by any of the discussants, even though both published papers in which similar problems were studied in a similar way. Greenwood knew of Ross's work and could have stimulated Soper to try and merge a statistical and a modelling approach. Greenwood (1916), in basically a review of the work by Ross and Brownlee, wrote (describing only the model of Ross with constant happenings, not the mass-action-type model):

> The advantage of Sir Ronald Ross's method, apart from its simplicity and elegance—advantages which are, however, no mean matters—lies in its generality, so that it may be possible to include the case hypothesized by Brownlee as a particular example . . . As restrictions are relaxed, the analysis will inevitably become more intricate, and, having devised an *a priori* law, one must devise . . . a way of applying the law to statistical data.
>
> It is high time that epidemiology was extricated from its present humiliating position as the plaything of bacteriologists and public health officials . . . The work of Sir Ronald Ross, of Dr. Brownlee, and of a few others should at least elevate epidemiology to the rank of a distinct science.

At this time, Greenwood showed no knowledge of the equally relevant work by McKendrick, nor did he do so in later publications (such as in Greenwood 1932). Ross did not publish on mathematical epidemiology in the period (he died in 1932), so we do not know whether Ross read Soper's paper. McKendrick shows knowledge of the paper in later work. He referred to it in an article in 1940 but only as the source of a graph of the Glasgow measles periodic curve. (It is something of an honour, however, because Soper is the only reference of 20 that is not a work by McKendrick!) Although the situation has improved, it is still the case that statistical and mathematical modellers of epidemic phenomena form two groups with far too little overlap and far too little understanding and knowledge of each other's results, theories, and problems. Clearly, the languages of the two groups and the problems they address have diverged, but strangely enough even in the early twentieth century, when the material was still similar, bifurcation occurs. One explanation might be that the approach by Ross and McKendrick was *a priori*, starting with assumptions, principles, and causes and without immediate need for statistical techniques, whereas the approach of Brownlee, Greenwood, and co-workers was *a posteriori*, starting from data of past outbreaks (epidemic curves) and working backwards to underlying principles (Kingsland 1985), where statistics is the natural and immediately necessary tool. In short, the difference might be explained by drawing on the difference between Snow's and Farr's hypothesis of epidemic decline. It is a pity that Soper had no opportunity to expand his work, because his ideas clearly aimed at a mix of both approaches.

5.6 | A SCIENCE TAKING FLIGHT

One aspect of old paradigms is that the assumptions underlying the principle tend to be obscure or missing in most work that adopts the paradigm in the early years. These assumptions are not fully realized by the first advocates of a paradigm. The first authors do not know that they are introducing a paradigm or even a metaphor, because for this realization they need a context in which the assumption must be replaced by something more complex. This context needs to grow with applications and with developing theoretical insight. With the context comes exceptions to the rule, and with the exceptions comes a clearer understanding of the underlying principles. An example is the homogeneous assumption as a global descriptor of interaction: any two randomly picked individuals have the same probability per unit of time to come into contact with each other. Systems adhering to mass-action are, in a sense, *well stirred*. Another principle is that the contact rate is proportional to a product of *densities*, that is, numbers of individuals of the various types per unit area. Both assumptions are clear if you bring to mind the origin of mass-action in chemical reactions, where molecules in a well-stirred reaction vessel collide with a rate proportional to the product of the respective concentrations of the various types of molecule. A likely positive exception to the ignorance of assumptions in the case of mass-action is the work of Ross because he did not introduce mass action out of context; Ross introduced mass-action as a natural step in the theory of happenings that he created.

There are also a few *consequences* of mass-action that were realized only slowly. One is that it is linear in the density of susceptibles. This implies that doubling the susceptible density would also double the density of contacts made per unit of time. For many types of contact this is clearly not justified, certainly not over a large range of densities. Another aspect is that in the discrete-time formulation of Hamer and Soper, researchers can overestimate the number of susceptibles that become infected, in the sense that in some time steps more susceptibles can be expected to succumb than are available, thus leading to negative population sizes. Researchers should therefore be careful in formulating discrete-time models with mass-action (see Diekmann & Heesterbeek 2000).

Of these aspects, the first to give rise to new descriptions of interaction is the theory of Wade Hampton Frost (now known unfairly as the Reed-Frost model) from 1928. This model does not suffer from overestimation of new victims, at least not in comparison with the discrete-time mass-action model. Frost never published his model, but the text of a lecture from 1928, in which he describes the idea, was eventually published (as Frost 1976). There is substantial literature on the Frost model and its relations to what is often called the *Soper model* (e.g., see Jacquez 1987, Dietz & Schenzle 1985, and the references given there). The first time these models and the Ross-McKendrick approaches appeared together in a detailed analysis of their differences and similarities was in work by Wilson and Burke (1942, 1943); they later appeared in the papers by Wilson and Worcester (1945). Unknown to all these authors was the work by En'ko in Russian in the late nineteenth century who published, also in connection to measles, another alternative mathematical model for discrete time (described earlier). In contrast to the similar Reed-Frost model, En'ko did not provide a mechanistic basis for his approach. Reed and Frost, however, went so far as constructing a mechanical analogue for their model, where black and white balls were mixed according to certain well-defined rules in a one-dimensional trough (see Fine 1977 for a detailed account).

Mass-action is still in high demand (e.g., it is recognized as making an important contribution to a recent theory for modelling the dynamics of structured populations; Diekmann, Gyllenberg, & Metz 2003). That it is still a relevant paradigm partly stems from the fact that the particular question a researcher wants to study with a model dictates the assumptions and ingredients that will underlie it. When results turn out not to be very sensitive to the precise metaphor used to describe interaction, then it makes sense to stick with the simplest descriptor. This does not imply that it is easy to come to that conclusion or that anybody goes to the effort of testing these assumptions within the model, but certainly not all criticism of mass-action descriptions in present-day models is justified: The nature of the problem studied influences this heavily. Surely Occam's razor principle has a large hand in the popularity of mass action in ecology and epidemiology. Researchers can interpret it as a valid *null model*, for which they realize in many situations that it is not a good mechanistic description of interactions but against which they can test the influence of more detailed mechanisms.

In recent decades, increasingly more involved mechanisms of interaction are required, because various types of heterogeneity—for example, social structure, spatial structure, and age, sex, and behaviour differences—have been shown to influence contact pattern, susceptibility, and infectivity. Also, more detailed information about these differences is now available and can be analyzed. The differences influence the mechanisms that operate on the level of the individuals and can be important in shaping phenomena at the population level where many such individuals interact. This has given rise to the modelling of a large variety of more heterogeneous types of interaction and more local mixing structures. Some of these models can be put on a strong mechanistic base, but many are heuristic and others lack even that. Especially in the last decade, however, more sophisticated methods have been developed. These methods will be treated in the next chapter by Matt Keeling.

Mass-action stood at the basis of the emerging study of epidemic phenomena as a science. One can wonder whether the many descriptions of the contact process that came later in the development of epidemiology and ecology can ever again have such a far-reaching influence and whether any of these paradigms will remain to influence the way contacts are modelled in these sciences as much as mass-action still does.

ACKNOWLEDGEMENTS

I would like to thank Klaus Dietz for detailed comments on an earlier version of this chapter.

REFERENCES

Aitchison, J., & Watson, G.S. (1988). A not-so-plain tale from the Raj. *In* "The Influence of Scottish Medicine: A Historical Assessment of Its International Impact" (D.A. Dow, Ed.). Parthenon Publishing Group, Carnforth, UK, pp. 113–128.

Bastiansen, O., Ed. (1964). "The Law of Mass Action: A Centenary Volume, 1864–1964." Det Norske Videnskaps-Akademi i Oslo.

Diekmann, O., Gyllenberg, M., & Metz, J.A.J. (2003). Steady-state analysis of structured population models. *Theor. Popul. Biol.* **63**, 309–338.

Diekmann, O., & Heesterbeek, J.A.P. (2000). "Mathematical Epidemiology of Infectious Diseases: model building, analysis and interpretation." John Wiley & Sons, Chichester, UK.

Dietz, K. (1988). The first epidemic model: A historical note on P.D. En'ko. *Austral. J. Stat.* **30A**, 56–65.

Dietz, K., & Heesterbeek, J.A.P. (2005). "Epidemics: The Discovery of Their Dynamics." In preparation.

Dietz, K., & Schenzle, D. (1985). Mathematical models for infectious disease statistics. *In* "A Celebration of Statistics: The ISI Centenary Volume" (A.C. Atkinson & S.E. Fienberg, Eds.). Springer-Verlag, New York, pp. 167–204.

En'ko, P.D. (1989). On the course of epidemics of some infectious diseases. *Int. J. Epidemiol.* **18**, 749–755.

Farr, W. (1866). On the cattle plague. *J. Soc. Sci.* March 20.

Fine, P.E.M. (1975). Ross's *a priori* pathometry: A perspective. *Proc. Roy. Soc. Med.* **68**, 547–551.

Fine, P.E.M. (1977). A commentary on the mechanical analogue to the Reed–Frost epidemic model. *Am. J. Epidem.* **106**, 87–100.

Fine, P.E.M. (1979). John Brownlee and the measurement of infectiousness: A historical study in epidemic theory. *J. Roy. Stat. Soc. A* **142**, 347–362.

Frost, W.H. (1976). Some conceptions of epidemics. *Am. J. Epidem.* **103**, 141–151.

Gillispie, C.C. (1972). "Dictionary of Scientific Biography." Charles Scribner & Sons, New York.

Greenwood, M. (1916). The application of mathematics to epidemiology. *Nature* **97**, 243–244.

Greenwood, M. (1931). Obituary of Herbert Edward Soper. *J. Roy. Stat. Soc.* **94**, 135–141.

Greenwood, M. (1932). "Epidemiology: Historical and Experimental: The Herter Lectures for 1931." Johns Hopkins Press, Baltimore.

Greenwood, M. (1936). Obituary of Sir William Hamer. *Brit. Med. J.* July 18, 154–155.

Guldberg, C.M., & Waage, P. (1864). Studier over Affiniteten. Avhandl. Norske Videnskaps. Akad. Oslo, I. Mat. Naturv. Kl.

Hamer, W.H. (1896). Age-incidence in relation with cycles of disease prevalence. *Trans. Epidem. Soc. Lond.* **16**, 64–77.

Hamer, W.H. (1906). Epidemic disease in England: The evidence of variability and of persistency of type. *Lancet* March 3rd, 569–574 (Lecture I), March 10th, 655–662 (Lecture II), March 17th, 733–739 (Lecture III).

Harvey, W.F. (1944). Anderson Gray McKendrick 1876–1943. *Edinburgh Med. J.* **50**, 500–506.

Heesterbeek, J.A.P. (2002). A brief history of R_0 and a recipe for its calculation. *Acta Biotheoretica* **50**, 189–204.

Jacquez, J.A. (1987). A note on chain-binomial models of epidemic spread: What is wrong with the Reed–Frost formulation? *Math. Biosci.* **87**, 73–82.

Kermack, W.O., & McKendrick, A.G. (1927). Contributions to the mathematical theory of epidemics, Part I. *Proc. Roy. Soc. Lond. A* **115**, 700–721. Reprinted in *Bull. Math. Biol.* (1991), **53**, 33–55.

Kingsland, S.E. (1985). "Modeling Nature: Episodes in the History of Population Ecology." University of Chicago Press, Chicago.

Kotz, S., & Johnson, N.L. (1997). "Breakthroughs in Statistics," Volume 3. Springer-Verlag, New York.

Lotka, A.J. (1923). Martini's equations for the epidemiology of immunizing diseases. *Nature* **111**, 633–634.

Lotka, A.J. (1925). "Elements of Physical Biology." Williams & Wilkins, Baltimore. Reprinted as "Elements of Mathematical Biology" (1956), Dover, New York.

Lund, E.W., & Hassel, O. (1964). Guldberg and Waage and the law of mass action. *In* "The Law of Mass Action: A Centenary Volume, 1864–1964" (O. Bastiansen, Ed.). Det Norske Videnskaps-Akademi i Oslo, pp. 37–46.

Martini, E. (1921). "Berechnungen und Beobachtungen zur Epidemiologie und Bekämpfung der Malaria." Gente, Hamburg.

McKendrick, A.G. (1911). The chemical dynamics of serum reactions. *Proc. Roy. Soc. Lond. B* **83**, 493–497.

McKendrick, A.G. (1912). The rise and fall of epidemics. *Paludism* **1**, 54–66 (*Transactions of the Committee for the Study of Malaria in India*).

McKendrick, A.G. (1926). Applications of mathematics to medical problems. *Proc. Edinburgh Math. Soc.* **44**, 98–130.

McKendrick, A.G. (1940). The dynamics of crowd infection. *Edinburgh Med. J.* **47**, 117–136.

Mellor, J.W. (1905). "Higher Mathematics for Students of Chemistry and Physics: With Special Reference to Practical Work," 2nd edition. Longmans, Green & Co., London.

Nye, E.R., & Gibson, M.E. (1997). "Ronald Ross, Malariologist and Polymath: A Biography." Macmillan Press, Houndmills, UK.

Ransome, A. (1880). On epidemic cycles. *Proc. Manchester Lit. Phil. Soc.* **19**, 75–96.

Ross, R. (1911). "The Prevention of Malaria," 2nd edition. John Murray, London.

Ross, R. (1916). An application of the theory of probabilities to the study of *a priori* pathometry: Part I. *Proc. Roy. Soc. Lond. A* **92**, 204–230.

Ross, R., & Hudson, H.P. (1917). An application of the theory of probabilities to the study of *a priori* pathometry. *Proc. Roy. Soc. Lond. A*, **93**, 212–225 (Part II), 225–240 (Part III).

Snow, J. (1853). "On Continuous Molecular Changes, More Particularly in Their Relation to Epidemic Diseases." J. Churchill, London.

Soper, H.E. (1929). The interpretation of periodicity in disease prevalence. *J. Roy. Stat. Soc.* **92**, 34–61 (followed by discussion: 62–73).

Wilson, E.B., & Burke, M.H. (1942). The epidemic curve. *Proc. Natl. Acad. Sci. USA* **28**, 361–367.

Wilson, E.B., & Burke, M.H. (1943). The epidemic curve, Part II. *Proc. Natl. Acad. Sci. USA* **29**, 43–48.

Wilson, E.B., & Worcester, J. (1945). The law of mass action in epidemiology. *Proc. Natl. Acad. Sci. USA* **31**, 24–34 (part I), 109–116 (part II).

6 | EXTENSIONS TO MASS-ACTION MIXING

Matt J. Keeling

6.1 | INTRODUCTION

Epidemiological models have been one of the major successes of quantitative modelling of biological phenomena (Anderson & May 1991). Such models have been able to accurately predict the spread of infectious agents and are at the heart of optimal public-health control measures (Ferguson *et al.* 2003). Fundamental to disease models is the transmission of infection, which is frequently captured as either mass-action or pseudo mass-action (Chapter 5). More recent developments, which have considered the individual, mechanistic nature of transmission, have shown that both mass-action and pseudo mass-action are approximations to the true dynamics of transmission. Here a range of modifications and improvements to these basic forms of modelling transmission are described. Although many of these improvements necessitate a significant increase in model complexity and computing power, recent advances have developed analytical approaches that provide many insights into when improvements to mass-action assumptions may be necessary.

Almost all epidemiological models operate by separating the population into a discrete set of classes (Anderson & May 1991). The most natural compartmentalisation is given by the *SIR* model, in which individuals are classed as susceptible to the disease, *S*; infected (and therefore infectious) with the disease, *I*; or recovered and therefore immune, *R*. In the simplest situation, births (*B*) are placed only into the susceptible class, and deaths (*d*) occur to individuals with an equal risk independent of class (Anderson & May 1979). Recovery from infection is also assumed to occur at a constant rate, g, such that the average infectious period (ignoring deaths) is $1/g$. This leads to the standard *SIR* equations:

$$\frac{dS}{dt} = B - \lambda S - dS$$
$$\frac{dI}{dt} = \lambda S - gI - dI \qquad (6.1)$$
$$\frac{dR}{dt} = gI - dR$$

where λ is the *force of infection,* which measures the rate at which susceptible individuals become infected and is controlled by the mixing between infected and susceptible individuals. The type of mixing experienced, and whether mass-action is a good approximation, depends on how λ scales with the level of infection, with susceptibles, and with population size N (Chapter 14). Thus, for pure mass-action:

$$\lambda(S, I, N) = \beta I \qquad (6.2)$$

For pseudo mass-action, another equation is used:

$$\lambda(S, I, N) = \beta \frac{I}{N}. \qquad (6.3)$$

Here, mass-action comes from the basic assumption that susceptible and infectious individuals meet one another at random and can spread the disease. Thus, paralleling work in the physical sciences on chemical reactions, we assume that when there are more infectious individuals there is a higher encounter rate between them and susceptible individuals, leading to a greater force of infection. The alternative pseudo mass-action assumption focuses on the proportion of the population that are infectious (I/N), essentially assuming that each susceptible interacts with a fixed number of individuals, and the proportion of these that are infectious determines the force of infection. When the population size, N, remains constant, there is no real difference between the two models because, for pseudo mass-action, β can be rescaled to absorb the $1/N$ term. It is only when the population size varies that the differences can be clearly observed; β is therefore a proportionality constant, which incorporates mixing rates and transmission probabilities.

For the standard disease models the force of infection is proportional to the level of infection within the population; we will now consider several modifications that change the basic form of this quantity and show how these are related to the mass-action assumption. We will consider three major types of modifications: changes in the functional form of the interaction term related to biological and social considerations, increased compartmentalisation because of spatial structure with non-random mixing, and methodological changes whereby considering the transmission at a more individual-based level alternative model equations are derived. Throughout this chapter, the type of mixing experienced will be related to the force of infection experienced by susceptibles.

6.2 | FUNCTIONAL FORMS

Functional forms offer one of the earliest and most appealing types of modification to the standard form of enemy-victim interactions. Their formulation was often based upon the interaction of parasitoids and hosts, or predators and prey; therefore, the concepts are frequently difficult to translate in a mechanistic manner to the spread of diseases. Under the assumption of mass-action mixing, such host-parasitoid equations are inherently unstable leading to biologically unrealistic extinctions; much interest has therefore been focused on means of stabilising these interactions (Hassell 2000). In this section, the concepts will be illustrated using the predator-prey and host-parasitoid systems that these functional forms were designed to stabilize, although attention will be given to when such modifications may be applicable to disease models.

A phenomenological modification to the host-parasitoid dynamics is to include a power-law mixing term, such that the rate of parasitism is given by:

$$\text{Rate of parasitism} = aHP^{1-m}, \tag{6.4}$$

where H and P are the number of hosts and parasitoids, m measures the degree of interference between parasitoids, and a is a proportionality constant similar to β in the disease models. This change to mass-action mixing originates from the observation that parasitoids may suffer from intraspecific competition such that as the number of parasitoids increases the rate of parasitism (per parasitoid) decreases. There is no clear mechanistic reason a power-law exponent is introduced, but this simple modification to the model has been supported by lab-based observations of host-parasitoid interactions and dynamics (Hassell 2000). More recent advances using high-powered analytical and computational techniques (described later) have shown that such power-law

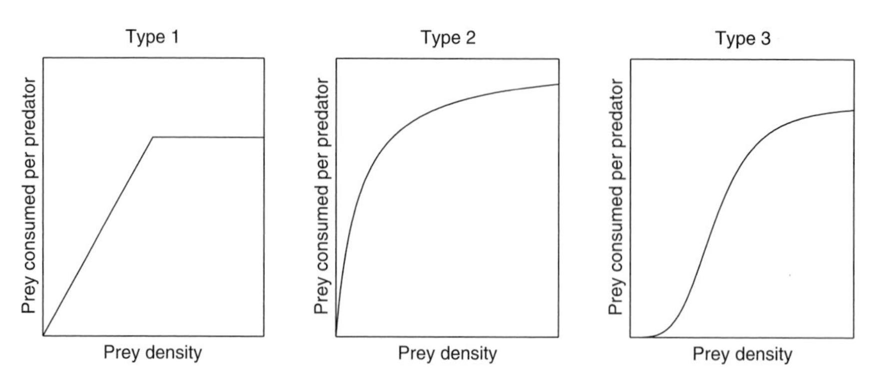

FIGURE 6.1 | The three functional forms classified by Holling (1959), which illustrate how the number of prey consumed per predator scales with prey density.

modifications to the mass-action assumption may be rooted in the effects of spatial patterns and aggregation and hence may be applicable to disease models.

Holling (1959) classified three basic types of interaction between predators and prey or parasites and hosts (Fig. 6.1). Although all three of these functional forms originate from the assumption of a mass-action encounter rate, once the encounter has occurred the interaction between the two organisms is modified by biological considerations, leading to a different term in the equations. The standard assumptions for the interaction between predatory and prey lead to a multiplicative term in the equations such that the consumption rate per predator increases linearly with the number of prey, whereas the Holling functional forms all show saturation as the number of prey increases. Type 2 functional responses are usually attributed to a handling time, T_h, such that items of prey require a fixed amount of time to be captured and consumed; therefore, even when prey are plentiful there is a clear upper limit to how many can be consumed in 1 day. The predation rate between predators (P) and prey (H) is approximated by:

$$\text{Rate of predation} = \frac{aHP}{1 + aT_h H}. \tag{6.5}$$

Clearly, from the perspective of a disease there is no handling time. Although there may be a delay between infection and being infectious (the incubation period), this does not limit the number of secondary cases produced by an infectious individual. However, the behavioural characteristics of the host may lead to handling-time effects. Consider a sexually transmitted disease. Irrespective of the number of available partners, there are theoretical limits, and in human society social limits (Johnson *et al.* 2001), to the number of possible partners within any given time frame and therefore a saturation to the mixing. A suitable modification to the epidemiological force of infection is therefore:

$$\lambda(S, I, N) = \beta \frac{I}{1 + T_h N}. \tag{6.6}$$

This handling-time approach effectively scales between pure mass-action when the population is small and the number of individuals limits transmission, and pseudo mass-action when the population is large and transmission is limited by the number of contacts it is possible to make. It is interesting to note that sexually transmitted diseases, which effectively have a long social handling time, have a long infectious period compared with most airborne infections. Although many plausible explanations for this exist, it is clear that a sexually transmitted disease with a short infectious period is unlikely to be successful because an individual will generally recover before finding a new sexual partner.

Another situation in which transmission is limited by time constraints rather than susceptible numbers is in vector-borne diseases. Worldwide, many infections are spread by mosquitoes. Examples include such high-impact diseases as malaria, West Nile virus, and dengue fever (Dye 1992). Transmission of these infections can only occur when a female mosquito takes a blood meal before laying eggs; this happens approximately every 3 days or so for adult mosquitoes. Therefore, irrespective of the number of suitable human hosts, the spread of infection is limited by the number of mosquitoes and their biting frequency. The force of infection to humans becomes:

$$\lambda(S, I, N) = \beta_M \frac{I_M}{1 + T_h N}, \tag{6.7}$$

where I_M is the number of infected mosquitoes. Interestingly, if the number of humans is large compared to the number of mosquitoes, such vector-borne diseases may fail to spread; successful transmission from mosquito to human and back to mosquito requires two mosquito bites and therefore a sufficiently high rate of an individual being bitten.

6.3 | METAPOPULATION MODELS

Metapopulation-like mixing may be an appropriate replacement for (pseudo) mass-action mixing whenever the environment of susceptibles is sufficiently patchy such that (partially) isolated subpopulations of hosts exist. Traditionally, metapopulation models originated from the study of insects in patchy environments, where a patch can be considered either occupied or empty (Levins 1969, Hanski & Gilpin 1997, Hanski & Gaggiotti 2004). At the most local scale, and from the perspective of the disease, each host is an isolated patch of resource, but this degree of patchiness is naturally incorporated in the partitioning of individuals into discrete classes (e.g., *S*, *I*, and *R*). Metapopulation models are usually focused at macroscale dynamics (Hanski & Gilpin 1997, Grenfell & Harwood 1997, Hanski 1999), where hosts (especially human hosts) are aggregated into communities within which mixing and hence transmission is relatively frequent and between which infection spreads at a lower rate (Grenfell & Bolker 1998).

Two formulations of the metapopulation are used in epidemiological modelling. The patch-occupancy or Levins assumption is to classify subpopulations into a limited set of states. As such, a subpopulation will be described as infected if disease is present—the precise prevalence of infection within the population is not explicitly considered (Hess 1996; Gog, Woodroffe, & Swinton 2002; McCallum & Dobson 2002). A more comprehensive but complex formulation is to model the precise state of each subpopulation in terms of the number of susceptible, infected, and recovered individuals. In such metapopulation models an extra layer of heterogeneity has been added such that individuals are differentiated by both their disease status (*S*, *I*, or *R*) and by the subpopulation or community to which they belong (Hanski 1994). Hence, S_i refers to the number (or proportion) of individuals in community i that are susceptible. Although it is intuitively obvious that the transmission within communities should be greater than the mixing between them, the precise form of the force of infection depends on the nature of the disease and the host.

Transmission of human diseases between communities usually occurs because of the short duration transitory movement of either susceptibles or infected individuals (Sattenspiel & Dietz 1995, Keeling & Rohani 2002). If P_{ij} is the proportion of time an individual from community i spends in community j, then for pseudo mass-action the force of infection for susceptible from community i becomes:

$$\lambda_i = \sum_k \beta_k \left(\frac{\sum_j I_j P_{ik} P_{jk}}{\sum_j P_{jk} N_j} \right). \tag{6.8}$$

This formulation considers the transmission from infectious individuals that live in community j to a susceptible from community i when both are in location k. This transmission is dominated by transmission within the home environment ($i = j = k$). Although in theory the movement parameters, P_{ij}, can be measured from detailed road, rail, and air traffic surveys, this can be a time-consuming and costly exercise (Hawkes & Hart 1993, Wilson 1995, Donnelly *et al.* 2003). Instead the coupling is often estimated in terms of the two population sizes and the distance between them, D_{ij}. The coupling is frequently assumed to obey a power-law relationship with distance, with the exponent to be determined from the available data. One of the two most common forms is pure distance:

$$P_{ij} \propto D_{ij}^{-\alpha} \tag{6.9}$$

or gravity, where movement into a population increases with the population size:

$$P_{ij} \propto N_j D_{ij}^{-\alpha}. \tag{6.10}$$

Most studies have found that the gravity model with an exponent, α, of around two works well for most human populations (Finkenstädt & Grenfell 1998, Xia *et al.* 2004), although this cannot hope to mimic all the complex movement patterns driven by social and economic forces.

For simplicity, and to understand the fundamental effects of partitioning the population, it is often assumed that all populations are of equal size and that interaction with all other communities is equal. The force of infection for susceptibles in community i is then (Keeling & Rohani 2002):

$$\lambda_i = \beta\left(\gamma^2 + [1-\gamma]^2\right)I_i + \beta\sum_{j \neq i}\left(\frac{2\gamma(1-\gamma)}{n-1} + \frac{n-2}{(n-1)^2}\gamma^2\right)I_j, \tag{6.11}$$

where γ is the proportion of time spent away from the home community, and n is the number of communities or patches. This formula again comes from calculating the probability that a susceptible individual from patch i will meet an infectious individual either in the home communities or when they are both visiting another location. Thus the first γ^2 term is the probability that a susceptible and an infectious individual from patch i both move to another patch where they can meet, and $[1-\gamma^2]$ is the probability that neither moves and they can meet in patch i. Hence, we are effectively assuming mass-action mixing within a community but that individuals can move between them. Note that using this force of infection, the basic reproductive ratio, R_0, (which measures the average number of secondary cases an infectious individual produces in a totally susceptible population) is independent of γ and this formulation therefore corresponds to when potential transmission contacts occur at a constant rate. This equation is an accurate approximation for commuter-like movements, for which the duration of a trip is short compared with the length of the infectious period. For epizootics (animal-based diseases), for which movement between patches tends to be permanent or of extremely long duration, only the movement of infected individuals contributes to interaction between patches. This form of movement has to be modelled explicitly and cannot be readily captured by a modified force of infection.

For sessile organisms, such as plants, a movement-based approach is clearly not valid; however, the metapopulation approach may still be applicable to the dynamics of diseases within fields or other isolated habitats (Swinton & Gilligan 1996). Considering the situation in which pathogen particles from one habitat have a low probability of spreading to other habitats, the force of infection is then given by:

$$\lambda_i = \beta_i \left(I_i + \sum_j P_{ij} I_j \right), \tag{6.12}$$

where P_{ij} measures the dispersal of pathogen between habitats. In this formulation, the coupling between habitats acts differently. Whereas in human populations the presence of another nearby community will affect the movement of individuals but not R_0, for plant pathogens dispersal is independent of the surrounding environment; so R_0 varies with the number and location of surrounding habitat.

The partitioning of the population into discrete habitats or communities can lead to heterogeneity (or asynchrony) between the various components (Hanski & Gilpin 1997, Grenfell & Harwood 1997, Lloyd & May 1996). In deterministic models, we generally find that all populations tend to an asymptotic state and that, if the populations have equal mixing patterns, these asymptotic states are identical. Therefore, in deterministic models the long-term dynamics are homogeneous. However, for stochastic (and therefore more realistic) models each community experiences a different set of random events; therefore, their trajectories diverge (Renshaw 1991). The degree of heterogeneity is controlled by the amount of stochasticity, which pushes population trajectories apart; the deterministic dynamics, which generally draw trajectories to the same fixed point or attractor; and the amount of coupling, which can lead to the populations experiencing similar forces of infection. From this we can conclude that small isolated populations should show the greatest heterogeneity, whereas large interconnected communities should be more homogeneous (Keeling & Grenfell 1999; Rohani, Keeling, & Grenfell 2002) (Fig. 6.2).

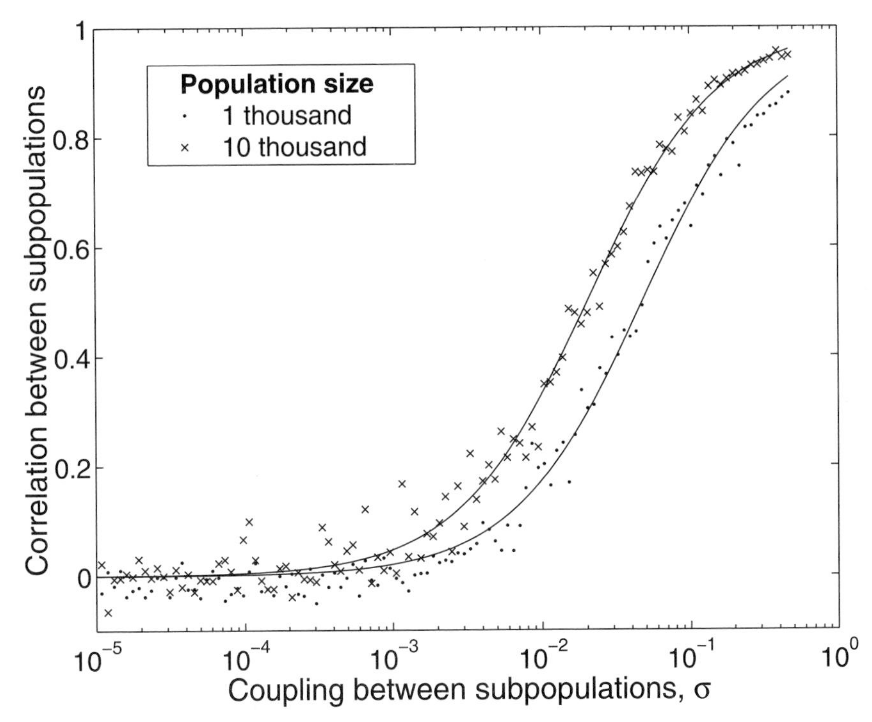

FIGURE 6.2 | The correlation (level of homogeneity) between disease levels in two stochastic *SEIR* (susceptible–exposed–infectious–recovered) subpopulations, which include a latent or exposed class. Two different population sizes are simulated, 1000 individuals (*dots*) and 10,000 individuals (*crosses*). Fitted curves are of the form $\sigma/(\xi + \sigma)$, which is based on theoretical predictions (Keeling & Rohani 2002), where σ is the coupling and ξ is a free parameter to be determined. Disease parameters match those of whooping cough, although seasonal forcing is ignored.

This heterogeneity has several implications for the global dynamics of diseases. The different levels of infection in different populations can lead to rescue effects—in which the low level of disease in one population can be rescued from extinction by imports from another community. The classic example of this is the fade-out pattern of measles, in which the disease frequently disappears from small communities only to be reintroduced from those populations in which the disease is endemic (Bartlett 1956, 1957; Grenfell, Bolker, & Kleczkowski 1995). This well-studied pattern is caused by two separate mechanisms. The first and most obvious is the stochastic behaviour coupled with the oscillatory dynamics that cause the extinctions—small populations that have lower numbers of infected individuals and experience more stochasticity are more prone to extinctions (Renshaw 1991; Fig. 6.3). The second factor is the import of new infections, which is governed by the non-

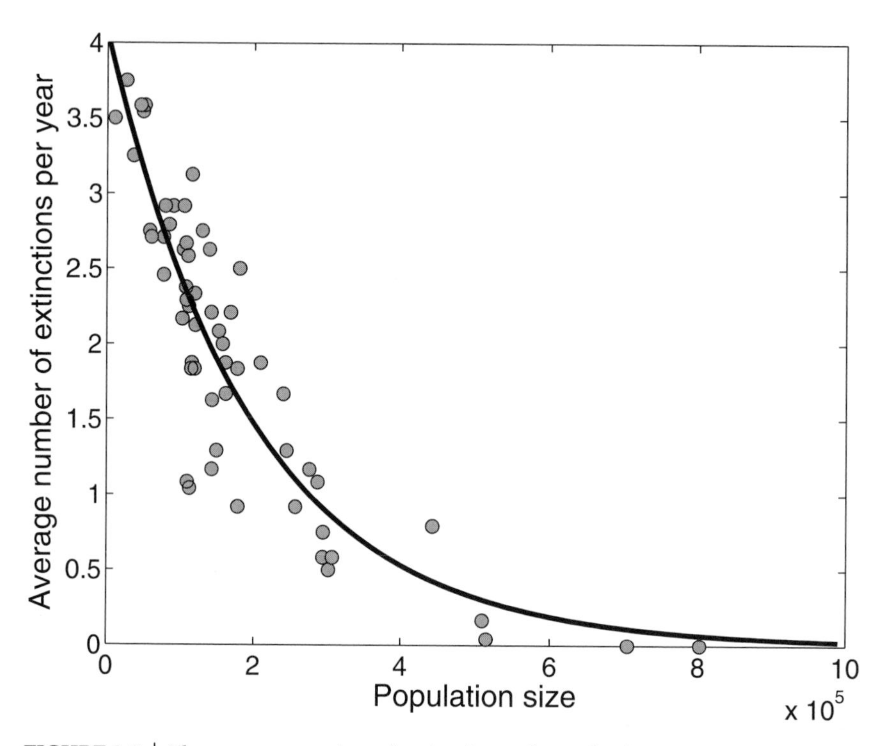

FIGURE 6.3 | The average number of extinctions of measles from 60 towns and cities in England and Wales from 1944 to 1968. An extinction is classed as 2 consecutive weeks without any reports of measles cases. Data are taken from the Registrar General's Weekly Reports (see http://www.zoo.cam.ac.uk/zoostaff/grenfell/measles.htm). The solid line is the best-fit exponential to the data.

mass-action mixing. Frequent imports, for example, because of the proximity of a large city, may be sufficient to prevent extinctions in small surrounding communities (Finkenstädt & Grenfell 1998). However, if imports are rare, they allow the proportion of susceptibles to build up during the extinct phase such that each new import triggers a large epidemic followed by an inevitable extinction (Bjørnstad & Grenfell 2001).

Associated with this phenomenon are the strong negative correlations that arise between the proportion of susceptible and infectious individuals within subpopulations (Keeling 2000). Intuitively, those communities or habitats that have high levels of infection have few susceptibles, whereas in those communities where the disease has become extinct the level of susceptibles will increase. This negative correlation, which arises because of the localized nature of enemy-victim interactions, influences disease persistence (Keeling, Wilson, & Pacala 2000). Localized extinctions may not necessarily be bad for a disease because this allows the host population to recover and be exploited later (Keeling 2000). It is therefore possible for a disease to persist at some global scale while the local dynamics exhibit continual cycles of extinctions and recolonizations (Grenfell & Harwood 1997). Whether the limited coupling between communities enhances or lessens persistence is crucially dependent on the within-community dynamics and the trade-offs between increased heterogeneity and increased isolation (Grenfell, Bolker, & Kleczkowski 1995; Grenfell & Bolker 1998, Keeling 2000, Hagenear *et al.* 2004).

Metapopulations are based upon the localized mass-action mixing of distinct populations or communities. As such, they present a reliable and robust representation of disease dynamics at a regional or national level. Although metapopulations that assume identical populations and global coupling are now relatively common and provide good insights into the role of heterogeneity, the use of more realistic models based on the location and movement of human populations is still in its infancy. Human populations are naturally partitioned into discrete towns, cities, and villages, and the inclusion of this extra degree of realism holds great promise for the future of disease modelling.

6.4 | CELLULAR AUTOMATA

Cellular automata are caricatures of epidemiological processes and are designed to illustrate novel features rather than reflect real situations (Mollison 1977; Yakowitz, Gani, & Hayes 1990; Johansen 1996; Durrett & Levin 1994). These models have developed from work in statistical physics and are predominantly based on a square lattice of cells, with each cell in one of a small number of states. For disease modelling it is often convenient to let each cell represent an individual and hence be classified as susceptible, infected, or recovered. Under such conditions diseases are modelled as *SIRS* (susceptible-infections-recovered-susceptible); this can be viewed either as recovered individuals dying and being replaced by newborn susceptibles or as a consequence of waning immunity. The infection process is assumed to occur locally such that an infected cell can spread the disease, at rate τ, to its nearest four or eight neighbours. Hence, the force of infection for a susceptible at cell (i, j) is:

$$\lambda_{(i,j)} = \tau \times \text{number of infected cells within neighbourhood}. \quad (6.13)$$

Therefore, once again infection operates through a localized form of mass-action.

The most popular form of cellular automata model, which approximates disease dynamics, is the so-called forest fire model (Bak, Chen, &

Tang 1990). In this model infection is assumed to spread relatively quickly, and the infection is short-lived. Loss of immunity is a far slower process, and at an even slower timescale there are rare imports of infection that can restart an epidemic. This leads to some interesting mathematical results, which I describe later; however, from a biological and epidemiological perspective several key elements of disease transmission are captured by forest fire models (Rhodes & Anderson 1998). The first is that the population is composed of individuals and therefore, the dynamics are stochastic with the inevitable risk of extinction. The second is that cases are highly clustered (Fig. 6.4) such that infection is highly aggregated, and there is often a strong negative spatial correlation between infected and susceptible individuals.

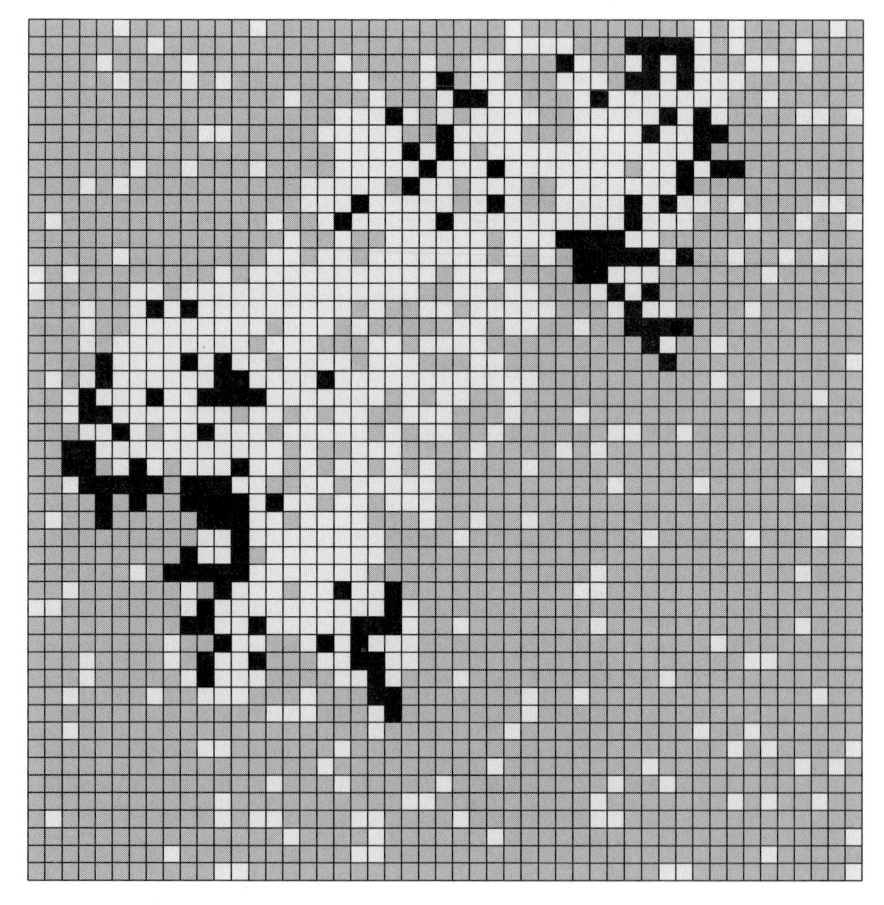

FIGURE 6.4 | Snapshot of a forest-fire cellular automata; infected cells (*black*), susceptible cells (*dark grey*), recovered cells (*light grey*). The average infectious period is 10 iterations, the average recovered period is 100 iterations, the neighbourhood is set at the four nearest cells, and the infection rate between neighbouring cells is $\tau = 0.2$.

The local nature of the interactions and local-scale correlations that develop have some major effects on the dynamics of diseases. The most pronounced is that infection is self-limiting because it rapidly depletes the level of susceptibles in the neighbouring environment (Rand, Keeling, & Wilson 1995). The spread of a SIR epidemic through the cellular automata lattice is therefore strongly related to ideas in percolation theory. In a simple two-dimensional lattice, a single infectious individual can infect any one of its four neighbours; hence, it would seem intuitive that for the disease to propagate the probability of transmitting infection to a neighbour, p_τ must be greater than $1/4$. However, each new secondary infection only has three susceptible neighbours (because one neighbour is already infected); therefore, for the disease to spread each of these secondary cases must infect at least one neighbour, so $p_\tau > 1/3$. This argument can be continued to third, fourth, and higher-generation cases, in which case we find $p_\tau > 0.593$ for the epidemic to spread indefinitely or percolate through the lattice.

The situation becomes far more complex when replenishment of susceptibles (births) is included. Although this reduces the critical threshold slightly, it also introduces notions of persistence. It also leads to the idea that although a large transmission probability is good in the short term, it may lead to such a rapid repletion of susceptibles that future generations cannot survive (Read & Keeling 2003). Therefore, there may be critical upper bounds to the transmission probability above which long-term persistence is impossible (Rand, Keeling, & Wilson 1995).

One interesting feature of cellular automata that endears them to physicists and mathematicians is that they are examples of self-organized critical phenomena (Bak, Chen, & Tang 1990; Tang & Bak 1988; Rhodes, Jensen, & Anderson 1997). In simple terms a set of patterns and properties arise that are robust to moderate changes in the parameters and generally depend only on the model containing very different timescales. These self-organized phenomena are frequently associated with power-law scaling such that two properties are related by a non-integer exponent. A classic example of this is the work of Rhodes and Anderson (1996a, 1996b), who used lattice models to explain the observed epidemic properties of measles in the Faroe Islands. Detailed case reports from these islands in the North Atlantic show that measles epidemics occur at irregular intervals but that the size, s, and duration, t, of these epidemics follow a power-law pattern:

$$\text{Prob}(\text{epidemic size} > s) \propto s^{-b}, \quad \text{Prob}(\text{epidemic duration} > t) \propto t^{-c}, \tag{6.14}$$

where $b \approx 0.28$ and $c \approx 0.8$ (Rhodes & Anderson 1996b).

Similar scaling exponents are obtained from cellular automata models, indicating that to some degree the concept of self-organized critical behaviour can be transferred from such simple models to real epidemiological situations (Rhodes & Anderson 1996a). Similar results hold for cases from a small population in England before vaccination, medium size and duration epidemics obey a power-law relationship; however, for large epidemics the depletion of susceptibles brings in density-dependent behaviour and causes deviations from the ideal (Fig. 6.5).

6.5 | NETWORK MODELS

Network models are closely related to cellular automata but have broken free from the rigid lattice structure. Formally, a network is a set of nodes (generally thought of as individuals), some of which are linked by connections—it is only through such connections that diseases can pass. Therefore, cellular automata models are a narrow class of networks in which only nearest-neighbour individuals are connected. Networks were first studied as part of the social sciences; it has only been in recent years, when modelling has shifted to a more individual-based framework, that networks have been seen as important epidemiological tools. This is particularly true for sexually transmitted infections (STIs), for which the network of sexual contacts may be sparse and readily defined. Mathematical networks are described by their graph–theoretical properties, in particular the matrix $\mathbf{G}$, where $\mathbf{G}_{ij}$ is one if there is a connection between nodes i and j and otherwise is zero (Keeling 1999a). Epidemiologically, we tend to concentrate on more observable quantities, such as the number of contacts per individual and the degree of clustering.

Four distinct types of network are considered here (Fig. 6.6). Lattices are the networks associated with cellular automata; they have a rigid structure and a fixed number of connections per individual, generally four or eight (Mollison 1977; Yakowitz, Gani, & Hayes 1990; Johansen 1996; Durrett & Levin 1994). All connections are localized, so strong correlations can develop; in this sense, the lattice networks are highly clustered. Small worlds are a simple extension to the lattice models (Watts & Strogatz 1998, Watts 1999). Individuals are distributed evenly around a circle, and connections are made to the n nearest neighbours. In addition, there is a low density of global connections between randomly chosen individuals. These long-range random connections determine the small-world nature of the network, break the stranglehold of local correlations, and allow the rapid spread of infection to the entire population. Spatial

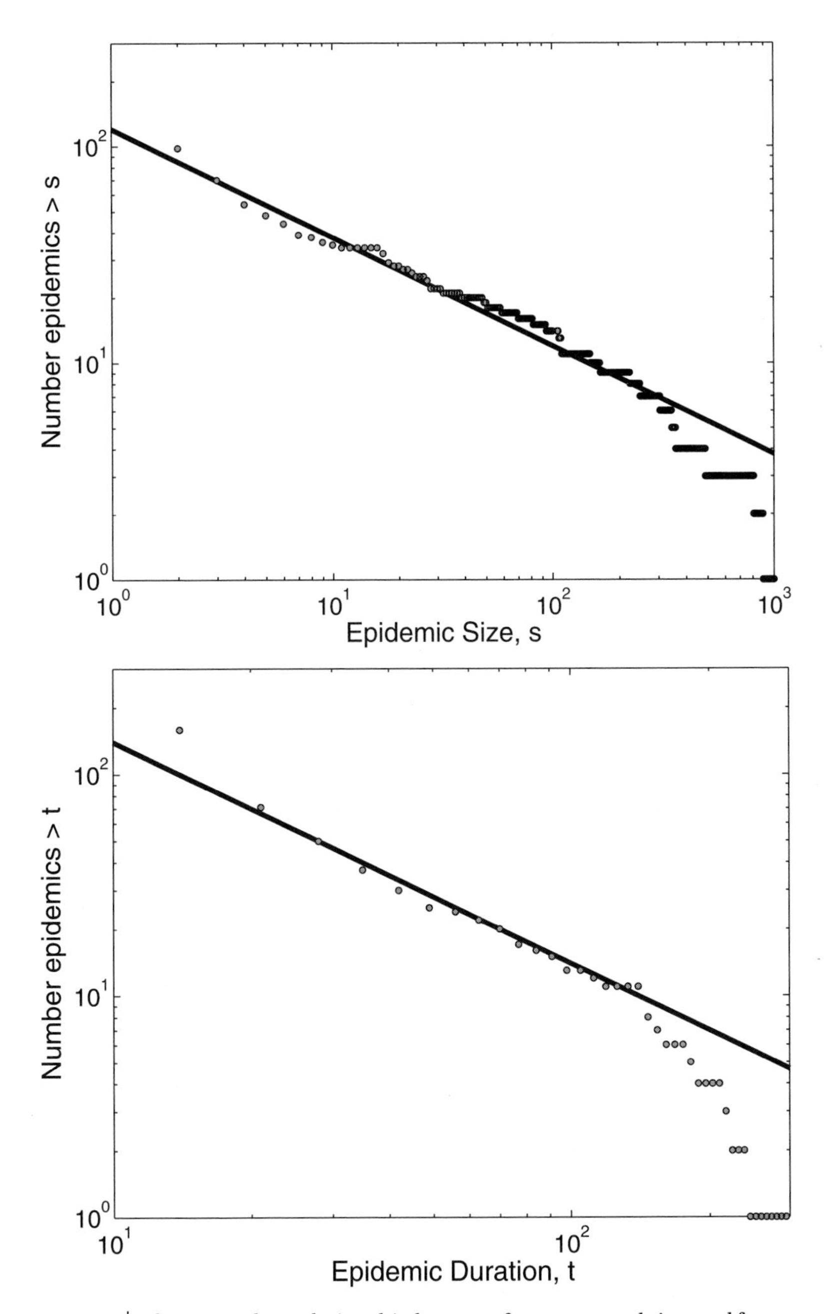

FIGURE 6.5 | The power-law relationship between frequency and size, and frequency and duration, for measles cases in Kings-Lynn—a small fairly isolated population in East Anglia, UK, from 1944 to 1968. Data are taken from the Registrar General's Weekly Reports. To help with the presentation of the data, the number of epidemics of greater size or greater duration is plotted. The calculated best-fit power laws to the data are Prob(epidemic size $> s$) $\propto s^{-0.5}$ and Prob(epidemic duration $> t$) $\propto t^{-0.9}$.

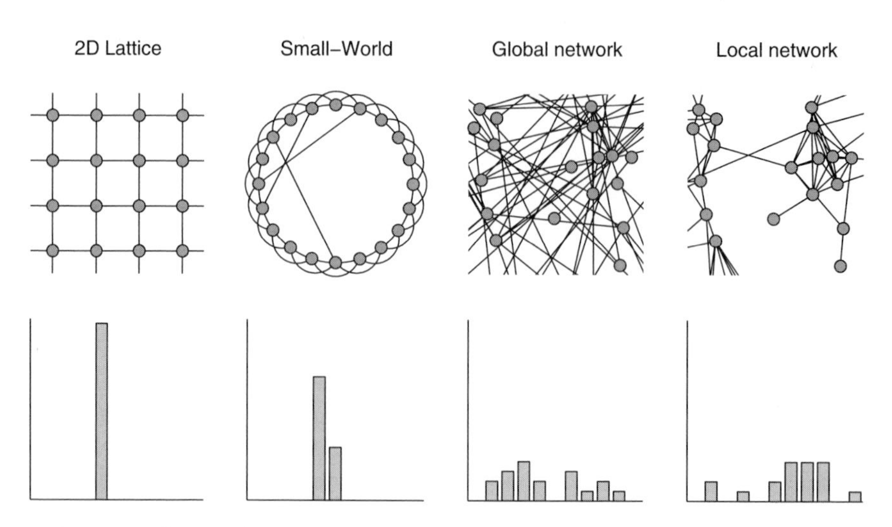

FIGURE 6.6 | Four distinct types of network: a two-dimensional lattice, a one-dimensional small world, a spatial network with global connections, and a spatial network with local connections. The graphs below each figure show the distribution of connections per individual—lattices and small worlds are homogeneous.

networks are formed by first randomly positioning individuals within a given area (or, in some cases, volume). Individuals are then connected randomly with a probability determined by the distances between them (Watts 1999, Read & Keeling 2003). Global networks assume that connections occur randomly irrespective of the separation; such models are a plausible representation of sexual contact networks in which clustering and triangular connections between three individuals are rare. Local networks assume that connections are more likely between nearby individuals, and often the connection probability is Gaussian with distance; such networks are reasonable approximations for the spread of airborne diseases in which triangular loops are the norm. The degree of clustering can be measured by the proportion of three connected individuals that form triangles—small worlds and local networks are highly clustered, whereas global networks have few triangular connections. Such clustering tends to reduce the speed at which infection spreads because fewer new contacts are reached in each generation.

The force of infection in such models can be related to the matrix **G**, which describes the set of connections. The force of infection to a susceptible individual, i, is given by:

$$\lambda_i = \beta \sum_j \mathbf{G}_{ij} I_j, \qquad (6.15)$$

where I_j is one if j is infected and otherwise is zero. Again this looks like mass-action transmission within the local environment described by the contact matrix. From this formulation it is clear that the disease dynamics are tightly slaved to the properties of the matrix **G** and hence to the properties of the network. There are three components to the network that primarily affect the dynamics: the average number of connections, the distribution of connections, and the degree of clustering.

1. As the average number of connections, $\bar{n}$, increases, there is an ever-greater potential for the infection to be spread. Although the expected number of secondary cases caused by a primary infection must increase linearly with $\bar{n}$, the growth rate of the epidemic in later generations may have a more complex dependence influenced by both the distribution and the clustering of contacts.
2. Lattice models and small worlds are homogeneous in the number of contacts per individual; in contrast, the spatial models show a much wider variance. Even when two networks have the same average number of contacts, the presence of a significant variance can have substantial effects. Individuals, which by chance have many contacts, are most likely to become infected; however, these are also the individuals that have the greatest opportunity to spread the infection. Hence, the rate at which infection can spread is increased by a heterogeneous network structure (Anderson & May 1991). This effect is even stronger if the network is assortative, such that highly connected individuals are more likely to connect to other highly connected individuals; such assortative mixing is fundamental to high-risk core groups, which play a pivotal role in the spread of sexually transmitted diseases (Hethcote & van Ark 1987, Anderson & May 1991).
3. Clustering, or transitivity, is defined by the presence of short loops within the network; the simplest definition is therefore:

$$\text{clustering } \phi = \frac{\text{number of triangular loops}}{\text{number of all triples}}, \tag{6.16}$$

which can be rewritten in terms of the connection matrix (Keeling 1999a):

$$\phi = \frac{\text{trace}(\mathbf{G}^3)}{\|\mathbf{G}^2\| - \text{trace}(\mathbf{G}^2)}, \tag{6.17}$$

where $\|\mathbf{G}\| = \sum_{i,j} \mathbf{G}_{ij}$ and $\text{trace}(\mathbf{G}) = \sum_i \mathbf{G}_{ii}$. A high degree of clustering means that strong local correlations can quickly develop; in this way clustering retards the spread of an epidemic because of the local depletion of susceptibles. In contrast, the presence of a few long-range connections allows the infection to escape from this restrictive local structure (Watts & Strogatz 1998). This is the principal driving force behind the small-world models; although the vast majority of their structure is regular and lattice-like, the few random connections drastically reduces the average path length between individuals and changes the patterns of spread.

Network-based models offer a fantastic opportunity to explore the dynamics of diseases at the most local scale. However, there are many difficulties in determining the correct network structure. For sexually transmitted diseases, information from contact-tracing and large-scale surveys provides some insights into the structure of the network (Johnson *et al.* 2001, Potterat *et al.* 2002), but such information is notoriously full of sampling and social biases. For airborne infections the problem is made all the more difficult by the uncertain nature of a connection and the many more connections per individual, although some progress is being made in situations in which close contact is needed for infection to be transmitted (Edmunds, O'Callaghan, & Nokes 1997; Wallinga, Edmunds, & Kretzschmar 1999).

6.6 | ANALYTICAL APPROXIMATIONS: POWER-LAW EXPONENTS

The adaptations to standard mass-action mixing that have been described so far (metapopulation, cellular automata, and network) are all highly computational, requiring extensive simulations to reveal their emergent phenomena. More recently, attention has returned to more analytical methods of improving the accuracy of the interaction (transmission) term. In this section, I will concentrate on these modifications to mass-action and explain how they act as a bridge between the standard mass-action models and the computationally intensive approaches. For all these types of analytical approaches, research motivated by epidemiological applications is at the forefront of development of this field.

Power-law exponents on the interaction terms between susceptible and infected individuals have been postulated as a plausible modification to mass-action mixing that could mimic the effects of spatial aggregation (Filipe & Gibson 1998, 2001; Finkenstädt & Grenfell 2000; Bjørnstad, Finkenstädt, & Grenfell 2002). In particular, the force of infection would become:

$$\lambda = \beta I^{1+\alpha} S^{1+\Psi}. \qquad (6.18)$$

More generally we could consider the mixing parameter β to be a function of susceptible and infectious levels. Intuitively, if α is less than zero, then this modified form acts to dampen the initial exponential growth phase of the epidemic:

$$I(t) \sim (\alpha\beta t)^{-1/\alpha}. \tag{6.19}$$

The other exponent, ψ, operates in the reverse manner: when ψ is greater than zero the density-dependent damping because of a decrease in the level of susceptibles is increased. These results are reflected in the equilibrium level of susceptibles (Fig. 6.7); when the disease is extant the equilibrium level of infection does not vary significantly. It is interesting to note that small changes in the exponents can have significant implications for the dynamics and that for values of α much above zero, extinction of the disease occurs.

Until recently, such power-law exponent models were mathematical curiosities with little firm grounding in observed behaviour. However, in recent years new work into childhood diseases has focused attention on this modification to the standard mass-action assumptions. Research by

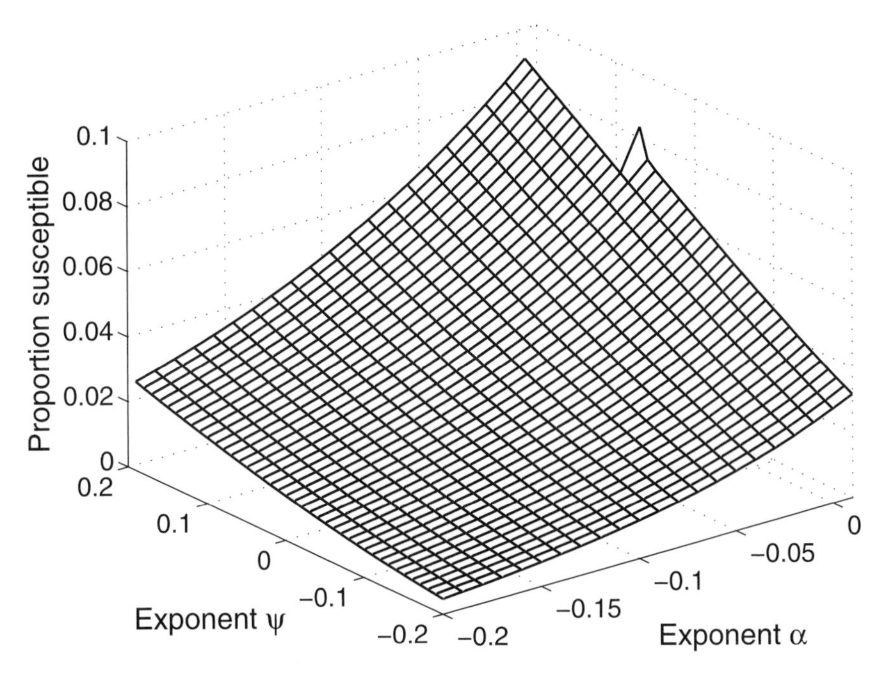

FIGURE 6.7 | An example of how the equilibrium level of susceptible varies with α and ψ. Parameters are measles-like, $\beta = 17/13$ and $g = 1/13$, with a constant population size and an average life expectancy of 50 years, $B = d = 5.5 \times 10^{-5}$. Positive values of α are usually associated with extinction of the disease.

Grenfell and co-workers (Finkenstädt & Grenfell 2000; Bjørnstad, Finkenstädt, & Grenfell 2002) has focused on a discrete-time model of measles infection (the *TSIR* model), which uses a 2-week updating interval. The number of biweekly cases, C_t, and the number of susceptible individuals, S_t, are given by:

$$\begin{aligned} C_{t+1} &= \beta_t C_t^{1+\alpha} S_t \\ S_{t+1} &= B_t - \beta_t C_t^{1+\alpha} S_t, \end{aligned} \tag{6.20}$$

where B_t is the known birth rate from the period of interest and β_t is a seasonally varying contact rate that mimics the opening and closing of school and the changes in the age profile of infected individuals. The 26 seasonal contact parameters (one for each 2-week period of the year), as well as the reporting rate, can be estimated from the known case-report data. However, of more interest is the exponent α, which was estimated to be −0.0311 (Finkenstädt & Grenfell 2000); the exponent ψ, which applies to the level of susceptibles, was ignored for technical reasons. Further work (Bjørnstad, Finkenstädt, & Grenfell 2002) has refined this approach by estimating a value of α for measles dynamics within 60 communities in England and Wales. All the estimated values of α are close to 0, lying between −0.2 and 0.1 (Fig. 6.8).

Figure 6.8 clearly illustrates that although the observed dynamics are close to those predicted by standard mass-action mixing (α is close to zero), there are statistically significant differences that can be captured by slight modifications to transmission terms. This work provides the strongest evidence to date that such power-law corrections may be a more accurate description of the dynamics. Although some of this may be attributed to the move from differential to discrete-time modelling, there is still a substantive change to the predictive dynamics.

6.7 | ANALYTICAL APPROXIMATIONS: PAIR-WISE MODELS

Both cellular automata and network models are based on the following premise: infection can only spread between pairs of connected individuals. Pair-wise models use this observation and base their calculations on the dynamics of pairs of connected individuals rather than simply the dynamics of individuals (Dietz & Hadeler 1988; Sato, Matsuda, & Sasaki 1994; Altmann 1995; Keeling, Rand, & Morris 1997; Bolker & Pacala 1997; van Baalen & Rand 1998; Bolker 1999; Keeling 1999a, 1999b; Bauch & Rand 2000; Eames & Keeling 2002).

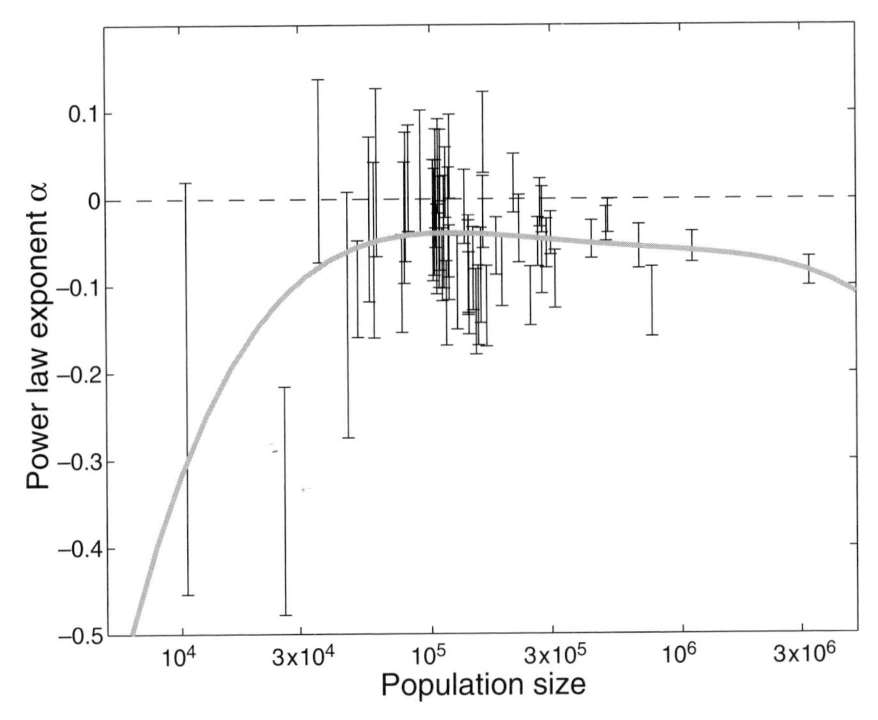

FIGURE 6.8 | The estimated values of the exponent α for measles in 60 cities in England and Wales from 1944 to 1968. Here, the dashed line corresponds to the standard mass-action assumption of $\alpha = 0$, and the grey line is a smoothed fit through the α values. The data are taken from Bjørnstad, Finkenstädt, & Grenfell 2002.

The force of infection, λ, can be considered as the product of two terms: the rate of transmission across a contact and the expected number of infectious individuals in contact with the susceptible in question. In the standard mass-action models it is assumed that every individual is weakly connected to every other in the population; hence, the expected number of infectious individuals in contact with the susceptible is simply proportional to the level of infection. In contrast, pair-wise models lose the assumption of random mixing and explicitly model the correlation between connected pairs (Rand 1999, Keeling 1999a, Bolker 1999).

In the simplest homogeneous networks, often called *random graphs*, in which each individual has exactly n contacts, the pair-wise models give analytical results. Denote the number of S-I connected pairs to be $[SI] = [IS]$, and for consistency of notation write the number of susceptible and infected individuals as $[S]$ and $[I]$, respectively. Under the mass-action assumptions of random mixing, the number of S-I pairs would be approximated by:

$$[SI] \approx n[S][I]/N. \tag{6.21}$$

However, for pair-wise modelling, we treat $[SI]$ as a variable. The force of infection is therefore:

$$\lambda = \tau \frac{[SI]}{[S]} = \tau n \frac{[S][I] + C_{SI}}{[S]N}, \tag{6.22}$$

where τ is the rate of transmission across a contact and C_{SI} is the spatial correlation between susceptible and infected individuals. If we know the number of S-I pairs, then this form of the force of infection (and therefore the standard differential equations) are exact. However, $[SI]$ is a variable and changes during the course of an epidemic as susceptibles are converted to infected individuals; we must therefore formulate equations for its behaviour.

In addition to the standard equations for the number of individuals of each type, we need to consider how the pairs evolve. To be able to iterate the standard equations, we only need to know $[SI]$; I will therefore concentrate on the dynamics of this quantity first:

$$\frac{d[SI]}{dt} = \tau([SS^{\leftarrow}I] - [I^{\rightarrow}SI] - [S^{\leftarrow}I]) - g[SI]. \tag{6.23}$$

The arrows have been added to indicate the transmission of infection involved. The first term is a triple (i.e., S connected to S connected to I) and accounts for the creation of an S-I pair caused by an S-S pair being infected. The next term refers to the loss of an S-I pair because the S is infected from outside the pair. The third term again refers to the loss of an S-I pair, but here infection is transmitted within the pair. Finally, there is recovery of the infected individual.

We now can begin to see a pattern: to accurately predict the dynamics of the number of individuals we need to know the number of pairs. However, to accurately predict the dynamics of pairs we need to know the number of triples. Although mathematical methods exist to deal with this infinite cascade of ever-longer chains of connected individuals, it is biologically plausible and computationally convenient to truncate this process at the pair-wise level. We approximate the number of triples in terms of the number of pairs, which is termed a *moment closure approximation* because it allows us to close the equations in terms of components that we know. We do this assuming that triples are composed of two sets of pairs that share a common central member:

$$[ABC] \approx \frac{n-1}{n} \frac{[AB][BC]}{[B]}. \tag{6.24}$$

This is a reasonable approximation when the network is not clustered; therefore, the ends of the triple (A and C) are unlikely to be connected (Keeling 1999a, Eames & Keeling 2002). When there is a significant amount of clustering, this triple approximation can be modified by taking the correlation between A and C into account (Keeling 1999a).

From the triple approximation it is clear that the dynamics of $[SI]$ depend on the number of $[SS]$ pairs, and for clustered networks the number of $[II]$ pairs also plays a role. Therefore, for completeness, we give the equations for all possible pairs:

$$
\begin{aligned}
\frac{d[SS]}{dt} &= -2\tau[SS^{\leftarrow}I] \\
\frac{d[SI]}{dt} &= \tau([SS^{\leftarrow}I]-[I^{\rightarrow}SI]-[S^{\leftarrow}I])-g[SI] \\
\frac{d[SR]}{dt} &= -\tau[I^{\rightarrow}SR]+g[SI] \\
\frac{d[II]}{dt} &= 2\tau([IS^{\leftarrow}I]-[S^{\leftarrow}I])-2g[II] \\
\frac{d[IR]}{dt} &= \tau[I^{\rightarrow}SR]+g[II]-g[IR] \\
\frac{d[RR]}{dt} &= 2g[IR]
\end{aligned}
\tag{6.25}
$$

We note that from these equations we can regain the dynamics of individuals ($[SI] + [II] + [IR] = n[I]$, etc.). Here we have only considered the simple *SIR* epidemic without births and deaths; these demographic events perturb the structure of the network and therefore are difficult to include in a natural manner (Keeling 1999b).

For the *SIR* epidemic in a random (nonclustered) network, the complete dynamics can be expressed with the following equations:

$$
\begin{aligned}
\frac{d[S]}{dt} &= -\tau[SI] \\
\frac{d[I]}{dt} &= \tau[SI]-g[I] \\
\frac{d[SI]}{dt} &= \frac{\tau[SI]}{n[S]}((n-1)[SS]-(n-1)[SI]-n[S])-g[SI] \\
\frac{d[SS]}{dt} &= -2\frac{\tau[SI]}{n[S]}(n-1)[SS]
\end{aligned}
\tag{6.26}
$$

Hence, in moving from a random-mixing model to an approximation of the effects of network structure we have only doubled the number of equations from two to four.

We note that for the same basic disease parameters, the pair-wise model predicts a far less severe epidemic with a slower rate of increase and a lower maximum (Fig. 6.9). This is because the pair-wise model can capture the local depletion of susceptibles, which limited the spread of infection in network and cellular automata models. The strength of this correlation effect can be seen in the infection modifier M, which is defined as the ratio of transmission rates under the pair-wise and random-mixing assumptions. Initially $M = 1$, because when there is only a single infectious individual in a sea of susceptibles there are no spatial correlations and the two models are in agreement. However, during the first few generations of the epidemic, as the local correlations build up, the modifier drops to a lower level $M \to 1 - 2/n$ (Keeling 1999a), where it remains until the epidemic becomes large and nonlinear effects begin to act. The period during which $M = 1$ is too short to be seen in Figure 6.9, and the asymptotic level of $M = 1 - 2/n = 0.6$ determines the initial dynamics. This asymptotic level of M allows us to estimate the basic reproductive ratio (calculated from the early growth rate) in such models (Keeling 1999a):

$$R_0^{\text{random mixing}} = n\tau \quad R_0^{\text{pair-wise}} = (n-2)\tau. \tag{6.27}$$

We therefore see that, in agreement with cellular automaton and network models, the pair-wise models need a higher transmission rate to persist.

The discrepancy between pair-wise and random-mixing models is largest when the number of connections per individual is small; hence, although the local correlations that develop may only have a limited effect on the dynamics of airborne diseases, for STIs—where most individuals have few contacts—the differences are far more significant. Hence, recent applications have focused on the use of pair-wise models for predicting the dynamics of STIs (Bauch & Rand 2000, Eames & Keeling 2002); STIs also have the considerable advantage that the associated networks are not highly clustered. To validate these methods it is useful to compare the results of direct simulation of infection on a computer-generated network with the solution to the pair-wise equations (Fig. 6.10).

An alternative form of pair-wise model, which is highly applicable to plant pathogens, is provided by considering transmission in two-dimensional space rather than on a network (Bolker & Pacala 1997, 1999;

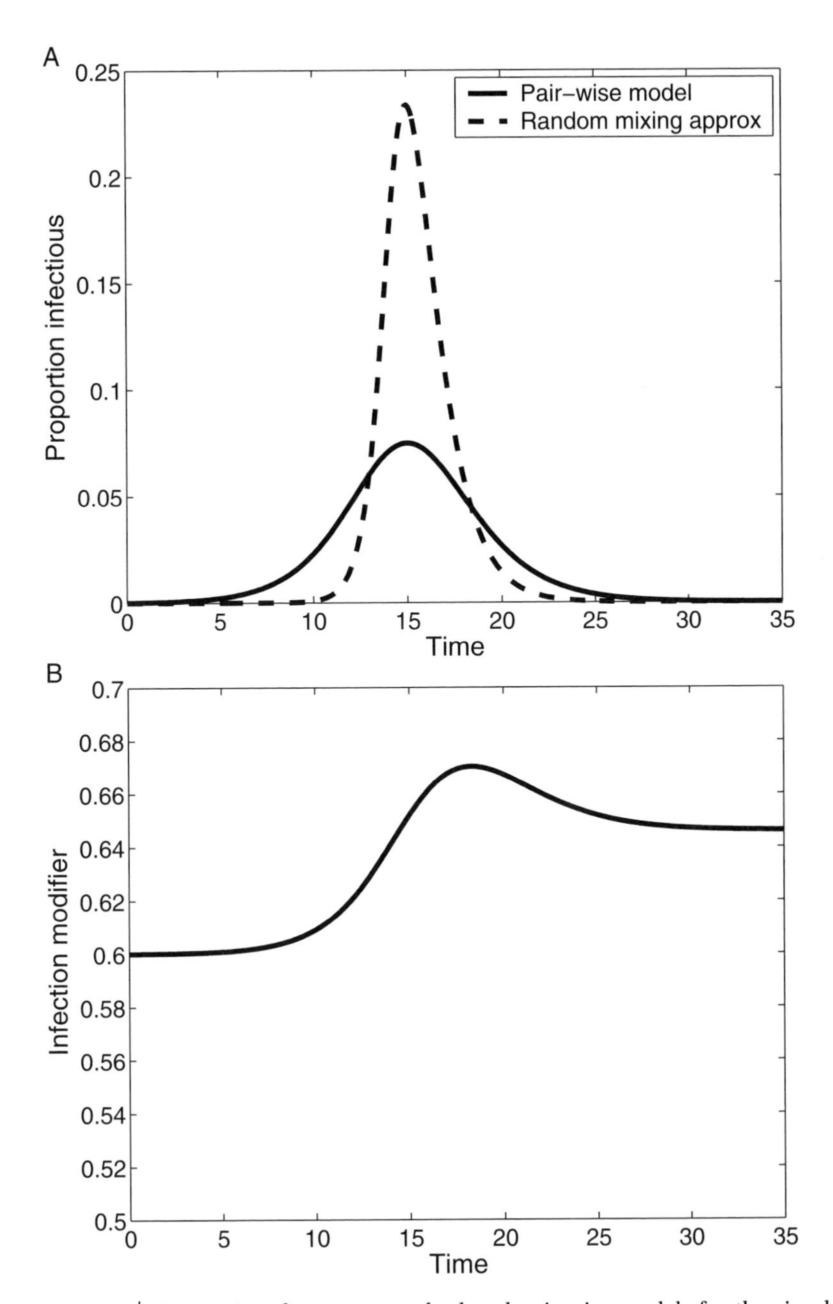

FIGURE 6.9 | Comparison between standard and pair-wise models for the simple *SIR* epidemic without births and deaths. The standard mass-action model uses the approximation $[SI] \approx n[S][I]/N$. For this example, $n = 5$, $\tau = 0.5$, and $g = 1$. In panel A, the two epidemic curves have been aligned such that their maxima occur at the same time. Panel B shows the infection modifier $M = [SI]N/(n[S][I])$, which measures the ratio between the true force of infection and that derived from the random-mixing approximation.

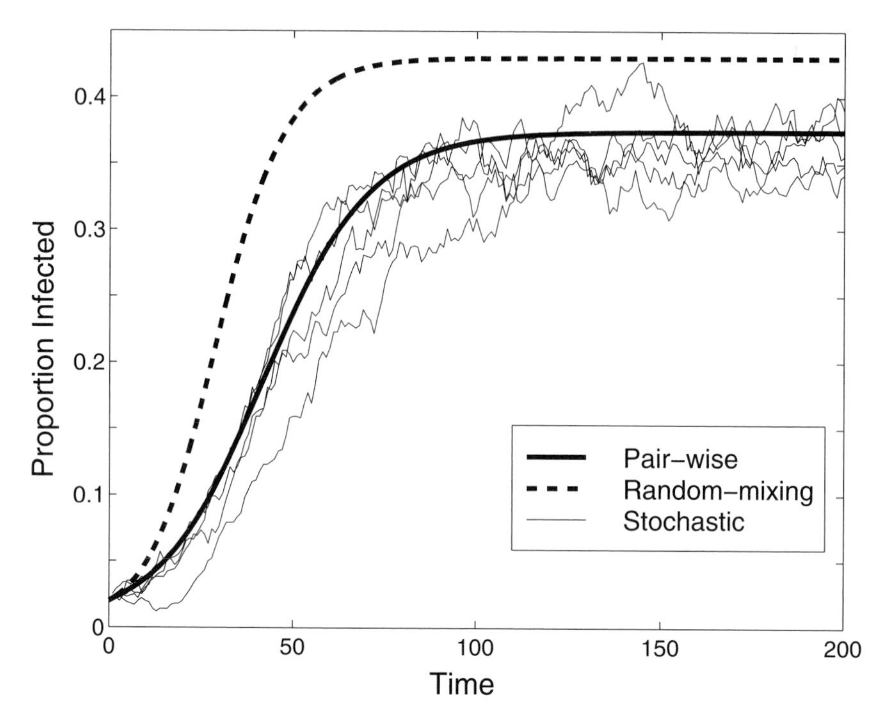

FIGURE 6.10 | Comparison of a random-mixing model, a pair-wise model, and stochastic simulations of an sexually transmitted disease epidemic in a heterogeneous spatial network with global connections. Both the random-mixing model and the pair-wise model are modified to account for the heterogeneity and assortivity in contact numbers.

Law, Murrell, & Dieckmann 2003). In this formulation the basic equations become:

$$\frac{dS}{dt} = B - \int_0^\infty \tau(r)[SI](r)dr - dS$$
$$\frac{dI}{dt} = \int_0^\infty \tau(r)[SI](r)dr - gI - dI \tag{6.28}$$

where $[SI](r)$ refers to the number of S-I pairs separated by a distance r, and $\tau(r)$ measures how infectivity changes with distance. Hence, the force of infection is given by:

$$\lambda = \frac{\int_0^\infty \tau(r)[SI](r)dr}{S} = \left(\int_0^\infty 2\pi r\tau(r)dr\right)I + \frac{\int_0^\infty 2\pi r C_{SI}(r)}{S}dr, \tag{6.29}$$

where the first term is proportional to the level of infection and is therefore the familiar mass-action component, and the second term is the modification caused by the spatial covariance C_{SI}. The presence of an integral in the equations makes the mathematics more formidable, although many of the concepts are similar to the network-based approach. Great care is needed with the triple approximations because any three individuals automatically form a triangular connection. Despite these difficulties, great progress has been made with this technique, which has provided good agreement with, and a deeper understand of, the results of full spatial simulations (Bolker & Pacala 1997, 1999; Law, Murrell, & Dieckmann 2003).

Pair-wise models offer some exciting possibilities for the future of disease modelling, especially of sexually transmitted diseases. Although full network simulations are difficult to parameterize and only provide information about one specific case, pair-wise models can be parameterized from readily attainable contact-tracing data and offer generic and robust predictions (Eames & Keeling 2002). The only limiting factor is the large number of differential equations involved for pair-wise models of heterogeneous networks; however, the iteration of such equations is still far quicker than the stochastic simulation of large networks.

6.8 | ANALYTICAL APPROXIMATIONS: MOMENT CLOSURE

The pair-wise models described previously deal with the buildup of correlations at the local scale because of the finite number of connections per individual. Correlations can also occur at the larger metapopulation scale, and these are primarily driven by the stochastic nature of the within-community dynamics. Again, we can go beyond the standard mass-action models by considering higher-order terms and making a moment closure approximation (Isham 1995; Keeling 2000; Keeling, Wilson, & Pacala 2000; Nasell 1999, 2002, 2003).

Consider a situation in which there is a large (or infinite) set of independent populations, all of which act independently and hence have differing levels of infection. Define $\langle\cdot\rangle$ to be the average of any given quantity over all the populations. In many situations this will be the same as the long-term time average of just one population. Now, consider the average dynamics across all populations and therefore across all stochastic realisations. First note that:

$$\frac{d\langle\cdot\rangle}{dt} = \left\langle \frac{d}{dt} \cdot \right\rangle. \tag{6.30}$$

That is, the rate of change of the mean is the mean of all the rates of change; hence, assuming mass-action mixing at the population level:

$$\begin{aligned}\frac{d\langle S\rangle}{dt} &= \langle B - \beta SI - dS\rangle = B - \beta\langle SI\rangle - d\langle S\rangle \\ \frac{d\langle I\rangle}{dt} &= \langle \beta SI - gI - dI\rangle = \beta\langle SI\rangle - (g+d)\langle I\rangle.\end{aligned} \tag{6.31}$$

Therefore, at the aggregate level the force of infection must take into account both the average levels of susceptible and infectious individuals within the population and the covariance between them:

$$\lambda = \beta\frac{\langle SI\rangle}{\langle S\rangle} = \beta\langle I\rangle + \beta\frac{C_{SI}}{\langle S\rangle}, \tag{6.32}$$

where C_{SI} is the covariance between S and I, and, as we have previously seen for metapopulations, this is usually negative. Thus, the aggregate model has a reduced force of infection, and therefore more susceptibles, than the standard mass-action models.

Once again we are in a situation where exact knowledge of $\langle SI\rangle$ would allow us to accurately predict the dynamics of the mean population levels. Following the previous formulation, it seems appropriate to treat $\langle SI\rangle$ as a variable and try to calculate its dynamic behaviour. In practice, it is often simpler and somewhat more biologically intuitive to model the covariance, C_{SI}.

$$\frac{dC_{SI}}{dt} = \frac{d}{dt}[\langle SI\rangle - \langle S\rangle\langle I\rangle] = \frac{d\langle SI\rangle}{dt} - S\frac{d\langle I\rangle}{dt} - I\frac{d\langle S\rangle}{dt}. \tag{6.33}$$

The rate of change of terms such as $\langle SI\rangle$ are calculated as follows:

$$\frac{d\langle SI\rangle}{dt} = \left\langle \sum_{\text{events}} \text{Rate of event} \times [\text{Change to } \langle SI\rangle] \right\rangle. \tag{6.34}$$

Consider this for the five events that can occur (birth, death of susceptible, death of infected, recovery, and transmission):

$$\begin{aligned}\frac{d\langle SI\rangle}{dt} = \langle & B\times[(S+1)I - SI] && \text{birth} \\ &+ dS\times[(S-1)I - SI] && \text{susceptible death} \\ &+ dI\times[S(I-1) - SI] && \text{infected death} \\ &+ gI\times[S(I-1) - SI] && \text{recovered death} \\ &+ \beta SI\times[(S-1)(I+1) - SI]\rangle && \text{transmission} \\ = \langle & BI - dSI - dIS - gIS + \beta SI(S - I + 1)\rangle.\end{aligned} \tag{6.35}$$

Once again, such a calculation involves higher-order terms ($\langle S^2 I\rangle$ and $\langle SI^2\rangle$) that need approximating, which can be done in terms of lower-order components by looking at the various combinations of singletons and pairs contained within the expression:

$$\langle SSI\rangle = \langle S\rangle\langle S\rangle\langle I\rangle + C_{SS}\langle I\rangle + 2C_{SI}S + T_{SSI}, \tag{6.36}$$

where T_{SSI} is the third-order correction and C_{SS} is naturally the variance in S. One standard closure assumption to make is that the third-order corrections are all zero. This leads to the following equation for the covariance:

$$\begin{aligned}\frac{dC_{SI}}{dt} = {} & C_{SI}(\beta\langle S\rangle + \beta - \beta\langle I\rangle - 2d - g) + \beta\langle S\rangle\langle I\rangle \\ & + \beta C_{SS}\langle I\rangle - \beta C_{II}\langle S\rangle.\end{aligned} \tag{6.37}$$

To complete the equations, a similar procedure must be followed to find equations for the variance of S and I.

$$\begin{aligned}\frac{dC_{II}}{dt} &= 2C_{II}(\beta\langle S\rangle - d - g) + 2\beta\langle I\rangle C_{SI} + \beta\langle S\rangle\langle I\rangle + \beta C_{SI} + (d+g)\langle I\rangle \\ \frac{dC_{SS}}{dt} &= -2C_{SS}(\beta\langle I\rangle + d) - 2\beta\langle S\rangle C_{SI} + \beta\langle S\rangle\langle I\rangle + \beta C_{SI} + B + d\langle S\rangle\end{aligned} \tag{6.38}$$

We have moved from the standard set of two differential equations to a set of five equations; however, this extra increase allows us to readily capture the effects of stochasticity and population-level correlations (Keeling 2000). Figure 6.11 shows the equilibrium level of the various parameters as the population size, and therefore the level of stochasticity, varies.

There are several features to notice in these graphs. The first is that, as expected, the relative effect of the covariance, C_{SI}, becomes more negative as the population size becomes smaller. This is reflected as a decrease in the force of infection and hence as an increase in the level of susceptibles. We also observe an increase in the standard error as the population size decreases and, therefore, stochasticity plays a larger role. Finally, for populations below 225,000 the equations break down; this is because the assumption that all third-order corrections are zero (which equates to assuming all distributions are normal) is not valid. Work using other, more biologically plausible assumptions (such as negative binomial distributions) can extend the range over which the equations work (Keeling 2000, Nasell 2003).

Moment closure approximations, such as those developed here, are an extremely powerful tool for understanding the roles of stochasticity and spatial segregation. They are increasingly being used to provide an analytical understanding of disease extinction, which is a stochastically driven phenomenon. Despite the variety of approaches, all models agree with the basic phenomena of a strong negative correlation between infected and susceptible levels within a population, which decreases the force of infection experienced. The challenge for using such models in the future will be to find robust and biologically appropriate approximations for the third-order corrections. This can only come from an interaction between theory and simulation.

6.9 | CONCLUSIONS

The standard principles of mass-action mixing still dominate the fields of ecology and epidemiology. Even the more complex examples described here employ this form of mixing at the most local of scales. It is only when we attempt to derive global population-level dynamics that the mixing rates are modified. Mass-action mixing is therefore not a paradigm lost but one that requires reinterpretation when examining problems at a broader scale.

There are now two directions that epidemiological modelling can take in the future; we can either focus on high-power computational techniques or on improved analytical approximations. With the recent increases in computer power it is feasible to model all individuals within a large population and to use some basic rules to simulate their local interactions. Such techniques may reveal unexpected emergent phenomenon and should provide the most accurate means of modelling a particular disease scenario. However, these models can only be as good

FIGURE 6.11 | Equilibrium values for the moment closure model with measles-like parameters ($R_0 = 17$, $g = 1/13$, $d = 5.5 \times 10^{-5}$, and $B = dN$) for various population sizes, N. Panel A shows the level of susceptible compared with the theoretical level predicted by the standard *SIR* model—note that there is negligible difference in the level of infection. Panel B gives the relative change in the force of infection because of the correlations; this is $\frac{C_{SI}}{\langle S \rangle \langle I \rangle}$. Finally, panels C and D show the standard error for susceptibles and infected individuals, $\frac{\sqrt{C_{SS}}}{\langle S \rangle}$ and $\frac{\sqrt{C_{II}}}{\langle I \rangle}$, respectively.

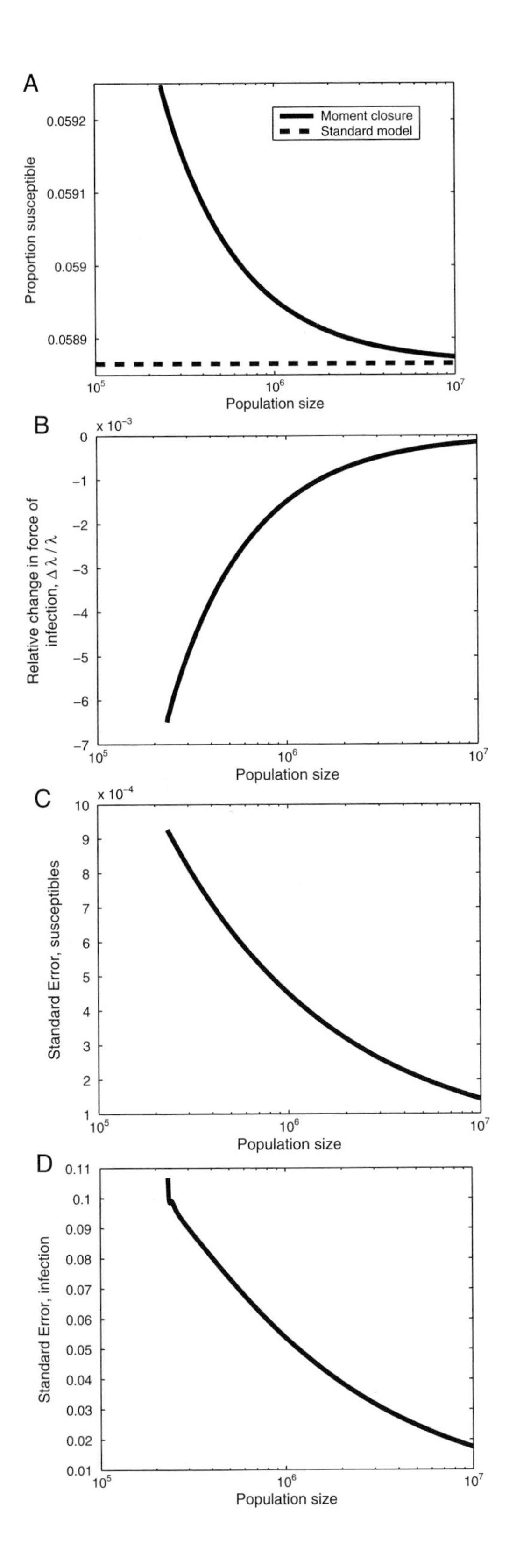
A
Moment closure
Standard model
Proportion susceptible
0.0592
0.0591
0.059
0.0589
Population size
B
x 10^-3
Relative change in force of infection, Δλ / λ
Population size
C
x 10^-4
Standard Error, susceptibles
Population size
D
Standard Error, infection
Population size

as the data used to parameterize them, and our general understanding of human mixing patterns is still extremely limited. These computer models also suffer from a lack of generality, such that the results are specific to the situation under scrutiny. In contrast, the analytical approaches offer a more robust approach and provide the opportunity to formulate general results. Often, these analytical approaches can be recast into a modification to the basic mixing terms and hence provide a more intuitive understanding of the problem.

Clearly, both computational and analytical methods have their merits and disadvantages. It can be hoped that by exploring these two distinct avenues in parallel, both approaches will benefit, leading to a better understanding of disease dynamics and more accurate predictions of given situations with the ultimate aim of more effective disease control and eradication.

REFERENCES

Altmann, M. (1995). Susceptible–infectious–recovered epidemic models with dynamic partnerships. *J. Math. Biol.* **33**, 661–675.

Anderson, R.M., & May, R.M. (1979). Population biology of infectious diseases, Part I. *Nature* **280**, 361–366.

Anderson, R.M., & May, R.M. (1991). “Infectious Diseases of Humans.” Oxford University Press, Oxford.

Bak, P., Chen, K., & Tang, C. (1990). A forest fire model and some thoughts on turbulence. *Phys. Lett. A* **147**, 297–300.

Bartlett, M.S. (1956). Deterministic and stochastic models for recurrent epidemics. *Proc. Third Berkeley Symp. Math. Stats. Prob.* **4**, 81–108.

Bartlett, M.S. (1957). Measles periodicity and community size. *J. Roy. Stat. Soc. A* **120**, 48–70.

Bauch, C., & Rand, D.A. (2000). A moment closure model for sexually transmitted disease transmission through a concurrent partnership network. *Proc. Roy. Soc. Lond. B* **267**, 2019–2027.

Bjørnstad, O.N., & Grenfell, B.T. (2001). Noisy clockwork: Time series analysis of population fluctuations in animals. *Science* **293**, 638–643.

Bjørnstad, O.N., Finkenstädt, B.F., & Grenfell, B.T. (2002). Dynamics of measles epidemics: Estimating scaling of transmission rates using a time series *SIR* model. *Ecol. Monogr.* **72**, 169–184.

Bolker, B. (1999). Analytic models for the patchy spread of plant disease. *Bull. Math. Biol.* **61**, 849–874.

Bolker, B., & Pacala, S.W. (1997). Using moment equations to understand stochastically driven spatial pattern formation in ecological systems. *Theor. Popul. Biol.* **52**, 179–197.

Bolker, B., & Pacala, S.W. (1999). Spatial moment equations for plant competition: Understanding spatial strategies and the advantages of short dispersal. *Am. Nat.* **153**, 575–602.

Dietz, K., & Hadeler, K.P. (1988). Epidemiological models for sexually transmitted diseases. *J. Math. Biol.* **26**, 1–25.

Donnelly, C.A., Ghani, A.C., Leung, G.M., Hedley, A.J., Fraser, C., Riley, S., Abu-Raddad, L.J., Ho, L.M., Thach, T.Q., Chau, P., Chan, K.P., Lam, T.H., Tse, L.Y., Tsang, T., Liu, S.H., Kong, J.H.B., Lau, E.M.C., Ferguson, N.M., & Anderson, R.M. (2003). Epidemiological determinants of spread of causal agent of severe acute respiratory syndrome in Hong Kong. *Lancet* **361**, 1761–1766.

Durrett, R., & Levin, S.A. (1994). The importance of being discrete (and spatial). *Theor. Popul. Biol.* **46**, 363–394.

Dye, C. (1992). The analysis of parasite transmission by bloodsucking insects. *Annu. Rev. Entomol.* **37**, 1–19.

Eames, K.T.D., & Keeling, M.J. (2002). Modelling dynamic and network heterogeneities in the spread of sexually transmitted diseases. *Proc. Natl. Acad. Sci. USA* **99**, 13,330–13,335.

Edmunds, W.J., O'Callaghan, C.J., & Nokes, D.J. (1997). Who mixes with whom? A method to determine the contact patterns of adults that may lead to the spread of airborne infections *Proc. Roy. Soc. Lond. B* **1384**, 949–957.

Ferguson, N.M., Keeling, M.J., Edmunds, W.J., Gani, R., Anderson, R.M., & Leach, S. (2003). Planning for smallpox outbreaks. *Nature* **425**, 681–685.

Filipe, J.A.N., & Gibson, G.J. (1998). Studying and approximating spatiotemporal models for epidemic spread and control. *Phil. Trans. Roy. Soc. B* **353**, 2153–2162.

Filipe, J.A.N., & Gibson, G.J. (2001). Comparing approximations to spatiotemporal models for epidemics with local spread. *Bull. Math. Biol.* **63**, 603–624.

Finkenstädt, B.F., & Grenfell, B.T. (1998). Empirical determinants of measles metapopulation dynamics in England and Wales. *Proc. Roy. Soc. Lond. B* **265**, 211–220.

Finkenstädt, B.F., & Grenfell, B.T. (2000). Time series modelling of childhood diseases: A dynamical systems approach. *J. Roy. Stat. Soc. C* **49**, 187–205.

Gog, J., Woodroffe, R., & Swinton, J. (2002). Disease in endangered metapopulations: The importance of alternative hosts. *Proc. Roy. Soc. Lond. B* **269**, 671–676.

Grenfell, B.T., & Bolker, B.M. (1998). Cities and villages: Infection hierarchies in a measles metapopulation. *Ecol. Lett.* **1**, 63–70.

Grenfell, B.T., & Harwood, J. (1997). (Meta)population dynamics of infectious diseases. *TREE* **12**, 395–399.

Grenfell, B.T., Bolker, B.M., & Kleczkowski, A. (1995). Seasonality and extinction in chaotic metapopulations. *Proc. Roy. Soc. Lond. B* **259**, 97–103.

Hanski, I. (1994). A practical model of metapopulation dynamics. *J. Anim. Ecol.* **63**, 151–162.

Hanski, I. (1999). "Metapopulation Ecology." Oxford University Press, Oxford.

Hanski, I., & Gilpin, M.E. (1997). "Metapopulation Biology: Ecology, Genetics, and Evolution." Academic Press, London.

Hanski, I., & Gaggiotti, O.E. (eds.) (2004) Ecology, Genetics, and Evolution of Metapopulations. Elsevier Academic Press.

Hassell, M.P. (2000). "The Spatial and Temporal Dynamics of Host–Parasitoid Interactions." Oxford University Press, Oxford.

Hawkes, S.J., & Hart, G.J. (1993). Travel, migration, and HIV. *AIDS Care* **5**, 207–214.

Hethcote, H.W., & van Ark, J.W. (1987). Epidemologic models for heterogeneous populations: Proportionate mixing, parameter estimations, and immunization programs. *Math. Biosci.* **84**, 85–118.

Hess, G. (1996). Disease in metapopulation models: Implications for conservation. *Ecology* **77**, 1617–1632.

Holling, C.S. (1959). Some characteristics of simple types of predation and parasitism. *Can. Entomol.* **91**, 385–298.

Isham, V. (1995). Stochastic models of host–macroparasite interaction. *Ann. Appl. Prob.* **5**, 720–740.

Johansen, A. (1996). A simple model of recurrent epidemics. *J. Theor. Biol.* **17**, 45–51.

Johnson, A.M., Mercer, C.H., Erens, B., Copas, A.J., McManus, S., Wellings, K., Fenton, K.A., Korovessis, C., Macdowall, W., Nanchahal, K., Purdon, S., & Field, J. (2001). Sexual-risk behaviour in Britain: Partnerships, practices, and HIV-risk behaviours. *Lancet* **358**, 1835–1842.

Keeling, M.J. (1999a). The effects of local spatial structure on epidemiological invasions. *Proc. Roy. Soc. Lond. B* **266**, 859–869.

Keeling, M.J. (1999b). Correlation equations for endemic diseases. *Proc. Roy. Soc. Lond. B* **266**, 953–961.

Keeling, M.J. (2000). Metapopulation moments: Coupling, stochasticity, and persistence. *J. Anim. Ecol.* **69**, 725–736.

Keeling, M.J., & Grenfell, B.T. (1999). Stochastic dynamics and a power law for measles variability. *Phil. Trans. Roy. Soc. Lond. B* **354**, 769–776.

Keeling, M.J., & Rohani, P. (2002). Estimating spatial coupling in epidemiological systems: A mechanistic approach. *Ecol. Lett.* **5**, 20–29.

Keeling, M.J., Rand, D.A., & Morris, A.J. (1997). Correlation models for childhood diseases. *Proc. Roy. Soc. Lond. B* **264**, 1149–1156.

Keeling, M.J., Wilson, H.B., & Pacala, S.W. (2000). Reinterpreting space, time lags, and functional responses in ecological models. *Science* **290**, 1758–1761.

Law, R., Murrell, D.J., & Dieckmann, U. (2003). Population growth in space and time: Spatial logistic equations. *Ecology* **84**, 252–262.

Levins, R. (1969). Some demographic and genetic consequences of environmental heterogeneity for biological control. *Bull. Ent. Soc. Am.* **15**, 237–240.

Lloyd, A.L., & May, R.M. (1996). Spatial heterogeneity in epidemic models. *J. Theor. Biol.* **179**, 1–11.

McCallum, H., & Dobson, A. (2002). Disease, habitat fragmentation, and conservation. *Proc. Roy. Soc. Lond. B* **269**, 2041–2049.

Mollison, D. (1977). Spatial contact models for ecological and epidemic spread. *J. Roy. Stat. Soc. B* **39**, 283–326.

Nasell, I. (1999). On the time to extinction in recurrent epidemics. *J. Roy. Stat. Soc.* **61**, 309–330.

Nasell, I. (2002). Stochastic models of some endemic infections. *Math. Biosci.* **179**, 1–79.

Nasell, I. (2003). An extension of the moment closure method. *Theor. Popul. Biol.* **64**, 233–239.

Potterat, J.J., Phillips Plummer, L., Muth, S.Q., Rothenberg, R.B., Woodhouse, D.E., Maldonado Long, T.S., Zimmerman, H.P., & Muth, J.B. (2002). Risk network structure in the early epidemic phase of HIV transmission in Colorado Springs. *Sex. Transm. Infect.* **78** (Suppl. I), i159 i163.

Rand, D.A. (1999). Correlation equations for spatial ecologies. *In* "Advanced Ecological Theory" (J. McGlade, Ed.). Blackwell Science, London, pp. 99–143.

Rand, D.A., Keeling, M.J., & Wilson, H.B. (1995). Invasion, stability, and evolution to criticality in spatially extended, artificial host–pathogen ecologies. *Proc. Roy. Soc. Lond. B* **259**, 55–63.

Read, J.M., & Keeling, M.J. (2003). Disease evolution on networks: The role of contact structure. *Proc. Roy. Soc. Lond. B* **270**, 699–708.

Renshaw, E. (1991). "Modelling Biological Populations in Space and Time." Cambridge University Press, Cambridge.

Rhodes, C.J., & Anderson, R.M. (1996a). Persistence and dynamics in lattice models of epidemic spread. *J. Theor. Biol.* **180**, 125–133.

Rhodes, C.J., & Anderson, R.M. (1996b). Power laws governing epidemics in isolated populations. *Nature* **381**, 600–602.

Rhodes, C.J., & Anderson, R.M. (1998). Forest fire as a model for the dynamics of disease epidemics. *J. Franklin I* **335B**, 199–211.

Rhodes, C.J., Jensen, H.J., & Anderson, R.M. (1997). On the critical behaviour of simple epidemics. *Proc. Roy. Soc. Lond. B.* **264**, 1639–1646.

Rohani, P., Keeling, M.J., & Grenfell, B.T. (2002). The interplay between determinism and stochasticity in childhood diseases. *Am. Nat.* **159**, 469–481.

Sato, K., Matsuda, H., & Sasaki, A. (1994). Pathogen invasion and host extinction in lattice structured populations. *J. Math. Biol.* **32**, 251–268.

Sattenspiel, L., & Dietz, K. (1995). A structured epidemic model incorporating geographic mobility among regions. *Math. Biosci.* **128**, 71–91.

Swinton, J., & Gilligan, C.A. (1996). Dutch elm disease and the future of the elm in the UK: A quantitative analysis. *Phil. Trans. Roy. Soc. Lond. B* **351**, 605–615.

Tang, C., & Bak, P. (1988). Critical exponents and scaling relations for self-organized critical phenomena. *Phys. Rev. Lett.* **60**, 2347–2350.

van Baalen, M., & Rand, D.A. (1998). The unit of selection in viscous populations and the evolution of altruism. *J. Theor. Biol.* **193**, 631–648.

Wallinga, J., Edmunds, W.J., & Kretzschmar, M. (1999). Perspective: Human contact patterns and the spread of airborne infectious diseases. *Trends Microbiol.* **7**, 372–377.

Watts, D.J. (1999). "Small Worlds: The Dynamics of Networks Between Order and Randomness." Princeton University Press, Princeton, NJ.

Watts, D.J., & Strogatz, S.H. (1998). Collective dynamics of "small-world" networks *Nature* **393**, 440–442.

Wilson, M.E. (1995). Travel and the emergence of infectious diseases. *Emerg. Infect. Dis.* **1**, 39–46.

Yakowitz, S., Gani, J., & Hayes, R. (1990). Cellular automaton modeling of epidemics. *Appl. Math. Comput.* **40**, 41–54.

7 | MASS-ACTION AND SYSTEM ANALYSIS OF INFECTION TRANSMISSION

James S. Koopman

7.1 | INTRODUCTION

In my 32 years as an epidemiologist with public health responsibility and as an infection transmission modeller, I have noted that epidemiologists and other infection scientists are rarely familiar with any history or theory of mathematical transmission models. Only a tiny fraction use transmission models. One reason for this disconnect is the mass-action paradigm and the ethos of using models to extract the essence of a system rather than to develop ever-more detailed and realistic theory that can guide infection control. The chapters by Heesterbeek (Chapter 5) and Keeling (Chapter 6) in this volume are relevant to how population infection analysis arrived at its current level of theory development and where it is going. I will describe these chapters in the context of theory change advancing a science that can analyze population transmission in ever-more detailed and informative ways. Transmission theory is increasingly addressing the great complexity of biological diversity affecting agent-host interactions and of social interaction patterns through which infection spreads. New model forms that allow analysis of such complexity are advancing theory change in this direction. I will expand upon a point in Keeling's chapter to enumerate ways in which the directions that theory change can take are determined by the model forms used to analyze infection transmission through populations. I will carry this further to argue that assessing the robustness of model inferences to realistic violation of simplifying model assumptions is a good way both to advance the population science of infection transmission and to serve public health. Robustness assessment requires relaxation of model assumptions that are intrinsic to model form. Thus, a central part of my presentation consists of the relationships among model forms, model assumptions, and how the directions that theory change can take are determined by model form.

Heesterbeek's entertaining history of mass action (Chapter 5) presents scientists with strong field experiences such as Hamer, Ross, and Soper groping for a handle on analyzing how infection levels rise and decline and how infection spreads in populations. Heesterbeek narrates how the mass-action paradigm arose spontaneously in epidemiology on more than one occasion and then was strengthened and made more useful by mathematics and analytic thinking adopted from chemistry. We can see in Heesterbeek's narrative an interaction between theory and perceptions of population patterns of infection in which theory takes a leading role and proves more productive than the approach that allows data to be the guiding light. But it is not the theory of mass action that guides and motivates these early researchers. They are motivated by infection control needs rather than by what will advance a scientific paradigm.

The theory of mass action became central to analyzing infection transmission systems throughout the middle and late twentieth century. But it had limited influence on those who took responsibility for the control of infection in populations. The highly abstract and unrealistic nature of mass-action models left most epidemiologists with public health responsibility uninspired either to gather the data to test the theory or to modify the theory in a way that could make the theory work for them. The science of infection transmission system analysis during this period was driven more by mathematicians than by epidemiologists seeking to control infection. The quantitatively trained individuals to whom epidemiologists related in the last half of the twentieth century were statisticians whose theoretical foundations were sampling theory. That is still true. Both masters and doctoral students in epidemiology take numerous statistics courses but are almost never exposed to the art of using mathematics to express causal theory.

But lately, public health epidemiologists are paying more attention to models of infection transmission. This comes partly from greater relevance and realism in population transmission models. Keeling (in Chapter 6) shows paths to more realistic model elaborations that retain some kernel of mass-action thinking but that relax unrealistic mass-action assumptions. The following are specific assumptions intrinsic to mass-action formulations:

1. encounters are generated by a random process in which individuals have no power to influence who they encounter,
2. infection transmission occurs during instantaneous one-off encounters,
3. encounters are simultaneous and symmetrical so that if individual A encounters B, then B encounters A during the same instantaneous encounter,
4. instantly after an encounter the population is thoroughly remixed.

In addition, differential equation models of continuous populations that formulate mass-action must assume that each population segment is infinitely divisible. That translates into assuming that an infinite number of individuals are in each population compartment.

Keeling in Chapter 6 presents a variety of model forms that relax these assumptions and add new dimensions to contact patterns in transmission models. Analysis of these dimensions is becoming a central activity in analyzing systems that circulate and disseminate infection. To categorize these dimensions, it is possible to distinguish encounter processes, linkage processes, and transmission processes. Encounter processes lead to instantaneous events bringing two individuals together. Linkage processes follow encounter processes and determine whether the two individuals in an encounter will become linked and, if so, for how long. Transmission processes carry infectious agents from one individual to another. An additional model form not described by Keeling involves transmission through media such as air, surfaces, hands, water, food, and fomites. This form does not make any of the listed mass-action assumptions and does not have elements corresponding to encounter or linkage processes.

Mass-action is a particular formulation of the encounter process. In the form described by Heesterbeek, it assumes there is no linkage process and it models the transmission process as a probability of transmission given an encounter. It assumes that neither the environment nor the needs of the actors in the transmission system lead to sustained or repeated contacts. This is one reason epidemiologists with public health responsibility have largely ignored mass-action models. Analyses of linkages between individuals have intuitively guided the infection control decisions of public health epidemiologists. Public health actions often flow from concepts of enduring contact structure, such as who sustains circulation of an infection, who bridges different populations that can sustain infection, who generates the most transmissions in low-risk populations, who amplifies transmission chains, and who forms immunity barriers that can stop transmission. Mass-action assumptions assume away the basis for answering these questions.

7.2 | MODEL FORMS AS PARADIGMS FOR THEORY CHANGE

Keeling's presentation points to more productive analyses of the systems that circulate and disseminate infection by suggesting that different model forms addressing similar issues have different things to offer by relaxing different assumptions in the mass-action paradigm. I will elaborate the classification of model forms presented by Keeling and show how these forms fit into a broader context of theory change.

The following seven model forms present distinct opportunities for theory change within a science of infection transmission. Each of the following model forms has distinct paths for model elaborations relevant to public health:

1. universal mass-action models,
2. structured mass-action models,
3. transmission media models,
4. fixed network models,
5. dynamic network models,
6. pair correlation models,
7. agent-based models.

We can view each model category as defining positive heuristics for a research programme as described by Lakatos (1980). They guide different groups of scientists in their analysis of infection transmission systems. Infection transmission science does not have entrenched camps that seek dominance for their theories. Most infection transmission scientists use one or more of these forms. There are, however, distinct sets of individuals from different disciplines pursuing each of the preceding model forms. During the mid-twentieth century, scientists with a mathematics background were the dominant users of the first two model forms. Then software became available that made numerical solution of differential equations a simple task, and more scientists with public health epidemiology and especially with biology or ecology backgrounds began using these model forms. Environmental scientists have dominated the development of the third model form. They commonly greatly simplify or ignore population transmission dynamics in their models. Scientists with a physics background have lately taken strongly to using fixed network models. Scientists with a sociology background that employed social network analysis represent another large and distinct group using this model form. Mathematicians also have a long history of inquiry using this model form. The pair correlation models represent a more recent innovation and are mainly used by scientists with a mathematical background. It seems likely that easy-to-use software would quickly expand the number of scientists using this model form. Agent-based models have been promoted by "complexity" scientists. The natural context in which agent-based models formulate causal actions is pulling public health epidemiologists and others with a less mathematical background into this paradigm.

The first model form was the focus of the Heesterbeek chapter in this volume. The simplest universal mass-action models assume that encounter rates are constant across time, across contexts, and across individuals. The form Keeling identifies as *mass-action* is now commonly referred to as density-dependent mixing, and the form he calls *pseudo mass-action* is commonly called frequency-dependent mixing (McCallum, Barlow, & Hone 2001).

The second model form derives from the first and is probably used by more scientists addressing transmission system questions than any other model form. The mass-action assumptions in encounter processes can be relaxed in a variety of ways. Keeling addresses one approach to this model form called metapopulation models. In such models, density-dependent mixing holds within local settings (Grenfell & Harwood 1997). Another approach involves structured mixing (Jacquez, Koopman, & Simon 1989), in which frequency-dependent mixing holds within activity settings. Structured mixing takes a statistical mechanics approach to locality such that the location of population is not specified at each time point and movements are not specifically modelled. Only probabilities of being at a location are modelled. Metapopulation models more explicitly locate population. Thus, the structured mixing approach is more appropriate for repeated transient movements and the metapopulation approach is more appropriate for longer-term movement. In either formulation, proportionate mixing is usually used within metapopulations or activity settings. As Keeling comments, these model forms retain an essential core of mass-action within locality. Because they retain this mass-action core, the scale of mixing at the population level in these model forms can go from assortative to proportionate mixing but not to disassortative mixing. Symmetrical disassortative mixing entails either a rejection or a mutual acceptance process (Koopman *et al.* 1988).

The frequency and density formulations make different assumptions regarding the effects of population density and individual capacities on the encounter process. Intermediate formulations uniting these are possible using the theory of limited handling capacity for encounters that Keeling presents or using a limited linkage capacity formulation (Riggs & Koopman 2004).

Structured mass-action models relax the assumption of instantaneous remixing after an encounter and specify who encounters whom by allowing mixing only within, and not between, structures. Modelling who has contact with whom has been a focus of many modelling efforts since the mid-1980s (Koopman 2004). They thus open a large area for theory development beginning with core group definition (Yorke, Hethcote, & Nold 1978) and more recently including the implications of mixing in multiple contexts (Ball, Mollison, & Scalia-Tomba 1997; Koopman *et al.* 2002).

The instantaneous and thorough remixing assumption of mass-action is relaxed in different ways that allow different types of theory change by the last five model forms. Models that elaborate the transmission process by specifying the media through which infection travels do not assume that contact is symmetrical as do mass-action models. In fact, they relax all of the assumptions listed for mass-action models. Transmission media such as surfaces or air are formulated as separate compartments that are contaminated by infectious individuals and that, in turn, contaminate susceptible individuals. Transmission direction in this case is determined by the time when different individuals either contaminate the media or are contaminated by the media. History gives direction to the causal arrow.

Transmission media models have natural relationships to structured mass-action models. Transmission occurs through local media; thus, transmission media models must specify enough structure to ensure that local contamination of and from media is reasonable. One advantage of such models is that they can incorporate well-developed theory regarding agent excretion, survival in the environment, and biological processes when contamination of different anatomical sites occurs (Haas, Rose, & Gerba 1999). Thus we can see again that the direction of theory change relates to model form. Another advantage of this model form is that it allows incorporation of data on levels of environmental contamination. Progress in developing new methodology for such measurements has been dramatic, and it is likely that such measurements can reveal as much or more about a transmission system as infectious agent identification in humans can.

Network models have contact structures based on links between pairs of individuals. Network structures are formulated by specifying linkage contingencies. These are of two types: (1) contingencies specifying aspects of the individuals involved in the pair formation and (2) contingencies specifying characteristics of the population patterns of already formed pairs. The former type of contingencies are an effective way of generating disassortative contact patterns by specifying a linkage contingency with lower probabilities of linkage when individuals with characteristics similar to one's own are encountered. Heterosexual relationships, teacher-student relationships, and nurse-patient relationships are examples. Other contingencies can generate interesting and important patterns such as small-world networks (Watts & Strogatz 1998) and scale-free networks (Borguna, Pastor-Sattoras, & Vespignani 2002; Liljeros *et al.* 2001).

In fixed network models there is no population encounter process and linkages are assumed to be fixed so that both encounter and linkage processes are assumed away. In all network models, including the pair correlation models, a simple formulation of the transmission process is just to specify a rate of transmission. Such specification presents a new context for theory change that does not exist in the first two model forms.

Recent work has expanded mathematical analysis methods for network models (Newman 2002, 2003). Fixed network structures with specific desired characteristics, such as degree distribution, connectivity, clustering, and locality separation, can be generated by specifying a probability for a particular linkage being formed as a function of how many such linkages are already formed in the population. This has provided the basis for interesting analyses of contact structure effects (Sander *et al.* 2002). This is an increasingly common approach in the physics literature dealing with network structures. Analytic tractability using this approach is constrained to fixed networks. In simulations, however, these same approaches can be used in dynamic models. Perhaps because fixed network models have this constraint, they have succeeded in opening a new realm for theory change that is inaccessible from other model forms.

Dynamic network models incorporate both an encounter process and a linkage process (Koopman *et al.* 2000). They can be formulated by adding a linkage process to any of the model forms that have an encounter process. They can also be formulated by adding dynamics to fixed networks. In either case, after individuals encounter one another, they have a chance of forming a linkage. The encounter process may be completely mass-action. More commonly, however, the linkages an individual has formed will affect the encounter process. Again this opens new areas for theory change that are inaccessible through other model forms. New theory can be formulated by relaxing assumptions that pair formation upon an encounter and pair dissolution given an established linkage are both independent of anything else going on in the model.

Analytic tractability is a virtue that is given up for greater realism in dynamic network models. The pair correlation models, to which Keeling has made such valuable contributions, could be viewed as a reduced form of dynamic network models that retains analytical tractability. As presented by Keeling, pair linkage in the pair correlation models is an instantaneous and inevitable result of an encounter. The encounter process is formulated as mass-action. The linkage process formulated is thus the simplest possible and specifies no contingencies. It appears possible to elaborate more realistic linkage processes in this model form, but the consequences for analyzability are unclear. The linkage processes in pair correlation models affect linkages only at the ego-centred or individual network level. They assume away higher-level network structures by specifying moment closure approximations at the level of triplets or higher-order linkages. These model forms capture some aspects of relationships between discrete individuals that mass-action models do not. They do so at the level of pairs, triads, or higher-order groupings. But they do not relax the continuous population assumption that requires an infinite number of individuals to make populations infinitely divisible, and they cannot capture the stochastic aspects of discrete individual encounters.

Agent-based models address many more details of both the encounter and the linkage process. They provide rules for the evolution of individual actions, leading to contacts that transmit infection rather than defining population patterns for such contacts in the way that compartmental models do. In such models, behavioural rules based on agent perceptions can determine contact with random processes playing no role. Each individual can survey other individuals with whom they might make contact, and either unilateral or mutual decision rules can determine whether an encounter will occur. Thus, once again, you can see how model form determines the direction that theory change can take. Theory elaboration capturing the effects of decision rules is facilitated by the agent-based models. But agent-based models can retain some of the mass-action assumptions, such as instantaneous and thorough remixing after an encounter, in their encounter process formulations. A major difference between agent-based models and other forms is that the former seeks to elucidate how contact patterns emerge from individual interaction rules rather than to make assumptions about the shape of those patterns. Thus, the object of theory change differs between those scientists pursuing agent-based models and those pursuing the other model forms. What the agent-based scientists pursue as central issues are taken as arbitrarily given patterns to be realistically described rather than explained by those employing the other model forms.

These model forms are linked by the assumptions they have in common and distinguished by the assumptions that they relax to add greater realism. In an ideal analytical world you would be able to nest these different model forms so that you could assess when transiting from one model form to another results in significantly improved descriptions of observations. A more practical way to link these model forms is to use the more detailed model forms to assess the robustness of inferences made using the simpler model forms. This puts the focus on model-based inference. Each inference you seek to make from a model analysis will require a different model that is as simple as possible but no simpler. Robustness assessment is the process of assessing whether the model is simpler than it needs to be to make an inference.

7.3 | ROBUSTNESS ASSESSMENT

One categorization of purposes for modelling infection transmission is as follows:

1. gain insights that help develop theories, infection control policies, or both,
2. develop theory,
3. make policies,
4. analyze data,
5. design studies.

Specific inferences in these classes are, correspondingly, as follows:

1. whether some phenomenon can be expected under specified conditions,
2. whether specific formulations correspond to real-world events,
3. what can be expected from infection control decisions,
4. what the values are for parameters representing real-world processes in a model,
5. what study designs are most informative.

If we find that an inference is made from a more abstract, simple model but is not made from a more detailed and thus more realistic model, then that inference is not robust to the assumptions relaxed by the more detailed and realistic model.

All models require unrealistic assumptions. As models are elaborated with greater realism, they relax these assumptions. Robustness assessment evaluates how realistic violations of assumptions affect the uses of model analyses. Robustness assessments are purpose specific. A model that is not robust for making policy decisions may be robust for gaining insight. In other words, even though a policy decision made using a model might be reversed when an unrealistic assumption is relaxed, a particular insight gained from the model may not be reversed.

The differences in assumptions for the different model forms, as presented previously, provide the basis for assessing robustness by comparing inferences from model analyses using two or more models making different assumptions (Koopman 2004). These transitions are mostly within the structured mass-action class of models I have described. Hopefully, as the science of infection transmission system analysis advances, software will be elaborated to facilitate transitions among many other model forms.

The robustness of insights from deterministic continuous population models in the structured mass-action class may be assessed for infinite population-size assumptions by converting from deterministic compartmental models to stochastic compartmental models. Going from the deterministic compartmental model to an agent-based model would relax so many different assumptions simultaneously that little specific knowledge would be gained about robustness.

Big differences in assumptions may not always be a barrier for using two model forms in robustness assessment. Suppose you want to assess the robustness of an inference from a data analysis in which you have fitted a deterministic structured mass-action model to a set of collected data. One could repeat the analysis using many variants of the deterministic model to assess the robustness of inferences from such a data analysis. But that does not relax the assumptions intrinsic to the model form. One could use the output from an agent-based model, gather similar data from such a model, and see whether the inferences from the data analysis are consistent with the detailed output that can be attained from the agent-based model. This type of activity is being supported by a major modelling endeavour with the acronym MIDAS supported by the United States National Institute of General Medical Sciences.

7.4 | ADVANCING A SCIENCE OF INFECTION TRANSMISSION SYSTEM ANALYSIS

Robustness assessment disciplines model development and research programme development. It provides a framework for identifying caveats to the inferences made in pursuing any of the modelling purposes listed previously. Conversely, it identifies areas of research needed to solidify particular inferences.

A focus on caveats to an analysis may not be the best way to promote the development of a science. Lakatos (1980) has pointed out how *negative heuristics* contribute to the development of *scientific research programmes* by not allowing reality or criticism to inhibit the development of promising but problematic lines of theory. Negative heuristics allow a scientific research programme to ignore inconsistencies between their theories and their data or criticisms about their validity. For example, the adherents of mass-action modelling have not halted their research because of the reality that their models either fail to fit available epidemic data or do so in a tautological manner. The mantra that they are capturing something of essence and that differences between model output and data can be treated as noise even when it has strong patterns has sustained them by steeling them to the criticism that their models are unrealistic.

Conversely, the adherents of agent-based modelling face many discouraging words such as those at the end of Keeling's chapter. They steel themselves against such rhetoric with the conviction that they are not seeking the same sort of things that scientists performing mathematical analyses are seeking. They can ignore such remarks because they know that by pursuing the analysis of agent-based formulations they may discover how and why population patterns emerge from individual decision algorithms, and it appears to them that mathematical analyses have little hope of doing this.

Solid negative heuristics about which criticisms should be ignored can contribute to developing a field of inquiry. Worrying too much about problems that might invalidate insights, theories, policies, data analyses, or study designs might paralyze scientific discourse. There has to be some balance between Hamer's criticism of promising new formulations, printed in a smaller typeface within his paper as described by Heesterbeek, and the unrestrained elaborations of mathematical analyses based on mass-action assumptions that led to dramatic growth in mathematics papers dealing with infection transmission in the latter half of the twentieth century.

At the current stage of the development of a science of population infection, there may be even more to be gained if proper respect is accorded to others who have different negative heuristics. Each modeller cannot be expected to master all the intricacies of model forms used by other modellers. Teamwork among modellers using different model forms will facilitate robustness assessments that use one model form to relax the assumptions of other model forms. In such a collaborative environment, a broad education in modelling should make modellers familiar with many model forms and with tools and skills required to use those model forms to achieve any of the five modelling purposes. But if great insight and skill is to be acquired in any model form, negative heuristics are needed to steel modellers against the criticisms of those using different heuristics for their modelling endeavours.

The research support environment needed to promote separate development of different model forms and to integrate them in robustness assessments is not ideal. Modellers need to criticize one another to promote the growth of knowledge. But such criticism too often works against funding research programmes. Perhaps such criticism can be put in a positive context by recognizing the necessity for robustness assessments that use multiple model forms.

Robustness assessments focus model development on the pursuit of specific objectives. Finding a lack of robustness identifies issues inadequately addressed either by a model or by available data. Thus, robustness assessments move model development in the direction of ever-more detailed and ever-more pertinent theory. They provide a framework for discourse among modellers, environmental scientists, infection process scientists, public health officials, and others. That is because different disciplines can focus on the same inferences. This is especially true for policy inferences, such as whether a particular water treatment procedure is cost-beneficial and whether decontamination, school closure, contact tracing, or special case ascertainment should be pursued in an epidemic. It is not just that each discipline contributes a different piece to the overall model. Practitioners of diverse disciplines are often willing to make inferences about public health actions from their perspective. Robustness assessment provides a framework for integrating and evaluating different perspectives. The different model forms I have described provide different perspectives on infection transmission issues. If we are careful in specifying the assumptions made in each model form and the assumptions that can be relaxed by each model form, we will find better ways to use multiple model forms in making robust inferences.

REFERENCES

Ball, F., Mollison, D., & Scalia-Tomba, G. (1997). Epidemics with two levels of mixing. *Ann. Appl. Prob.* **7**, 46–89.

Borguna, M., Pastor-Sattoras, R., & Vespignani, A. (2003). Absence of epidemic threshold in scale-free networks with connectivity correlations. *Phys. Rev. E* **90** *028701*.

Grenfell, B., & Harwood, J. (1997). (Meta)population dynamics of infectious diseases. *TREE* **12**, 395–399.

Haas, C.N., Rose, J.B., & Gerba, C.P. (1999). "Quantitative Microbial Risk Assessment." Wiley, New York.

Jacquez, J.A., Koopman, J.S., & Simon, C.P. (1989). Structured mixing: Heterogeneous mixing by the definition of activity groups. In "Mathematical and Statistical Approaches to AIDS Epidemiology" (C. Castillo-Chavez, Ed.). Springer-Verlag, New York, pp. 301–315.

Koopman, J. (2004). Modeling infection transmission. *Annu. Rev. Publ. Health* **25**, 303–326.

Koopman, J., Simon, C., Jacquez, J., Joseph, J., Sattenspiel, L., & Park, T. (1988). Sexual partner selectiveness effects on homosexual HIV transmission dynamics. *J. Acqu. Imm. Def. Syndr.* **1**, 486–504.

Koopman, J.S., Chick, S.E., Riolo, C.S., Adams, A.L., Wilson, M.L., & Becker, M.P. (2000). Modeling contact networks and infection transmission in geographic and social space using GERMS. *Sex. Transm. Dis.* **27**, 617–626.

Koopman, J.S., Chick, S.E., Simon, C.P., Riolo, C.S., & Jacquez, G. (2002). Stochastic effects on endemic infection levels of disseminating versus local contacts. *Math. Biosci.* **180**, 49–71.

Lakatos, I. (1980). "The Methodology of Scientific Research Programmes." Cambridge University Press, New York.

Liljeros, F., Edling, C.R., Nunes-Amaral, L.A., Stanley, H.E., & Aberg, Y. (2001). The web of human sexual contacts. *Nature* **411**, 907–908.

McCallum, H., Barlow, N., & Hone, J. (2001). How should pathogen transmission be modelled? *TREE* **16**, 295–300.

Newman, M.E.J. (2002). Spread of epidemic disease on networks. *Phys. Rev. E* **66**, 016128.

Newman, M.E.J. (2003). Mixing patterns in networks. *Phys. Rev. E* **67**, 026126.

Riggs, T.J., & Koopman, J.S. (2004). A stochastic model for vaccine trials of endemic infections using group randomization. *Epidemiol. Infect.* **132**, 927–938.

Sander, L.M., Warren, C.P., Sokoloff, I.M., Simon, C.P., & Koopman, J.S. (2002). Percolation on heterogeneous networks as a model for epidemics. *Math. Biosci.* **180**, 293–305.

Watts, D.J., & Strogatz, S.H. (1998). Collective dynamics of "small-world" networks. *Nature* **393**, 440–442.

Yorke, J.A., Hethcote, H.W., & Nold, A. (1978). Dynamics and control of transmission of gonorrhea. *Sex. Transm. Dis.* **5**, 51–56.

III | COMMUNITY ECOLOGY

8 | COMMUNITY DIVERSITY AND STABILITY: CHANGING PERSPECTIVES AND CHANGING DEFINITIONS

Anthony R. Ives

8.1 | INTRODUCTION

Mathematical theory has been applied to numerous topics in community ecology. These include questions such as how many species can coexist within communities (MacArthur 1972), are there rules dictating the structure of food webs (Pimm 1982), and what explains the relative abundances of species (Preston 1962). Here I focus on the relationship between community diversity and community stability. Because perspectives on this topic have changed dramatically over the last 50 years, this provides a good topic to investigate if paradigm shifts have occurred in theoretical community ecology. In his companion chapter, Kevin McCann gives a summary of the history of this question and his view of current breakthroughs and future challenges to theory.* My goal is largely orthogonal to McCann's. Even though it appears that successive hypotheses about

* Although I have been designated the "senior ecologist" for this section. I was only elevated to senior status when a more senior, and more qualified, author had to withdraw. For the perspectives on similar issues by a truly senior ecologist, I suggest Robert May's introduction to the 2001 reprinting of *Stability and Complexity in Model Ecosystems.*

the relationship between diversity and stability have replaced each other, I want to show that it is not so much that the failure of one hypothesis led to its replacement by another. Instead, I think the dominant definition of stability has changed over the last 50 years, and these changes in definition have understandably led to changing conclusions about the relationship between diversity and stability. Although these changes have occurred within the theoretical literature, investigations of the relationship between diversity and stability have proceeded somewhat independently within the empirical literature. The changes within theoretical ecology and the contrast between theory and empiricism make it difficult to draw any conclusion about the general relationship between diversity and stability, if indeed there is one.

The chapter is organized into four parts. The first is a brief history of perspectives on the relationship between diversity and stability. This partly repeats the history in McCann's chapter, although my account of the more recent history expands beyond strict theory into experimental perspectives on diversity and stability. I then present a simple model that illustrates several different relationships between diversity and stability, or more properly, relationships between diversity and different definitions of stability. The goal here is to show that different perspectives on the relationship between diversity and stability are complementary rather than at odds. My intent is not to try to resolve conflict and end with a chorus of *Kumbaya*, but is instead to shift the debate towards selecting a suitable definition, or definitions, of stability for particular situations and needs of the researchers. Then, I describe the history of the sometimes rocky relationship between theory and empiricism stemming from different definitions of stability. Finally, I give some examples of specific ideas that I think can be tested, or at least explored, to derive insights into the relationship between diversity and stability, regardless of the preferred definition of stability.

8.2 | HISTORY

In the 1950s and 1960s a common view was that greater diversity generated greater stability because of the inherent redundancy of more diverse communities (Leigh 1965, Margalef 1969). This redundancy would allow some species to assume the role of other species if they declined in abundance or went extinct because of some environmental disturbance. The theory leading to this conclusion was largely qualitative. The archetypal example of this view is MacArthur (1955) who assumed that greater stability is synonymous with the number of pathways along which energy can flow through a food web and the equitability of energy flow among pathways. Thus, in more diverse systems, if one pathway or species is eliminated, there are other pathways that maintain energy flow from the bottom to the top of the food web.

In the 1970s theoreticians adopted the view that more diverse systems are less stable, because the chance that a randomly constructed model community is stable decreases with the size of the community (Gardner & Ashby 1970, Lawlor 1978, Pimm 1979). This occurs heuristically because, as the number of links within a food web increases, there is an increasing chance that at least one of them could drive instability, and instability of a single link within a food web will propagate through the entire community making the whole unstable. This conclusion was generated using explicitly mathematical approaches. For example, May (1972, 1973) randomly constructed model ecosystems and subjected them to stability analysis, a technique that has a long and rich history in mathematics and the "hard" sciences. This approach involves finding an equilibrium point for the model community and determining whether the densities of species within the community tend to return to this point if they are perturbed a small distance away.

Since the 1970s, more complex nonlinear model ecosystems have been studied, and cases have been found in which more diverse communities can be more stable than their simpler counterparts (McCann, Hastings, & Huxel 1998). Nonlinear models often have no stable equilibrium point, and persistence relies upon explicitly nonlinear dynamic behaviours of the models. Because nonlinear dynamic behaviours are difficult to study mathematically, it has been difficult to draw any broad conclusions about the relationship between diversity and stability in nonlinear systems. Nonetheless, the body of theory on nonlinear dynamic systems is best viewed as an extension and elaboration of the theoretical work started in the 1970s and exemplified by May (1973). The work has extended our knowledge of nonlinear dynamic systems and revealed cases of nonlinear phenomena creating diverse yet stable communities, but it has not overturned the general conclusion that more diverse communities are likely to be less stable.

In the 1990s, a third perspective about the relationship between diversity and stability came into vogue: although diversity may decrease the stability of individual species within a community, it may simultaneously increase the stability of the community as a whole (Tilman 1996). This perspective grew primarily from experimental studies on complex communities. Nonetheless, it has received some theoretical support from models that are explicitly stochastic and therefore capable of mimicking a randomly fluctuating environment (May 1974a, Doak *et al.* 1998, Tilman, Lehman, & Bristow 1998, Ives, Gross, & Klug 1999, Lehman & Tilman 2000).

The basic tenet of this perspective is that increasing the number of species within a food web may cause individual species to fluctuate more wildly. These fluctuations of individual species, however, will likely be asynchronous, so decreases in densities of some species will be compensated by increases in densities of other species. Thus, although diversity may increase the fluctuations of individual species, the aggregate density of all species in the community may nonetheless fluctuate less.

In summary, over the last 50 years views have changed from diversity causing communities to be stable, to diversity causing communities to be unstable (at least most of the time), to diversity causing communities to be stable while destabilizing the densities of the constituent species. These perspectives have different roots. The first derived from largely qualitative theory, which McCann aptly describes as an intuitive approach. The second was highly quantitative, employing the mathematics of dynamic systems analysis. The third was largely motivated by empirical studies, with theory playing a secondary role. This historical timeline leads to the following thesis about how perspectives about the relationship between diversity and stability have changed: intuition was overturned by rigorous mathematical theory, which itself was overturned by hard empirical evidence.

Although this thesis might sound appealing, I think it is mostly wrong. Changes in hypotheses were not driven by a cascade of failing paradigms. Instead, the definition of stability changed, and as the definition changed, so did the perceived relationship between diversity and stability. Thus, it is not that each successive hypothesis about the relationship between diversity and stability supplanted its predecessor; instead, each successive definition of stability supplanted its predecessor.

That changes in the definition of stability have driven changes in the perceived relationship between diversity and stability in no way trivializes the debate. However, recognizing how definitions of stability have changed shifts the debate towards selecting a definition of stability that best relates to the situation, system, or problem at hand. Comparing patterns of population variability among communities from the tropics to the arctic will likely require a different definition of stability than asking how a lake community will likely change as the pH drops because of acid rain.

8.3 | MULTIPLE TYPES OF STABILITY IN A MODEL ECOSYSTEM

Here I show that all three of the following can be simultaneously correct: (1) diversity increases community stability, (2) diversity decreases community stability, and (3) diversity decreases individual species stability and increases whole community stability. These patterns correspond, respectively, to the dominant views held in the 1950s and 1960s, in the 1970s and 1980s, and in the 1990s until the present, and they are represented, respectively, by MacArthur (1955), May (1973), and Tilman (1996). To illustrate that all three patterns can simultaneously be correct, I construct simple model ecosystems and analyze them according to the three different definitions of stability listed. As is traditional and necessary for any exercise such as this, I make the disclaimer that the goal is not to model any particular natural system but is instead to paint a broad picture of possibilities that could, in a crude way, be exhibited by some natural system somewhere.

The model communities are built around the following discrete-time, Lotka-Volterra-like equations:

$$x_i(t+1) = x_i(t)\exp\left(r_i + \sum_{j=1}^{n} b_{ij}x_j(t) + \varepsilon_i(t)\right). \tag{8.1}$$

Here, $x_i(t)$ is the density of species i at time t, r_i is the intrinsic rate of increase of species i, b_{ij} is the interaction coefficient giving the per capita effect of species j on the per capita population growth rate of species i, and $\varepsilon_i(t)$ is a random variable that represents the extrinsic effects of the environment on per capita population growth rates. I assume that $\varepsilon_i(t)$ has a mean of zero and is independent among species and independent through time so that species experience the environment without correlation among species and without temporal autocorrelation. This set of equations for n species is simple in that interspecific interactions are linear and additive. It is closely related to the continuous-time models analyzed by May (1973) and is somewhat generic, in the sense that other models with linear and additive interspecific interactions will lead to conclusions similar to those presented in this chapter.

To create random communities, I used an approach modified from that of May (1973). The defining characteristic of a model community is the pattern of interactions among species, given by the interaction coefficients b_{ij}. For intraspecific interactions, I set $b_{ii} = -1$ so that each species experiences intraspecific population regulation. For interspecific interactions b_{ij} $(i \neq j)$, I used two different procedures to create distinctly different types of communities. To create communities containing only competitors, I selected values randomly from a uniform distribution between –0.91 and 0. The minimum of –0.91 was used to give the same mean squared interaction strength, $s^2 = 0.5$, used by May (1973, Fig. 3.6). To create communities with interactions in addition to competition, I selected b_{ij} from a uniform distribution with minimum and maximum values of −0.91 and 0.91. This also gives a mean squared interaction strength of $s^2 = 0.5$ and includes predator-prey interactions when b_{ij} and b_{ji} have opposite signs (on average 50% of the time), competitive interactions when b_{ij} and b_{ji} are both negative (25% of the time), and mutualism when b_{ij} and b_{ji} are both positive (25% of the time).* I will refer to these two types of communities as *competitive* and *arbitrary* communities.

For the model analysis, I first use the dominant view of stability from the 1970s and 1980s and then use those from the 1950s and 1960s and the 1990s. Although this is not chronological, it is necessary for the presentation, as will be made clear below.

8.3.1 | The 1970s and 1980s

In the 1970s and 1980s, the stability of a community was equated with the property that all species in the community could persist indefinitely. Mathematically, for deterministic systems without environmental variability (i.e., with $\varepsilon_i(t) = 0$), this requires the existence of a stable equilibrium point or a stable but non-stationary mathematical structure, such as a stable limit cycle or a strange attractor (Hastings *et al.* 1993). Even if persistence does not involve a stable equilibrium point, it almost certainly requires an unstable equilibrium point that serves as the mathematical backbone for a stable but non-stationary structure.† Thus, determining whether a community is persistent is a two-step process, first to find a possible equilibrium point with all species having positive densities and second to figure out whether this equilibrium point is stable or whether there is another stable structure that maintains positive population densities indefinitely.

The equilibrium point of equation (8.1) consists of densities x_i^* that satisfy the following:

$$r_i + \sum_{j=1}^{n} b_{ij} x_j^* = 0. \qquad (8.2)$$

---

* In contrast to Fig. 3.6 in the book by May (1973), I assumed connectance (the probability that pairs of species interact) is 1 rather than 0.5. This is for the technical reason that, when connectance is 0.5, almost all feasible communities are stable (Fig. 8.1A, B). I used a connectance of 1 simply to make Fig. 8.1A, B more interesting.

† For equation (8.1), this is certainly true. I cannot think of a plausible biological model in which this is not true, although an exception may exist.

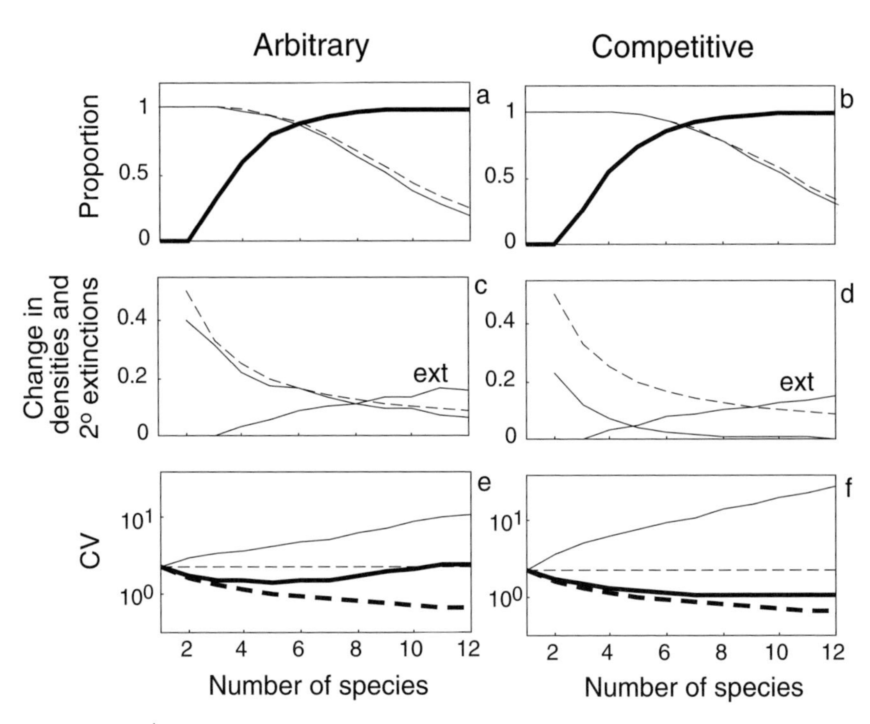

FIGURE 8.1 | For (a) arbitrary and (b) competitive communities, the relationship between community diversity (number of species) and the proportion of randomly constructed communities that do not have a feasible (positive) equilibrium point (*heavy line*). Of those communities that have a feasible equilibrium, the proportion that has a stable equilibrium is shown by a light line, and the proportion that is persistent (either with or without a stable equilibrium) is shown by a dashed line. Persistence for communities without a stable equilibrium was assessed by iterating the model 10,000 times and defining communities as persistent if all species densities remain greater than 10^{-6}. For (c) arbitrary and (d) competitive communities, the change in combined species densities following the removal of a randomly chosen species when species are allowed to interact (interactive communities, *solid line*) or not interact (noninteractive communities, *dashed line*). The proportion of the community lost because of secondary extinctions is given by the lines labelled "ext". For (e) arbitrary and (f) competitive communities, the CV of individual species densities for interactive (*thin solid lines*) and corresponding noninteractive communities (*thin dashed line*), and the CV of combined species densities for interactive (*thick solid line*) and noninteractive communities (*thick dashed line*). For all community sizes, communities were randomly assembled until 10,000 communities with feasible equilibria were found. Of these, only those with stable equilibria were used for panels c–f. All intraspecific interaction coefficients b_{ii} were set to −1, and interspecific coefficients were selected from uniform distributions between −0.9086 and 0.9086 for arbitrary communities and between −0.9086 and 0 for competitive communities. The other model parameters are r_i, which were set to 0.1 for all species.

If the solutions x_i^* are all positive, then the community is called feasible (Pimm 1982). The stability of a feasible equilibrium point is determined mathematically by asking whether population densities, when perturbed from the equilibrium point, will return towards equilibrium through time. If they instead move from equilibrium, then the equilibrium point is unstable. Mathematically, the stability of the equilibrium point depends on the *community matrix* **A**. For equation (8.1), the diagonal elements of **A** are $1 + b_{ii}x_i^*$ and the off-diagonal elements are $b_{ij}x_j^*$. If all of the eigenvalues of **A** have magnitudes less than one, then the equilibrium is stable.*

Much criticism and hand-wringing has centred on analyses that search for stable equilibria, with the charge that nature is not "equilibrial" because there is continuous variability created by environmental fluctuations (Wiens 1977, Strong 1986). For (8.1), however, this criticism would be misplaced. Because the equations have per capita population growth rates (the exponential terms) that are linear combinations of species densities, adding environmental variability through $\varepsilon_i(t)$ has little effect on the predictions about persistence made by the deterministic, equilibrial model. If the community is not persistent in its deterministic form without environmental variability, it will not persist with environmental variability either—environmental variability is most likely to decrease the chances of persistence. Non-theoreticians have sometimes attributed a mystical ability of variability and non-equilibrial dynamics to make species persist, although a mathematical evaluation reveals that there needs to be specific types of nonlinearities before environmental variability leads to persistence (Chesson 1994, 2000). I conveniently and intentionally excluded these specific types of nonlinearities from the models considered.

If a feasible equilibrium is found that is unstable, there are two alternatives: the community may not persist, with at least one species going extinct, or the community may persist in a non-stationary way. Generally, the easiest way to separate these alternatives is to iterate the model for a long time to see if all species persist.

Figure 8.1 gives results for simulated communities ranging in diversity from $n = 1$ to 12 species. For both communities with arbitrary interactions (with the sign of b_{ij} unconstrained) and competitive communities (with all $b_{ij} < 0$), the proportion of the randomly constructed

---

* In analyzing other models, the criterion from stability is often that the eigenvalues of the community matrix have real parts less than zero. The difference here is that (8.1) is posed in discrete time rather than with differential equations in continuous time. For continuous-time models, the real parts of the eigenvalues must be less than zero for stability.

communities without a feasible (positive) equilibrium point increases with increasing diversity (Fig. 8.1a, b). Furthermore, of those communities that have feasible equilibrium points, the proportion that is persistent decreases with increasing diversity. Finally, almost all persisting communities have a stable equilibrium point.

These results are closely related to those of May (1973), although they differ in details. Rather than randomly selecting values of b_{ij}, May followed Gardner and Ashby (1970) in constructing communities by randomly selecting elements of the community matrix **A** and calculating the corresponding eigenvalues. In effect, May's approach assumes that a feasible equilibrium exists and then asks whether this equilibrium is stable. Furthermore, he does not explicitly address the persistence of the community in the absence of a stable equilibrium, but given that (at least for equation (8.1)) most persistent communities have a stable equilibrium point, May's results are similar to those I have just presented.

In summary, when stability is measured in terms of the persistence of intact communities (i.e., communities that retain all species indefinitely), more diverse communities are doubly unlikely to be stable. First, more diverse, randomly constructed communities are less likely to have positive equilibrium densities for all species. Second, even when all species have positive equilibria, these equilibria are less likely to be stable in diverse communities. It is very rare that a randomly constructed community of many strongly interacting species can persist indefinitely without at least a few species going extinct.

8.3.2 | The 1950s and 1960s

MacArthur's (1955) concept of stability is different. He defines stability according to the suggestion made by Odum (1953); stability is measured by "the amount of choice which the energy has in following the paths up through the food web" (MacArthur 1955, p. 534). He uses the analogy of a network in which energy is passed among nodes (species). Because increasing the number of species in a community will increase the number of energy paths, more diverse systems will, all else being equal, be more stable.

MacArthur's view of energy flow is static in the sense that the energy flow is constant for a given food web network. One way to translate his idea about the network structure of food webs into the dynamic communities given by equation (8.1) is to ask how the removal of a random species from the community will change the flow of energy through the remaining species. Then, the more stable community is defined as the community that suffers the least change in energy flow with the removal of a random species. Although (8.1) does not explicitly include energy flow, MacArthur equates energy flow through a species with the density of that species, so I will ask how removing a species from the community affects the density of the remaining species. Finally, MacArthur addresses energy flow from bottom to top of the food web, so the correct response variable should be the density of species at the top of the food web. Nonetheless, this is difficult to model for randomly constructed food webs, and it is not very relevant for purely competitive communities. Therefore, I will assess the response of a community to the random removal of a species in terms of the combined density of all remaining species.

Figure 8.1c, d gives the consequences of removing a randomly selected species from those randomly constructed communities that have feasible, stable equilibrium points. It only makes sense to start with communities that have a stable equilibrium and are therefore stable in the sense of May (1973), because this ensures persisting communities with constant densities as envisioned by MacArthur (1955). To give insight into the mechanism underlying the response of communities to species removals, I compare the randomly constructed communities with *non-interactive* communities in which the intraspecific interaction terms, $b_{ii} = -1$, are the same as the randomly constructed communities but the interspecific interactions, b_{ij} $(i \neq j)$, are set to zero.

Not surprisingly, in the non-interactive communities the proportional change in combined species densities following the removal of a random species decreases with increasing size of the community, n. Specifically, if there are n species in the community, then the fraction of the combined density caused by removing one species is just $1/n$. For the community with arbitrary interactions, the same pattern holds, with the proportional change in combined density close to the case of the non-interacting communities (Fig. 8.1c). Furthermore, when a single species is removed from the arbitrary community, secondary extinctions occur among the remaining species; for the 12-species communities, on average about 20% of the 11 species remaining following the immediate removal of 1 species proceed to go extinct. For the competitive communities, comparable numbers of secondary extinctions occur, yet the change in combined species densities is considerably lower than that of the corresponding non-interacting community (Fig. 8.1d).

These results confirm, at least partly, the conclusion stated by MacArthur (1955): more diverse communities are more stable in the sense that the proportional change in combined densities is lower in more diverse communities. Somewhat surprisingly, in randomly constructed communities with arbitrary species interactions, stability of more diverse communities is not increased by interactions among species. In contrast, purely competitive interactions cause smaller changes in combined density relative to the case of no interactions. Finally, although more diverse communities are more stable against changes in combined densities, more diverse communities lose a greater proportion of their species through secondary extinctions. Thus, if stability were measured in terms of secondary extinctions, diverse communities would be less stable.

8.3.3 | The 1990s

The final definition of stability involves variation in species densities and variation in the combined densities of species (Tilman 1996). Out of necessity, I will again restrict analysis to communities that pass the test for stability *sensu* May (1973); they must have a feasible and stable equilibrium. Even though in the deterministic setting communities can persist without the equilibrium being stable, in the stochastic setting environmental fluctuations often drive at least one species extinct unless the underlying deterministic model has a stable equilibrium. To generate variability, the random variable $\varepsilon_i(t)$ is assumed to have nonzero variance. To assess the effects of species interactions, again I compare the cases of randomly constructed communities with those of noninteractive communities with $b_{ij} = 0$ $(i \neq j)$. Finally, to measure stability, I use the coefficient of variation (CV), which increases with variability and hence is inversely related to stability.

In the noninteractive communities, increasing diversity n has no effect on the CV of individual species densities because, in the absence of species interactions, the community size makes no difference for individual species. Nonetheless, the CV of combined densities decreases with n. This has been called *statistical averaging* (Doak *et al.* 1998) and the *portfolio effect* (Tilman, Lehman, & Bristow 1998). Because I assumed that $\varepsilon_i(t)$ fluctuate independently among species, when environmental fluctuations increase the densities of some species, they simultaneously decrease the densities of others. The lack of synchrony among species fluctuations causes the fluctuations in combined densities to decrease, leading to lower CVs in combined densities than in individual species densities.

When species interactions are included, increasing diversity increases the CV of individual species densities. This occurs because species experience environmental fluctuations not only directly but also through interactions with other species whose densities are fluctuating. Despite this increase in the CV of individual species densities, the CV in combined densities still decreases, at least over much of the range of diversity n. This is because, despite the increase in individual species CVs, species interactions create *compensatory dynamics,* in which decreases in the densities of some species are offset by increases in their competitors, prey, or both. For competitive communities, in the special case of symmetrical competition in which b_{ij} $(i \neq j)$ have the same value, compensatory dynamics cancel the increases in individual CVs so that the CV of combined species densities with competitive interactions is the same as the CV of combined densities in noninteractive communities (Ives, Gross, & Klug 1999). The result that interactions among species in purely competitive communities have no effect on the variance of the combined densities of all species was previously shown by May (1974a).*

For the randomly constructed competitive communities in Figure 8.1f, however, asymmetries among competitors mean that the increase in CVs of individual species are not completely balanced by compensatory dynamics, leading to higher CVs than when there are no interactions among species. The effect of asymmetries is greater in arbitrary communities (signs of b_{ij} unconstrained), with the CV in combined species densities increasing with n for more diverse communities (Fig. 8.1e) even though no such increase would occur if species interactions were symmetrical (Ives, Klug, & Gross 2000).

8.3.4 | Summary

The simulated communities produced by equation (8.1) demonstrate three distinct relationships between diversity and stability depending on the definition of stability selected. For randomly constructed communities, the chances of producing a community that is persistent (or has a stable, positive equilibrium point) decreases with increasing diversity. This suggests that communities in nature are not a random collection of possible communities, at least not if possible communities are created by the random selection of interaction coefficients. For those communities that have a stable equilibrium, more diverse communities are more stable to the removal of a random species in that the proportional change in combined species densities is less than for less-diverse communities. Nonetheless, more diverse communities are more likely to experience secondary extinctions. Finally, for persistent communities that experience environmental fluctuations, diversity can increase the

---

* I did not know this until doing reading to write this chapter; therefore, my previous papers on the topic did not properly acknowledge May's work. This is a particularly egregious oversight because May was my PhD advisor.

CV of individual species densities and decrease the CV of combined densities. Thus, for stability defined in terms of population variability, diversity has different effects on individual species than on the whole community.

Because all of these relationships between diversity and stability were generated by the same model, the variety of conclusions is not the result of different model assumptions. Instead, the variety is created by different definitions of stability. The three definitions of stability I have considered are only a small sample of the possible definitions of stability; in a recent review, Grimm and Wissel (1997) catalogued 167 definitions of 70 stability concepts used by ecologists. Also, I have adopted a narrow definition of diversity as simply the number of species in a community. Mixing and matching definitions of diversity with definitions of stability might lead to many different theoretical relationships between diversity and stability. I think a key goal in community ecology is to identify which definitions of diversity and stability are most relevant for the natural systems and research questions in hand.

8.4 | TESTING RELATIONSHIPS BETWEEN DIVERSITY AND STABILITY

Alongside theoretical work over the last 50 years, there have been numerous empirical studies of diversity and stability. Here, I describe the often unhappy relationship between theory and empiricism chronologically, focusing not on empirical studies per se but instead on the interactions between theory and empiricism. As in previous sections, but possibly even more so, my description is brief and simplistic, and it glosses over many ideas and studies. There is only space to give a caricature of the theory-empiricism divide. My objective is not to review empirical studies of stability but only to show that, although definitions of stability in the theoretical literature have changed through time, these changes have occurred largely independently from changes in the concepts of stability appearing in the empirical literature. Theory and empiricism have largely been disjoint.

8.4.1 | The 1950s and 1960s

As I explained previously, the theory of MacArthur (1955) matched the intuitive view that diversity begets stability. Thus, at least initially, theory and empiricists generally agreed. A common citation for empirical support of MacArthur (1955) is Elton (1958), who lists six arguments.

For example, one of his arguments is that simple mathematical and laboratory systems containing one predator and one prey species tend to be unstable. He implicitly assumes that more complex predator-prey systems, such as those found in nature, are more stable than these simple systems and that the stability of natural systems is the result of their diversity. To this day, there is neither theoretical nor experimental evidence to support the generality of these assumptions (May 2001). Another of Elton's arguments is that ecosystems in the diverse tropics are more stable than those in the depauperate temperate zone. This again requires the leap of faith that diversity is the causal reason for differences in stability between the tropics and the temperate zone, and some more recent analyses even question the basic premise that tropical systems are more stable (Wolda 1978). In concluding, Elton does not question the positive effect of diversity on stability, only how strong it is: "It is a question for future research, but an urgent one, how far one has to carry complexity in order to achieve any sort of equilibrium" (1958, p. 153).

This initial agreement between theory and empiricism did not last long because more quantitative empirical analyses gave a more complex picture. For example, Pimentel (1961) showed for the herbivorous insect community of cabbages, communities in a field with 300 other plant species were less likely to outbreak than those in fields with cabbages planted as monocultures, probably because the field with the diverse plant community had a greater abundance of predators relative to the herbivores attacking cabbages. Although this supports the general pattern that greater diversity is stabilizing, the explanation—that greater plant diversity causes greater predator abundance—is unrelated to energy flows through food webs. Watt (1964, 1965) analyzed patterns of forest moth dynamics, showing that polyphagous species (hence, those interacting with more diverse plant communities) were more likely to show outbreak dynamics than monophagous species, yet moth species that shared host plants with many other moth species (and hence experienced greater competition) were less likely to show outbreak dynamics (although a re-analysis of the data overturned some of these conclusions; see Redfearn & Pimm 1987). In a microcosm study, Nelson Hairston *et al.* (1968) found that increasing bacterial diversity increased the stability of *Paramecium* grazers, yet increasing *Paramecium* diversity (up to three species) decreased stability because one *Paramecium* species went extinct.

In short, despite the dominance of the intuitive view that diversity begets stability, ecologists empirically investigating diversity and stability did not always toe this party line. These ecologists thus were not satisfied with theory: "Appeal to MacArthur's (1955) formalism of this concept seems frequently to serve as a reassurance that the vague verbal argument does not entirely lack a theoretical basis" (Hairston *et al.* 1968, p. 1100).

8.4.2 | The 1970s and 1980s

The theory developed in the 1970s directly challenged the intuitive view that diversity leads to stability, and this fomented an active debate both among and between theoreticians and empiricists. On the more theoretically oriented side of empiricism, attempts were made to reconcile the small chance that a diverse model community is stable (i.e., persistent) with the observation that diverse communities occur abundantly in nature. Much of this work involved analyzing real food webs to look for patterns that might explain their stability. For example, the key result of May's (1973) work is that the chances of a randomly constructed community being stable decreases with both the number of species in the community and the connectance among species, where connectance is measured by the probability that a pair of species interacts. More diverse communities need not be less stable if their connectance is lower; specifically, if the number of species with which an average species interacts does not change with community diversity, then stability does not change with increasing diversity. Examinations of large collections of food webs (Rejmanek & Stary 1979, Briand 1983) found that connectance did decrease with increasing diversity. How much of this relationship is real, however, and how much just reflects the idiosyncrasies of empirical researchers drawing food webs can be, and has been, questioned (Paine 1988).

At the same time, empirical studies continued to show relationships between diversity and stability that were varied and complex, depending on the system under study, the experimental or natural disturbances affecting the system, and the concept of stability applied by the researchers (McNaughton 1978, Luckinbill 1979, van Voris *et al.* 1980, Pimm 1984b). This work proceeded somewhat independently of the theoretical work on diversity and stability. As in the 1950s and 1960s, some empiricists chafed at the broad conclusions drawn by theoreticians: "Continued assertions of the validity of one or another conclusion about diversity-stability, in the absence of empirical tests, are acts of faith, not science" (McNaughton 1977, p. 516). Ironically, even though this quote echoes the sentiment of Hairston and his colleagues (1968) (quoted in the previous subsection), in this case McNaughton is sceptical of theory showing that diversity is destabilizing, whereas Hairston and his colleagues were sceptical of theory showing that diversity is stabilizing.

8.4.3 | The 1990s

Since the beginning of the 1990s, there has been an increasing number of empirical studies addressing diversity and stability. An important recognition was the distinction between the effects of diversity on individual species and the effects of diversity on the community; diversity may increase the variability of individual species and simultaneously decrease the variability of the community as a whole. This realization is widely believed to reconcile the discrepancy between the theoretical results from May (1973)—that diversity creates instability—and those sometimes found in empirical studies in which more diverse communities are more stable. Although Tilman (1996) correctly points out that this distinction between population and community variability is made by May (1974b, p. 231), May suggests this distinction in a different context from his theoretical work showing that more diverse ecosystems are less likely to be stable (May 1974a). For the randomly constructed communities that I described in Section 8.3.1 (Fig. 8.1a, b), increasing diversity decreases the stability (variability) of both the entire community and all species within it; hence, these results cannot be reconciled to empirical studies by entreating the distinction between species-level and community-level processes.

In the 1990s, theory did not keep up with the growing number of definitions of stability used in empirical studies. Empirical studies measured stability in a variety of ways: variability of different community variables (Tilman 1996; Cottingham, Brown, & Lennon 2001; Petchey *et al.* 2002), magnitude of responses to particular perturbations (Pfisterer & Schmid 2002), rates of return to "normal" conditions following particular disturbances (Tilman & Downing 1994), susceptibility to invasions or extinctions (Hodgson, Rainey, & Buckling 2002), similarity among communities with the same number of species (Naeem & Li 1997), and many others (Grimm & Wissel 1997). This growing collection of measures of stability is to some extent understandable because researchers were trying to cope with the constraints of doing experiments with their systems. With each new study, often a new measure of stability was needed, and this measure de facto became a new definition of stability. Thus, empiricists were devising new definitions of stability to suit their needs, generally without recourse to theoretical ideas on stability. This is not a novel observation; numerous authors have called for restraint and precision in the use of the term stability (e.g., see Connell & Sousa

1983, Pimm 1984a, Berryman 1991, Pimm 1991, and Grimm & Wissel 1997). Despite these calls for order, however, definitions of stability keep multiplying.

8.4.4 | Summary

Over the last 50 years, there has been tension between theory and empiricism, with empiricists often sceptical of the broad conclusions drawn from theory and theoreticians frustrated by the narrow views of empiricists shaped by their particular study systems. Ironically, empiricists do not seem to have been happy whether theoreticians claim there is a positive or a negative relationship between diversity and stability. The disjoint between theory and empirical studies reflects how difficult it is to mesh theoretical concepts of stability and empirical measures of stability. On the one hand, theory often attempts to make broad conclusions about diversity and stability; on the other hand, empirical workers must accept the particular characteristics of their study systems.

Theory seems to be failing to give a unifying framework with which to summarize and assess numerous empirical studies. An alternative approach is to abandon the attempt to create a general theory of diversity and stability, and to instead use theory more precisely to understand particular systems. I am sympathetic with the observation by Hairston and his colleagues that "the frequent violation of mathematical theory by biological peculiarity suggests that models must make allowances for such effects" (1968, p. 1100).

8.5 | SUGGESTIONS FOR SPECIFIC "TESTS"

There is much to be gained by a closer tie between theoretical and empirical investigations into diversity and stability. As described previously, different definitions of stability lead to different theoretical relationships between diversity and stability, thereby ruling out the possibility of a general conclusion about diversity and stability without dealing, case by case, with the many possible definitions of stability. Rather than look for generalities in the relationship between diversity and stability, I think it will be more profitable to examine the mechanisms that underlie the multiple diversity-stability relationships. Building up case studies of systems whose dynamics are well understood might do more to reveal generalities about diversity and stability than studies that simply try to give a yes-or-no answer to whether the diversity of a particular system is positively associated with some measure of stability.

I will now address three questions about the mechanisms underlying the relationship between diversity and stability. These questions go "back to basics", addressing fundamental ecological issues that would be as familiar to ecologists 50 years ago as to ecologists today. Even though these are general ecological questions, I think they will be key to answering how diversity and stability are related in real ecosystems.

Q1. | What Is the Most Appropriate Measure of Diversity?

Species diversity necessarily implies that species are different; many identical species is no different ecologically from a single species. Variability among species may involve differences in species sensitivities to a gamut of environmental stressors, differences in trophic location, differences in relative abundances, etc. The diversity of a community depends, at least partly, on the diversity of its constituent species, so cataloguing differences among species is essential. Although both theoretical and empirical studies often equate species diversity with species richness (number of species), species richness is too crude a measure of diversity to understand a specific system.

The best measure of diversity will depend on the definition of stability. For example, when predicting the response of a community to a particular environmental stressor, a key component of diversity is differences among species in their sensitivity to the stressor. Similarly, a key component of the diversity of a community being invaded by an exotic species will be the trophic complexity of the community in relation to the trophic location of the invader. In general, there is likely no hard-and-fast rule as to what differences among species, and hence what type of diversity, will likely be important, and there is no substitute for good ecological knowledge about the system at hand.

Q2. | How Strong Are Species Interactions, and Are They Linear and Additive?

Some definitions of stability depend on the strength of interactions among species; for example, in the dominant 1970s definition of stability, the stronger the interactions among species, the less likely randomly constructed communities will be persistent. Furthermore, asymmetries in interaction strengths are likely important; for example, using the dominant 1990s definition, the variability (CV) of combined species densities is unaffected by the strength of species interactions when interaction strengths are symmetrical, but asymmetries increase the CV if there are no interactions among species (Fig. 8.1e, f). Quantifying the magnitude of species interactions seems essential for addressing the mechanisms underlying stability.

Quantifying species interactions will be easiest if per capita population growth rates of species are linear and additive functions of the densities of species with which they interact. When this is the case, not only can species interactions be quantified in a pairwise fashion, but models of community interactions are also simpler (e.g., equation (8.1)). Many studies, however, have documented nonadditive interactions in which, for example, the presence of one predator decreases the effectiveness of a second predator by scaring their shared prey into a refuge (Crowder, Squires, & Rice 1997). Non-additive interactions might lead to interesting and exotic properties of the type described by McCann in his companion chapter. Again, however, non-additive effects are likely properties of specific systems that are difficult to generalize.

Q3. | What Dictates the Structure of Communities?

Natural communities represent not random assemblages of species but rather collections of species that can coexist. Understanding how they coexist will likely reveal structural properties of the community that affect stability (May 2001, pp. xxiv–xxv). This applies not only to natural, pristine communities but also to degraded communities. Studies of diversity are often justified by appealing to the worldwide, anthropogenic extinction of species. Because these extinctions are not random, the resulting communities are not random; the surviving species have characteristics that allowed them to survive.

Understanding how species coexist and the resulting structure of the community is an old and difficult problem. Nonetheless, it directly relates to the relationship between diversity and stability. For example, if two competing species coexist indefinitely, they must differ somehow, whether in the resources they use, the suites of predators that attack them, or some other factor (Chesson 1991). These differences might be key to anticipating whether one species will increase in density and the other will decrease, thus leading to compensation that stabilizes the variation in total community biomass. Similarly, coexistence might be the result of broad-scale processes so that a given species in a geographically delimited community cannot coexist in the absence of subsidies through immigration. In this case, stability of the local community against species loss could only be understood by looking at regional mechanisms of coexistence. Processes that allow coexistence will likely be processes that affect multiple definitions of stability, and hence, studying both coexistence and stability will be mutually beneficial.

A research programme designed to investigate these three questions for a specific system might seem antitheoretical. Clearly, such a research programme will not be able to make sweeping claims about diversity and stability. Nonetheless, as a theoretician I would prefer a few case studies that document the mechanisms explaining the relationship between diversity and stability than many studies showing either a positive or a negative relationship between diversity and (some definition of) stability without any explanation of why.* Theory tailored to understanding particular systems might help quantify and substantiate the mechanisms underlying stability. Stability is a complex set of concepts that is best clarified through the lens of mathematics.

I do not want to leave the impression that Theory (with a capital 'T') is useless. Theoretical models such as those illustrated in Figure 8.1 can give a sense of what factors may be worth exploring in real systems. Nonetheless, I think theoretical models aimed at understanding broad relationships between diversity and stability should be treated as case studies themselves, rather than as oracles revealing truth. Just like the results of a detailed empirical case study, the results of a detailed theoretical model may suggest what factors might affect the diversity-stability relationship in other systems. But just like the match between two real systems, it is unlikely that the match between a theoretical and a real system will be one to one.

8.6 | SUMMARY

The main goal of this chapter was to demonstrate that the relationship between diversity and stability depends on the definition of stability selected. Thus, the debate about the effect of diversity on stability is primarily a debate about what is the most relevant definition of stability. There does not have to be one best definition of stability; different definitions may be more appropriate in different situations, and consider-

* Maybe it is a false claim that I am a theoretician. I have been involved in empirical research on pea aphids for the last 15 years, trying to understand how the diverse assemblage of natural enemies regulates pea aphid populations. The difficulty of defining stability, let alone understanding how it is affected by diversity, for the pea aphid system intimidates me and makes me sceptical of applying general theoretical models to my study system.

ing multiple definitions applied to the same situation may be instructive. What is clear, however, is that stability is not a single, simple, easily defined concept.

Empirical tests of the relationship between diversity and stability need to pay attention to mechanisms. I think it is unlikely that general patterns will emerge that explain diversity-stability patterns across many communities without numerous exceptions. Understanding mechanisms might be the best route to finding generalities because mechanisms will reveal which assumptions of different theoretical models hold and which do not hold for particular empirical systems.

The value of theory is that it provides inspiration for understanding nature. I do not think that theory involves assembling models based on laws of nature from which truth can then be found. Instead, just like a detailed empirical case study, a detailed theoretical study provides a comparative reference point from which to view other systems. If the theory is interesting, it will stimulate further studies not aimed to test the theory but instead aimed to investigate insights that the theory might spark.

ACKNOWLEDGEMENTS

Brad J. Cardinale, Kelley J. Tilmon, and an anonymous reviewer made exceptionally valuable comments on this chapter. I thank Bea Beisner and Kim Cuddington for thinking me senior enough to participate in this book.

REFERENCES

Berryman, A.A. (1991). Stabilization or regulation: What it all means. *Oecologia* **86**, 140–143.

Briand, F. (1983). Environmental control of food web structure. *Ecology* **64**, 253–263.

Chesson, P. (1991). A need for niches? *TREE* **6**, 26–28.

Chesson, P. (1994). Multispecies competition in variable environments. *Theor. Popul. Biol.* **45**, 227–276.

Chesson, P. (2000). Mechanisms of maintenance of species diversity. *Annu. Rev. Ecol. Syst.* **31**, 343–366.

Connell, J.H., & Sousa, W.P. (1983). On the evidence needed to judge ecological stability or persistence. *Am. Nat.* **121**, 789–824.

Cottingham, K.L., Brown, B.L., & Lennon, J.T. (2001). Biodiversity may regulate the temporal variability of ecological systems. *Ecol. Lett.* **4**, 72–85.

Crowder, L.B., Squires, D.D., & Rice, J.A. (1997). Nonadditive effects of terrestrial and aquatic predators on juvenile estuarine fish. *Ecology* **78**, 1796–1804.

Doak, D., Bigger, D., Harding, E., Marvier, M., O'Malley, R., & Thomson, D. (1998). The statistical inevitability of stability–diversity relationships in community ecology. *Am. Nat.* **151**, 264–276.

Elton, C.S. (1958). "The Ecology of Invasions by Animals and Plants." Methuen, London.

Gardner, M., & Ashby, W.R. (1970). Connectance of large dynamical (cybernetic) systems: A critical value for stability. *Nature* **228**, 784.

Grimm, V., & Wissel, C. (1997). Babel, or the ecological stability discussions: An inventory and analysis of terminology and a guide for avoiding confusion. *Oecologia* **109**, 323–334.

Hairston, N.G., Allan, J.D., Colwell, R.K., Futuyma, D.J., Howell, J., Lubin, M.D., Mathias, J., & Vandermeer, J.H. (1968). The relationship between species diversity and stability: An experimental approach with protozoa and bacteria. *Ecology* **49**, 1091–1101.

Hastings, A., Hom, C.L., Ellner, S., Turchin, P., & Godfray, H.C.J. (1993). Chaos in ecology: Is mother nature a strange attractor? *Annu. Rev. Ecol. Syst.* **24**, 1–33.

Hodgson, D.J., Rainey, P.B., & Buckling, A. (2002). Mechanisms linking diversity, productivity, and invasibility in experimental bacterial communities. *Proc. Roy. Soc. Lond. B* **269**, 2277–2283.

Ives, A.R., Gross, K., & Klug, J.L. (1999). Stability and variability in competitive communities. *Science* **286**, 542–544.

Ives, A.R., Klug, J.L., & Gross, K. (2000). Stability and species richness in complex communities. *Ecol. Lett.* **3**, 399–411.

Lawlor, L.R. (1978). A comment on randomly constructed ecosystem models. *Am. Nat.* **112**, 445–447.

Lehman, C.L., & Tilman, D. (2000). Biodiversity, stability, and productivity in competitive communities. *Am. Nat.* **156**, 534–552.

Leigh, E.G. (1965). On the relation between the productivity, biomass, diversity, and stability of a community. *Proc. Natl. Acad. Sci. USA* **53**, 777–783.

Luckinbill, L.S. (1979). Regulation, stability, and diversity in a model experimental microcosm. *Ecology* **60**, 1098–1102.

MacArthur, R.H. (1955). Fluctuations of animal populations and a measure of community stability. *Ecology* **36**, 533–536.

MacArthur, R.H. (1972). "Geographical Ecology." Harper & Row, New York.

Margalef, R. (1969). Diversity and stability: A practical proposal and a model of interdependence. In "Diversity and Stability in Ecological Systems" (G.W. Woodwell & H.H. Smith, Eds.), Brookhaven Symposium in Biology 22, Brookhaven National Laboratory, Upton, NY.

May, R.M. (1972). Will a large complex system be stable? *Nature* **238**, 413–414.

May, R.M. (1973). "Stability and Complexity in Model Ecosystems." Princeton University Press, Princeton, NJ.

May, R.M. (1974a). Ecosystem properties in randomly fluctuating environments. In "Progress in Theoretical Biology" (R. Rosen & F.M. Snell, Eds.). Academic Press, New York, pp. 1–50.

May, R.M. (1974b). "Stability and Complexity in Model Ecosystems," 2nd edition. Princeton University Press, Princeton, NJ.

May, R.M. (2001). "Stability and Complexity in Model Ecosystems." Princeton Landmarks in Biology. Princeton University Press, Princeton, NJ.

McCann, K., Hastings, A., & Huxel, G.R. (1998). Weak trophic interactions and the balance of nature. *Nature* **395**, 794–798.

McNaughton, S.J. (1977). Diversity and stability of ecological communities: A comment on the role of empiricism in ecology. *Am. Nat.* **111**, 515–525.

McNaughton, S.J. (1978). Stability and diversity of ecological communities. *Nature* **274**, 251–253.

Naeem, S., & Li, S. (1997). Biodiversity enhances ecosystem reliability. *Nature* **390**, 507–509.

Odum, E.P. (1953). "Fundamentals of Ecology." Saunders, Philadelphia.

Paine, R.T. (1988). Food webs: Road maps of interactions or grist for theoretical developments? *Ecology* **69**, 1648–1654.

Petchey, O.L., Casey, T., Jiang, L., McPhearson, P.T., & Price, J. (2002). Species richness, environmental fluctuations, and temporal change in total community biomass. *Oikos* **99**, 231–240.

Pfisterer, A.B., & Schmid, B. (2002). Diversity-dependent production can decrease the stability of ecosystem functioning. *Nature* **416**, 84–86.

Pimentel, D. (1961). Species diversity and insect population outbreaks. *Annu. Entomol. Soc. Am.* **54**, 76–86.

Pimm, S.L. (1979). Complexity and stability: Another look at MacArthur's original hypothesis. *Oikos* **33**, 351–357.

Pimm, S.L. (1982). "Food Webs." Chapman & Hall, London.

Pimm, S.L. (1984a). The complexity and stability of ecosystems. *Nature* **307**, 321–326.

Pimm, S.L. (1984b). Food chains and return times. In "Ecological Communities: Conceptual Issues and the Evidence" (D.R. Strong, D. Simberloff, L.G. Abele, & A.B. Thistle, Eds.). Princeton University Press, Princeton, NJ, pp. 397–412.

Pimm, S.L. (1991). "The Balance of Nature?" University of Chicago Press, Chicago.

Preston, F.W. (1962). The canonical distribution of commonness and rarity. *Ecology* **43**, 185–215.

Redfearn, A., & Pimm, S.L. (1987). Insect pest outbreaks and community structure. In "Insect Pests" (P. Barbosa & J.C. Schultz, Eds.). Academic Press, Orlando, FL, pp. 99–133.

Rejmanek, M., & Stary, P. (1979). Connectance in real biotic communities and critical values for stability in model ecosystems. *Nature* **280**, 311–313.

Strong, D.R. (1986). Density-vague population change. *TREE* **1**, 39–42.

Tilman, D. (1996). Biodiversity: Population versus ecosystem stability. *Ecology* **77**, 350–363.

Tilman, D., & Downing, J.A. (1994). Biodiversity and stability in grassland. *Nature* **367**, 363–365.

Tilman, D., Lehman, C.L., & Bristow, C.E. (1998). Diversity–stability relationships: Statistical inevitability or ecological consequences? *Am. Nat.* **151**, 277–282.

van Voris, P., O'Neill, R.V., Emanuel, W.R., & Shugart, H.H. (1980). Functional complexity and ecosystem stability. *Ecology* **61**, 1352–1360.

Watt, K.E.F. (1964). Comments on long-term fluctuations of animal populations and measures of community stability. *Can. Entomol.* **96**, 1437–1442.

Watt, K.E.F. (1965). Community stability and the strategy of biological control. *Can. Entomol.* **97**, 887–895.

Wiens, J.A. (1977). On competition and variable environments. *Am. Sci.* **65**, 590–597.

Wolda, H. (1978). Fluctuations in abundance of tropical insects. *Am. Nat.* **112**, 1017–1045.

9 | PERSPECTIVES ON DIVERSITY, STRUCTURE, AND STABILITY

Kevin S. McCann

9.1 | INTRODUCTION

The role of diversity, and structural complexity, in the dynamics and stability of ecosystems is a long-standing and unresolved issue in ecology (e.g., see Darwin 1859). This quest has left behind a long history that weaves together an array of personalities, many of whom significantly affected both ecological and mathematical development (e.g., see May 1976). It is also an area of investigation that has shown a dramatic increase in scientific activity lately (De Ruiter, Neutel, & Moore 1995; Naeem & Li 1997; van der Heijden *et al.* 1998; Berlow 1999; Neutel, Heesterbeek, & De Ruiter 2002), paralleling heightened scientific and public awareness of the heavily human-affected earth (Reid 1997, Levin 1999, Riciarddi & Rasmussen 2000) and advances in computational ability and understanding of complex dynamic systems (Strogatz 2001). As a result of the growing understanding of the enormity of the effect of humans on the biosphere, it is unlikely that the current interest in the diversity-stability question will fade.

The belief that nature's vast interconnectedness is important extends beyond a purely scientific discussion. The *wholeness* and *interconnectedness* of life are recurrent themes that transcend individual religions (Campbell & Moyers 1991). Implicit in these mythological themes is the notion that the interconnectedness of those that inhabit the earth is of vast importance for the day-to-day working of the earth. In some sense, this interconnected web of life is cast as a vast, impenetrable complexity, whose sheer existence and baroque beauty suggest the hand of God. One intriguing aspect of this religious view is that complexity, in a sense, is seen as a means to create an ordered functional universe (e.g., the web of life, Mother Earth, produces a consistently plentiful amount of deer to harvest each year). Here, then, is the notion that complexity generates a simple outcome (i.e., a relatively consistent harvest), an entirely intuitive belief but one that fascinates the human mind and lies at the heart of this often heated and intriguing debate.

Below, I describe the history of the diversity-stability debate. I argue that there have been three relatively distinct periods with two major transitions in thought. One period of mostly intuitive belief suggested nature's complexity gives rise to stability. A second period arose with the rigorous application of mathematics and dynamic systems theory that puts this intuitive belief to the test. And a current period of scientific attack is attempting to identify how specific attributes of structural complexity interact with spatial and temporal *noise* to stabilize naturally diverse and complex communities.

9.2 | A BRIEF HISTORY OF DIVERSITY AND STABILITY

9.2.1 | The Intuitive Years

The early ecological interest in diversity and stability revolved largely around intuitive interpretations. As explained earlier, both mythological and early scientific ideas seemed to suggest that diverse, highly interconnected systems were responsible for the persistence and consistency of natural systems on ecological timescales. Ecologists attempted to create a logical basis for this belief. Odum (1953), for example, simply came out and defined community stability as "the amount of choice which the energy has in following the paths up through the food web". The specific reason for this choice was never explicitly described and so suggests that it relies largely on an intuitive appeal that portioning the energetic pie tends to stabilize.

Not long after Odum's definition, Robert MacArthur (1955) published a well-known paper that attempted to give some scientific rigor to Odum's earlier statement. In this paper, MacArthur sketched a series of food webs and described the ramifications of energy partitioning for stability using information theory. His idea, although intriguing, plays more like an intuitive appeal masqueraded in informal mathematics.

The notion of stability here is somewhat vague but largely entails the response of the community to a perturbation that influences the density of at least one of the species. MacArthur's argument revolved around the notion that a perturbation that reduced the density of a population would be least felt by the community if other species consuming that organism were generalists feeding on other species as well (i.e., were embedded in a more complex web). In this way, when one prey was diminished the consumers would be buffered against this perturbation by turning their attention to the other prey species in the food web. At the heart of this idea is the realization that consumers have the ability to integrate across different prey involved in energetically diverse pathways. Curiously, one assumption is that of an equilibrium system, but this notion of a predator/consumer integrating across pathways, arguably, acts most effectively under non-equilibrium assumptions—an idea not easily testable without advances in dynamic systems theory and computation.

Shortly after MacArthur's semi-formal treatment, Charles Elton (1958) took an entire chapter in his famous book, *The Ecology of Invasions by Animals and Plants*, to explore the relationship between diversity and stability. His approach, like the others, was largely intuitive, although it drew from earlier theoretical models and anecdotal empirical evidence pertaining to the influence of invasions on ecological communities. Elton presented six lines of reasoning that I think can be summarized using three more broadly defined themes. Although his definition of stability vacillated, he went to great length to emphasize that he was generally considering dynamic instabilities that drove "destructive oscillations" and "population explosions" in food webs.

First, he felt that because systems of simple models were subject to extraordinary instabilities, increased stability would accompany increased model complexity and diversity. Here he was drawing from the early theoretical work of Lotka and Volterra, which produced dynamics defined by neutral stability, and from simple laboratory microcosms that have proved to be the definition of instability (e.g., microcosms are frequently so unstable that stability is measured as time to extinction). Second, he argued that simplified food webs were more vulnerable to invaders. In this case, he relied on evidence that suggests that pest outbreaks occur readily in monocultures and other habitats greatly simplified by the actions of humans. Finally, he argued that island food webs were notorious for the extensive effect incurred by invasive species. Here, the logic was that island food webs are less diverse than mainland webs. All these lines of reasoning are interesting and together compelling; however, they also include a suite of other factors that obfuscate the exact role of diversity. The mere fact that diversity correlates with all Elton's arguments does not necessarily suggest that diversity is the governing force behind his three generalizations. As an example, each of the three generalizations offered by Elton is also clearly related to spatial scale. Microcosms, monocultures, and models take place in spatially homogenized or unstructured arenas, and island food webs are clearly systems limited in spatial extent relative to their continental counterparts.

There were attempts to include dynamics in this early line of reasoning (e.g., Elton's destructive cycles and MacArthur's pathways distributing shocks or perturbations); however, they were not rigorous applications of systems science. The tools of dynamic systems were simply not a part of the ecologist's toolbox. Soon things were to change as a suite of mathematicians and physicists were entering ecology, clearing the way to challenge some of these ideas. They were to find that this early reasoning, although seemingly sound, had some logical holes.

9.2.2 | The Limits to Diversity

In the early 1970s mathematical ecologists began to wrestle with the diversity-stability problem for large model communities (e.g., see Gardner & Ashby 1970, May 1972, and May 1974). This was the start of a time that did much to point out the limits of diversity as a stabilizer. The development of computer technology was beginning to open the area of dynamic systems to new avenues previously unexplored by the pencil-and-paper mathematical techniques of the 1950s. Previously intractable systems, or intractable questions (e.g., global dynamic behaviour), were fair game and the analysis of large, nonlinear, or both types of systems became a focus of intense scientific interest. Scientists from numerous realms made significant contributions to these mathematical developments, including Robert May's finding that simple, discrete ecological systems beget chaotic dynamics (May 1976).

The hallmark of chaos is an extreme sensitivity to initial conditions (e.g., the butterfly flaps its wings and causes a storm elsewhere) and dynamics that twist out noisy patterns. The curious thing about chaos is that it contains elements of both pattern (e.g., a geometrically defined attractor) and random behaviour (unless the initial value is precisely known the dynamics are relatively unpredictable), yet this all arises from a deterministic system. Chaos gave scientists the sudden recognition that a nonlinear feedback process of a defined signature (e.g., May's difference equations have a delay of 1 year) interacts and mixes with other feedback processes of different signatures (e.g., the signature generated by the degree of density-dependent overcompensation in May's equations). They mix in a way that the addition of these feedbacks becomes greater than the individual feedbacks themselves. This can be seen upon examination of the spectral signature of a chaotic time series. Although there are spikes at some characteristic frequencies (the signatures of the major oscillatory drivers responsible for the pattern in the chaotic dynamics), there are also spikes almost everywhere else. Curiously, and almost magically, the signatures mix like a Jackson Pollack painting—a blend of pattern and noise. I believe that this fascinating result significantly influenced the perspective of scientists. If even simple systems can generate chaotic dynamics, what could complex systems generate? Would this mean that nature's complex palette would inspire an even more delirious and unpredictable form of chaos? It turns out that in an unconstrained and diverse world this is essentially true (May 1972, May 1974): unconstrained complexity readily and rapidly drives inspired amounts of instability.

From a diversity-stability perspective, the culmination of this intense period of activity can be seen in May's seminal book (1974). May, and others, found that diversity begets instability. The logic behind this is not far off that which I just suggested. In May's own words (1974):

> A variety of explicit counterexamples have demonstrated that a count of food web links is no guide to stability. This straightforward fact contradicts the intuitive verbal argument often invoked, to the effect that the greater the number of links and alternative pathways in the web, the greater the chance of absorbing environmental shocks, thus damping down incipient oscillations. The fallacy in this intuitive argument is that the greater the size and connectance of the web, the larger the number of characteristic modes of oscillation it possesses: since in general each mode is as likely to be unstable as stable (unless the increased complexity is of a highly special kind), the addition of more and more modes simply increases the chance for the total web to be unstable.

This is a point that seems absolutely true. In a world without some kind of structure, increased diversity means an increased potential for the coupling of oscillators, which drives increased dynamic instability.

The model experiment that produced these results can be interpreted as follows: May wished to test the way diversity and complexity (where this is equivalent to increased connectance) influenced stability. He therefore generated diversity and connectance using random statistical universes (i.e., unstructured). Thus, he clearly and correctly tested the hypothesis that diversity and increased connectance beget stability. The formal mathematics suggest that diversity does not influence stability.

Nonetheless, May (1972, 1974) was aware that this did not resonate with what most empiricists were finding—that diversity and complexity tended to promote with persistent communities (Goodman 1975, McNaughton 1978). May's work meant that ecologists must focus away from diversity and seek to understand what important biological structures wove the fabric upon which stability relied. Interestingly, although underplayed, these early investigations made some suggestions to the resolution of these problems. May and others (Gardner & Ashby 1970, DeAngelis 1975, Yodzis 1981) suggested that patterns in interaction strength and compartmentalization may play an important role in stabilizing complex dynamic systems. From a food web perspective, researchers were recommending that ecologists search for, and identify, important stabilizing structural attributes of food webs.

The theoretical work of the early 1970s thus inspired a factory-like production on the structure of food webs that carries on to this day in the form of network theory (e.g., see Dunne, Williams, & Martinez 2002). In this sense, I could argue that the contribution of the 1970s ignited a rather drastic change in the research agenda; a change that is still growing and thriving today.

9.2.2.1 | Revisiting the Data: Some Emerging Generalizations

In the quest for meaningful biological structure, one obvious starting place was structure within the food web. After the 1970s, a burgeoning food web statistics industry was created, inspired by Cohen, Briand, and Newman's (1990) synthesis of 33 food webs. The early work proved fascinating, suggesting a suite of patterns that were robust across several webs. Further research began to suggest that certain food web complexities (compartments and omnivory) were both dynamically unstable and infrequent in real food webs (Pimm 1979, Pimm & Lawton 1980). Nonetheless, some researchers (Winemiller 1990, Polis 1991, Hall & Raffaelli 1991, Martinez 1993, Cohen *et al.* 1993) began to question the resolution and meaning of these early food web statistics and their theories. Most webs were naturally biased towards specific taxa, and all webs were incomplete (Cohen *et al.* 1993).

A crowd of empiricists studying food webs began to find recurrent structures in food webs that had been overlooked (Winemiller 1990, Polis 1991, Martinez 1992, Hall & Raffaelli 1991). Gary Polis, for example, one of the central figures, pressed hard to show scientists that omnivory was replete in desert food webs and, with Don Strong, championed the role of spatial subsidies and multichannel pathways in food webs

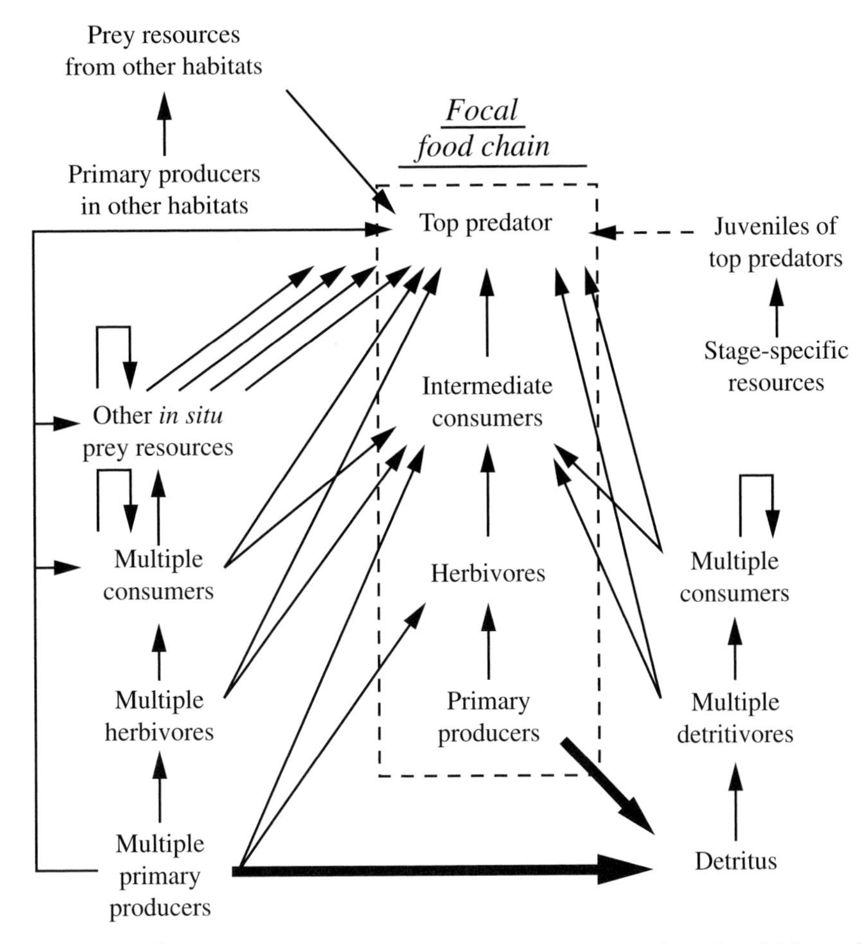

FIGURE 9.1 | Polis and Strong's (1996) schematic representation of multichannel omnivory highlighting factors that have been overlooked in food web ecology. Of note is the emphasis on the role of space. Specifically, they noted multiple producers existing in different habitats and detrital and grazing channels that tended to be coupled by higher-order consumers through movement and life history. Redrawn from Polis & Strong 1996.

(Fig. 9.1). Indeed, Polis and Strong (1996) argued that ecologists needed to expand their spatial scale and recognize that many of the focal webs we were studying were coupled through generalist consumers using both top-down (consumption) and bottom-up mechanisms (nutrient transfer). In Figure 9.1, such a generalized food web is schematically depicted, and some of the major empirical generalizations of more completely resolved food webs are outlined. These ideas resonated with others in that they emphasized that the scale that ecologists look at food webs must be fully reconsidered (reviewed in Polis & Winemiller 1996). From a dynamic systems perspective, these interactions that crossed different scales meant feedbacks that ultimately could modify the behaviour of any focal food web. In a sense, these results suggested that population, community, ecosystem, and landscape ecologists must begin to interact. At the same time, other researchers pointed out that food webs were not only variable in space but also highly variable in time (Warren 1989, Winemiller 1990; Reagan, Camilo, & Waide 1996). Winemiller's study system, Amazonian floodplain rivers, showed dramatic variation in food webs from season to season. Empirical studies were expanding the spatial and temporal scale of food web ecology in their search for important biological structure (reviewed in Winemiller & Polis 1993).

9.2.2.2 | Revisiting the Theory

There has long been the recognition that patterns in interaction strength might play a role in stabilizing food webs (May 1972, Yodzis 1981). Experimentally generated patterns in functional interaction strength (*sensu* Paine 1992), for example, have tended to find that most functional interactions are weak and only a few are strong. In addition, energy flows in food webs (e.g., per capita consumption rates) suggest that there is a tendency towards many weak linkages and a few strong linkages (Winemiller, in press). Shortly after the empiricists pointed out that certain food webs' structures were more common than originally thought (e.g., intraguild predation and omnivory), theorists began to explore such structure. One interesting approach was put forth by Holt (1996), who effectively suggested that an obvious extension to early Lotka-Volterra models of two-species interaction theories (e.g., competition, predation, and mutualisms) was to study relatively simple but ubiquitous food web structures (e.g., intraguild predation and omnivory). These common structures can be extracted out of whole webs and isolated as a *food web module*. In this way, theory could develop beyond the well-established theories for direct interactions (e.g., predator-prey and competition). Although some ecologists had pointed out the potential importance of a non-equilibrium perspective (DeAngelis & Waterhouse 1987), ecologists were suddenly becoming more willing and able (e.g., computationally intensive dynamic systems theory) to consider problems from a non-equilibrium perspective (Tilman 1996, Chesson & Huntley 1997, McCann 2000).

One result that followed out of these emerging perspectives was the averaging effect (Tilman 1996; Doak *et al.* 1998; Tilman, Lehman, & Bristow 1998). The basic idea behind the averaging effect is that variable population dynamics can be summed to give relatively stable community dynamics as long as the different populations show differential response to variable conditions. This differential response can be randomly driven (Doak *et al.* 1998) or induced by life history differences that drive differential response to varying abiotic conditions (Tilman, Lehman, & Bristow 1998). Thus, although each individual population can show significant variation through time, the sum of all these differentially responding organisms produces a relatively stable aggregate community biomass.

Another emerging theory, viewed from either an equilibrium or a non-equilibrium perspective, explored the implication of interaction strength within food web modules. Here, researchers asked how interaction strength influenced the dynamic behaviour and stability of some ubiquitous food web modules (McCann & Hastings 1997; Huxel & McCann 1998; McCann, Hastings, & Huxel 1998; Post, Conners, & Goldberg 2000). The results showed that interaction strength could play an enormously important role in stabilizing food webs. Specifically, weak interactions (in the sense of per capita energy flow) can be easily positioned within a food web such that they muted potentially strong and oscillatory interactions (Fig. 9.2A–C). Figure 9.2 illustrates the strategic placement of two weak interactions (Fig. 9.2B, C) and the subsequent stabilization of the time series. This sequence of stabilization appears to be because of the following two mechanisms (McCann 2000):

1. some weak interactions (e.g., competitive interactions) (Fig. 9.2B) can redirect energy from a potentially oscillatory consumer. In essence, the weak interaction reverses the paradox of enrichment as energy or productivity is shunted from a potentially oscillatory interaction (Rosenzweig 1971). Note, however, that a strong interaction placed similarly would only contribute to more intense and complex oscillations (*sensu* May 1974), and
2. generalist consumers (Fig. 9.2C) can, through preferential feeding, drive out-of-phase resource or prey dynamics (McCann 2000). Out-of-phase dynamics can be summed to give a relatively stable resource or prey community biomass and thus enable a relatively stable response of the consumer to the resource variability (i.e., out-of-phase dynamics give the consumer the option to respond to low resource densities by shifting its attention towards a resource that is not at low resource densities).

This latter mechanism again suggested that interesting stabilizing structures may be operating because of variability. This mechanism is a top-down (or combination of top-down and bottom-up) way to get a

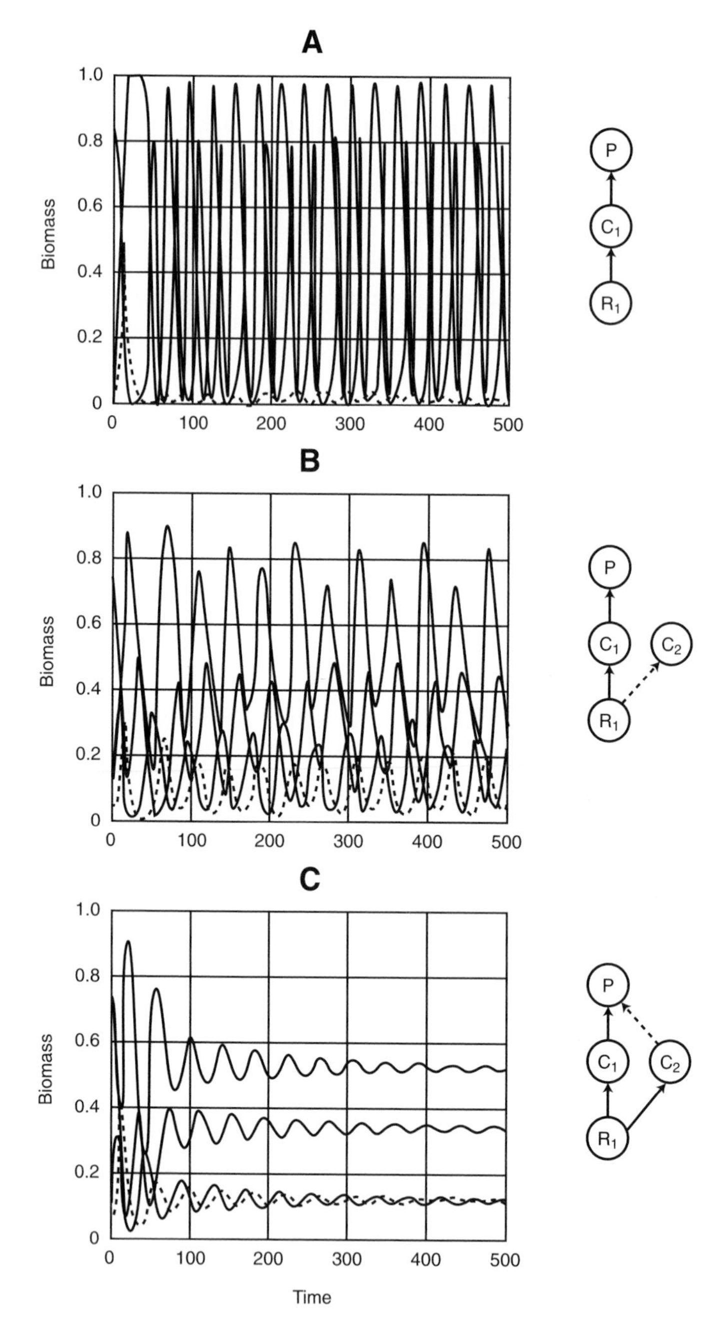

FIGURE 9.2 | An example of the successive addition of weak interactions to unstable food web modules. All three time series employ the models and parameters from McCann, Hastings, & Huxel 1998. (A) A food chain showing chaotic dynamics that are wildly unstable. (B) The same chain with an additional intermediate consumer (C_2) that is weakly consuming the resource (i.e., the C_2-R interaction is stable in isolation). The interaction is still oscillatory but less so than the food chain. (C) The same case again but with another weak interaction between the top predator (P) and C_2. The time series begins to become even more stable and approaches equilibrium.

differential response and is therefore closely aligned to the bottom-up ideas of the averaging effect. Certain trade-offs readily generate such a situation. For example, if organisms that are more competitive are less tolerant to predation, then this readily produces a situation with out-of-phase resource or prey responses. These results rely on consumers that forage on multiple resources with a preference. The results also appear to be consistent with theoretical efforts that have employed optimal foraging theory (Krivan 1996, Fryxell & Lundberg 1997, Krivan & Sikder 1999).

These new theories still assume a homogenized spatial world, and empiricists wrote repeatedly of the importance of space (Paine 1988, Peters 1983, Winemiller 1990, Polis 1991, Polis & Strong 1996). I now turn to some recent directions and briefly anticipate some future directions in this important area of research.

9.2.3 | Some Current and Future Considerations: Food Webs Across Space and Time

Although space has played a large role in population ecology and direct interactions (McCauley, Wilson, & deRoos 1996), the consideration of the role of space on food web dynamics is relatively recent (Holt 1996; Polis, Anderson, & Holt 1997; Nachman 2001; Callaway & Hastings 2002; McCann *et al.*, in press; Teng & McCann 2004). In a series of articles, Holt (1996, 2002) and others (Loreau, Mouquet, & Holt 2003; Holt & Hoopes, in press) have begun to tie metapopulation theory to community and ecosystem perspectives (dubbed metacommunity and metaecosystem, respectively). They have argued cogently that this larger perspective has the potential to unite population, community, and ecosystem perspectives. More specifically, they have argued that expanding the spatial scale of food webs may allow ecologists to more completely understand such long-standing issues as food chain length, trophic control (see also Polis, Anderson, & Holt 1997), island biogeography, and food web stability or instability.

Along a similar research theme, some ecologists have begun to consider empirical arguments to frame a more general spatial theory of food webs (McCann, Rasmussen, & Umbanhowar, in review). Polis and Strong (1996) emphasized that different habitats contained different primary producers and that these tended to be coupled by higher-ordered generalist consumers. This result is consistent with two empirical generalizations: (1) that generalist foraging tends to increase with higher-order consumers (Polis & Strong 1996; Cohen, Jonsson & Carpenter 2003) and (2) that higher-order organisms tend to be larger and more mobile than

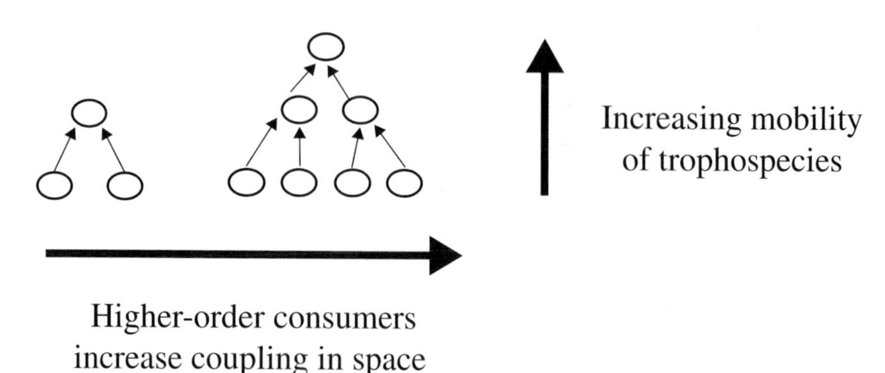

FIGURE 9.3 | A schematic representation of food webs in space. Higher-order organisms are increasingly more generalized in their foraging and increasingly more mobile. Thus, higher-order organisms couple lower-level habitat compartments.

their prey (Peters 1983; Brown, Stevens, & Kaufman 1996; McCann *et al.*, in press). These relationships are schematically summarized in Figure 9.3 and together create a simple framework for a general spatial theory. Some researchers (McCann *et al.*, in press) have begun to consider the implication of such spatial coupling on the dynamics and stability of coupled food webs. The results suggest that in spatially extended systems with differentially responding resources or prey, behaviour (i.e., movement) by the larger, more mobile organism can act as a potent stabilizing force, especially when considered in a non-equilibrium context.

The result is easily presented and consistent with earlier theory emerging from spatial population ecology (e.g., see McCauley, Wilson, & deRoos 1996 and Fryxell & Lundberg 1997). Effectively, larger organisms can respond to variation in space by moving from areas where prey or resource densities are low and towards areas where prey or resource densities are high. The outcome is the release of predatory pressure on prey when prey species are at low densities and increasing predatory pressure when prey species attain high densities—precisely the arrangement needed to reduce extreme variation in density. From the consumer perspective, its rapid behavioural response allows it to track variable resource or prey densities at a larger spatial scale. Clearly, the result relies on the underlying idea that resources in different habitats are responding differentially through time. It turns out that this variation can be abiotically driven or driven by the top-down predatory pressure of generalist consumers if the consumer tends to prefer one organism significantly more than other organisms (this is a manifestation of the weak interaction effect) (McCann, Hastings, & Huxel 1998). So again, like the averaging effect described for a single trophic level (Tilman, Lehman, & Bristow 1998), the notion of differential responses within a non-equilibrium perspective suggest that food web stability may unfold from variability in space and time.

Pimm and Lawton (1980) found little evidence for compartments in food webs except at huge spatial scales or if they considered the coupling of detrital webs to grazing webs. Recent analysis of food webs, using interaction strength or energy flow, found that compartments might be more ubiquitous than early investigations suggested (Krause *et al.* 2003). It is interesting to reconsider how the coupling of food webs within a spatial perspective will influence the food web compartments. In Figure 9.4A, a food web in which weak and strong interactions are essentially uniformly distributed throughout the food web is depicted. Such a configuration does not drive compartmented food web structure, and in light of the result from Krause and her colleagues' (2003), may not characterize natural systems. Figure 9.4B, on the other hand, shows a distribution of interaction strengths that generates strong compartmentalization. Another interesting potential distribution of interaction strengths that generates compartments of a slightly different kind is illustrated in Figure 9.4C. Here, one will find not only a compartmentalized web but also some compartments that may tend to contain stronger interactions than other compartments (i.e., there is the potential not only for weak interactions but also for weak compartments).

Soil ecologists have argued for such structure for some time in their underground food webs (Moore & Hunt 1988). They have suggested that bacterial energy channels tend to break down more labile detritus and turn over much more rapidly than fungal energy channels that tend to arise out of more recalcitrant detrital sources. Similarly, an argument can be made for littoral or benthic pathways in lakes versus pelagic pathways in lakes. Benthic invertebrates tend to turn over on a much longer timescale then the rapid turnover of zooplankton on phytoplankton. Finally, it has been suggested for some time that detrital webs are slower and more donor-controlled than grazing webs. Teng and McCann (2004) recently reconsidered the stabilizing role of compartments and found that compartments can be potent stabilizing forces. Again, particularly if compartments (like species) tend to respond differentially in time, behavioural responses by higher-order consumers can then average across these variable out-of-phase subsystems. Hence, strong and weak compartments could be an important form of food web structure that contributes to the persistence of ecological systems.

I offer these emerging ideas in hope that they outline the potential that may await us if we are bold enough to venture into the implications of

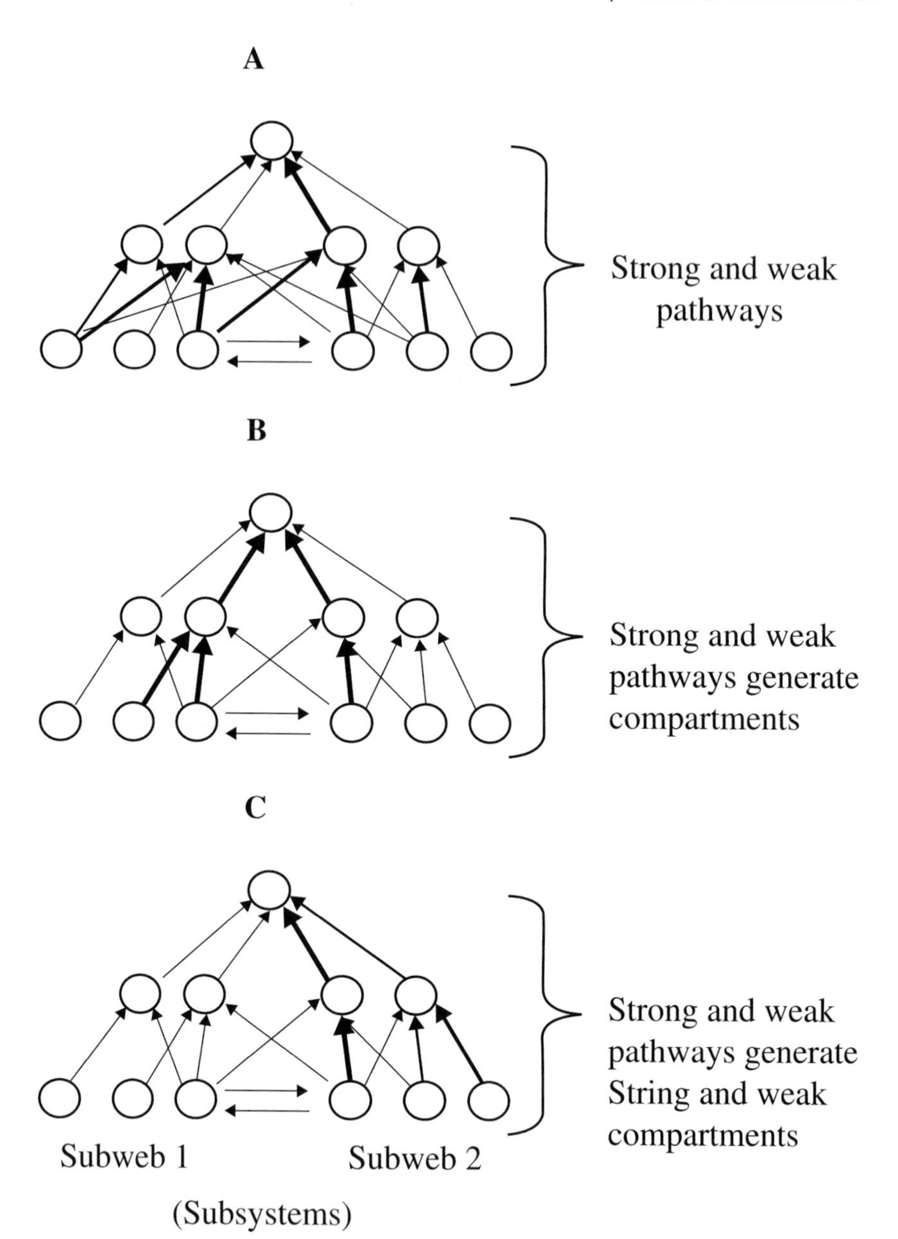

FIGURE 9.4 | Three examples of the distribution of weak interactions in a food web. (A) Uniformly distributed weak interactions will not tend to produce compartments even if weak interactions are ignored. (B) Weak interactions are distributed such that food webs have compartments, although weak and strong interactions still exist within individual compartments. (C) Weak interactions are distributed such that food webs have compartments, although weak and strong interactions are positioned such that there also exists the tendency for weak and strong compartment flows.

variability in space and time on food web dynamics and composition. It appears that this variability in biological structure may lead us to understand intriguing structural changes in food webs as a response to dynamic conditions. Thus, embracing the "noise" in space and time might contribute to significant empirical and theoretical advances in understanding. Importantly, several common threads (non-equilibrium perspective, differential response, and scale) in the preceding sections suggest that there may be some ways to unfold this vast complexity in a manageable way (e.g., compartments in space and compartments in time). Ecologists have begun to realize that we must cross long-standing scientific boundary lines.

REFERENCES

Berlow, E. (1999). Strong effects of weak interactions in ecological communities. *Nature* **398**, 330–334.

Brown, J.H., Stevens, G.C., & Kaufman, D.M. (1996). The geographic range: Size, shape, boundaries, and internal structure. *Annu. Rev. Ecol. Syst.* **27**, 597–623.

Callaway, D., & Hastings, A.M. (2002). Consumer movement through differentially subsidized habitats creates a spatial food web with unexpected results. *Ecol. Lett.* **5**, 467–470.

Campbell, J., & Moyers, B. (1991). "The Power of Myth." First Anchor Books, New York.

Cohen, J.E., Briand, F., & Newman, C.M. (1990). Community food webs: Data and theory. *Biomathematics* **20.** Springer-Verlag, New York.

Cohen, J.E., Jonsson T., & Carpenter, S.R. (2003). Ecological community description using the food web, species abundance, and body size. *Proc. Natl. Acad. Sci. USA* **100**, 1781–1786.

Cohen, J.E., & 23 others (1993). Improving food webs. *Ecology* **74**, 252–258.

Chesson, P., & Huntly, N. (1997). The role of harsh and fluctuating conditions in the dynamics of ecological communities. *Am. Nat.* **150**, 519–553.

Darwin, C. (1959). On the Origin of Species. London, John Murray.

DeAngelis, D.L. (1975). Stability and connectance in food web models. *Ecology* **56**, 238–243.

DeAngelis, D.L., & Waterhouse, J. (1987). Equilibrium and nonequilibrium concepts in ecological models. *Ecol. Monogr.* **57**, 1–21.

De Ruiter, P., Neutel, A.M., & Moore, J. (1995). Energetics, patterns of interaction strengths, and stability in real ecosystems. *Science* **269**, 1257–1260.

Doak, D.F., Bigger, D., Harding, E., Marvier, M., O'Malley, R., & Thompson, D. (1998). The statistical inevitability of stability–diversity relationships in community ecology. *Am. Nat.* **151**, 264–276.

Dunne, J.A., Williams, R.J., & Martinez, N.D. (2002). Network structure and biodiversity loss in food webs: Robustness increases with connectance. *Ecol. Lett.* **5**, 558–567.

Elton, C.S. (1958). "The Ecology of Invasions by Animals and Plants." Chapman & Hall, London.

Fryxell, J., & Lundberg, P. (1997). "Individual Behaviour and Community Dynamics." Chapman & Hall, Toronto.

Gardner, M.R., & Ashby, W.R. (1970). Connectance of large dynamical (cybernetic) systems: Critical value for stability. *Nature* **228**, 784.

Goodman, D. (1975). The theory of diversity–stability relationships in ecology. *Q. Rev. Biol.* **50**, 237–266.

Hall, S.J., & Raffaelli, D. (1991). Food web patterns: Lessons from a species-rich web. *J. Anim. Ecol.* **60**, 823–842.

Holt, R.D. (1996). Community modules. In "Multitrophic Interactions in Terrestrial Ecosystems" (M. Begon, A. Gange, & V. Brown, Eds.). Chapman & Hall, London.

Holt, R.D. (2002). Food webs in space: On the interplay of dynamic instability and spatial processes. *Ecol. Res.* **17**, 261–273.

Holt, R.D., & Hoopes, M.F. (in press). Food web dynamics in a metacommunity context: Modules and beyond.

Huxel, G., & McCann, K. (1998). Food web stability: The influence of trophic flow across habitats. *Am. Nat.* **152**, 460–469.

Krivan, V. (1996). Optimal foraging and predator–prey dynamics. *Theor. Popul. Biol.* **49**, 265–290.

Krivan, V., & Sikder, A. (1999). Optimal foraging and predator–prey dynamics, Part II. *Theor. Popul. Biol.* **55**, 111–126.

Krause, A., Frank, K., Doran, M., Ulanowicz, R., & Taylor, W. (2003). Compartments revealed in food web structure. *Nature* **426**, 284–285.

Levin, S. (1999). "Fragile Dominion: Complexity and the Commons." Helix Books, Reading, MA.

Loreau, M., Mouquet, N., & Holt, R.D. (2003). Metaecosystems: A theoretical framework for a spatial ecosystem ecology. *Ecol. Lett.* **6**, 673–679.

MacArthur, R.H. (1955). Fluctuations of animal populations, and a measure of community stability. *Ecology* **36**, 533–536.

Martinez, N.D. (1993). Effects of resolution on food web structure. *Oikos* **66**, 403–412.

May, R.M. (1972). Will a large complex system be stable? *Nature* **238**, 413–414.

May, R.M. (1974). "Stability and Complexity in Model Ecosystems," 2nd edition. Princeton University Press, Princeton, NJ.

May, R.M. (1976). Simple mathematical models with very complicated dynamics. *Nature* **261**, 459–467.

McCann, K. (2000). The diversity–stability debate. *Nature* **405**, 228–233.

McCann, K., & Hastings, A.M. (1997). Reevaluating the omnivory–stability relationship in food webs. *Proc. Roy. Soc. Lond. B* **264**, 1249–1254.

McCann, K., Hastings, A.M., & Huxel, G.R. (1998). Weak trophic interactions and the balance of nature. *Nature* **395**, 794–798.

McCann, K., Rasmussen, J., & Umbanhowar, J. (in press). The dynamics of spatially coupled food webs. Ecology Letters.

McCauley, E., Wilson, W.G., & deRoos, A.M. (1996). Dynamics of age-structured predator–prey populations in space: Asymmetrical effects of mobility in juvenile and adult predators. *Oikos* **76**, 485–497.

McNaughton, S.J. (1978). Stability and diversity in ecological communities. *Nature* **274**, 251–252.

Moore, J., & Hunt, W. (1988). Resource compartmentation and the stability of real ecosystems. *Nature* **333**, 261–263.

Nachman, G. (2001). Predator–prey interactions in a nonequilibrium context: The metapopulation approach to modeling "hide-and-seek" dynamics in a spatially explicit tritrophic system. *Oikos* **94**, 72–88.

Naeem, S., & Li, S. (1997). Biodiversity enhances ecosystem reliability. *Nature* **390**, 507–509.

Neutel, A.M., Heesterbeek, J.A.P., & De Ruiter, P.C. (2002). Stability in real food webs: Weak links in long loops. *Science* **296**, 1120–1123.

Odum, E.P. (1953). "Fundamentals of Ecology." Saunders, Philadelphia.

Paine, R.T. (1992). Food web analysis through field measurement of per-capita interaction strength. *Nature* **355**, 73–75.

Peters, R.H. (1983). "The Ecological Implications of Body Size." Cambridge University Press, Cambridge.

Pimm, S.L. (1979). "The Structure of Food Webs." *Theor. Popul. Biol.* **16**, 144–158.

Pimm, S.L., & Lawton, J.H. (1980). Are food webs divided into compartments. *J. Anim. Ecol.* **49**, 879–898.

Polis, G.A. (1991). Complex trophic interactions in deserts: An empirical critique of food web theory. *Am. Nat.* **138**, 123–155.

Polis, G.A., & Strong, D.R. (1996). Food web complexity and community dynamics. *Am. Nat.* **147**, 813–846.

Polis, G.A., & Winemiller, K.O. (1996). "Food Webs: Contemporary Perspectives." Chapman & Hall, London.

Polis, G.A., Anderson, W.B., & Holt, R.D. (1997). Toward an integration of landscape and food web ecology: The dynamics of spatially subsidized food webs. *Annu. Rev. Ecol. Syst.* **28**, 289–316.

Post, D.M., Conners, M.E., & Goldberg, D.S. (2000). Prey preference by a top predator and the stability of linked food chains. *Ecology* **81**, 8–14.

Reagan, D.P., Camilo, G.R., & Waide, R.B. (1996). The community food web: Major properties and patterns of organization. In "The Food Web of a Tropical Rain Forest" (D.P. Reagan & R.B. Waide, Eds.). University of Chicago Press, Chicago.

Reid, W.V. (1997). Strategies for conserving biodiversity. *Environment* **39**, 16–43.

Riciarddi, A., & Rasmussen, J. (2000). Extinction rates of North American freshwater fauna. *Cons. Biol.* **13**, 1220–1222.

Rosenzweig, M.L. (1971). Paradox of enrichment: Destabilization of exploitation ecosystems in ecological time. *Science* **171**, 385–387.

Strogatz, S.H. (2001). Exploring complex networks. *Nature* **410**, 268–275.

Teng, J., & McCann, K. (2004). The dynamics of compartmented and reticulate food webs. *Am. Nat.* **164**, 85–100.

Tilman, D. (1996). Biodiversity: Population versus ecosystem stability. *Ecology* **77**, 350–363.

Tilman, D., Lehman, C., & Bristow, C.E. (1998). Diversity–stability relationships: Statistical inevitability or ecological consequence? *Am. Nat.* **151**, 277–282.

van der Heijden, M.G.A., Klironomos, J.N., Ursic, M., Moutogolis, P., Streitwolf-Engel, R., Boller, T., Weimken, A., & Sanders, I.R. (1998). Mycorrhizal fungal diversity determines plant biodiversity, ecosystem variability, and productivity. *Nature* **396**, 69–72.

Warren, P.H. (1989). Spatial and temporal variation in the structure of a freshwater food web. *Oikos* **55**, 299–311.

Winemiller, K.O. (1990). Spatial and temporal variation in tropical fish trophic networks. *Ecol. Monogr.* **60**, 331–367.

Yodzis, P. (1981). The stability of real ecosystems. *Nature* **284**, 544–545.

10 | DIVERSITY AND STABILITY: THEORIES, MODELS, AND DATA

David Castle

> For a suggestive metaphor, the elements of scientific knowledge are not so much building stones laid at the foundation of edifices of facts but are more like mangrove trees, interpenetrating and rerooting themselves as their means of survival and growth.
>
> —RAVETZ 1989

10.1 | INTRODUCTION

Community ecology has undergone several upheavals in its theory and its core concepts. These have been ably described by Anthony Ives and Kevin McCann in their contributions to this volume. Although the two authors do not disagree that theoretical change has occurred, they do hold different interpretations about *what* has changed in community ecology theory and what it suggests about the work that has been done. For McCann, there has been a steady turnover of community ecologists' understanding of the diversity-stability relation, and for Ives, changed conceptions of stability are the drivers of the research agenda. This difference in interpretation could be resolved by digging deeper in the historical record and coming down on one side or another, but this would be an exercise in splitting hairs over positions that are not, in the main, incompatible. Instead, this chapter will take a backed-out and thus more synoptic view of Ives' and McCann's contributions to understand why and how theory change arises and what useful lessons can be derived from this analysis for community ecology.

Theorizing about conceptual change has been a major research programme for historians and philosophers of science. The focus has tended to be on the epochal transformations in theory rendered by major theoretical discoveries, an approach that tends to disregard significant if somewhat routine changes in theory (Castle 2001a). Routine theory change is in some ways more interesting, because unlike the big *eureka!* discoveries in science, which are often ineffable creative processes of individuals, routine theory change is laid bare in the to-and-fro of the debates that take place among scientists in a variety of *fora*. The first two sections of this chapter describe the varieties of change to ecological theories, models, and data that produce law-like generalizations at all three levels. Thereafter, a short rehearsal of Ives' and McCann's positions will be complemented with an analysis of their points of consensus and contention. The analysis will lead to some conclusions about the contemporary issues in community ecology before offering some ideas about the future of theory in community ecology.

10.2 | WHY CARE ABOUT THEORY CHANGE?

Like all scientists, biologists constantly strive to make their science as rigorous and robust as possible. Biologists face special challenges in their pursuit of the related goals of disciplinary excellence and scientific truth. Given the relatively short period in which the field can be thought of as unified by evolutionary theory, this is to be expected. After all, just a century and a half has elapsed since *On the Origin of Species* was published. Some time has been necessary to allow biology to balance the assessment of relative contributions to knowledge from observation studies and experimentation. Observation studies are often wrongly treated as being merely descriptive, as if no hypothesis could possibly be at play in such work. Experimentation, by contrast, is often given more credit for its explanatory value than it is perhaps due (Brandon 1994). Biology has been somewhat less conducive to mathematization than physics and chemistry, an association that is being newly cemented as the various -omic sciences brand biology as primarily an information-managed science and perhaps secondarily a mathematically formal science (Ideker, Galitski, & Hood 2001). Meanwhile, the many recognized subspecializations within the biological sciences have their own priorities, methodological commitments, and constraints imposed by the phenomena they undertake to study.

These admittedly broad reflections provide a context in which to consider theory change in biological subspecializations. Major theoretical change in biology, for example, the current revolution in genomics, has effects that ripple throughout biological subspecializations. Equally, biologists working within a subspecialization must adapt to theory change that arises uniquely within their subspecialty. The practicing biologist must keep track of a dynamic inventory of theories, models, and data that run deep in detail and broad in their interconnectedness. In recent accounts of theory change in science, a common observation has been that, in the normal course of things, not all of this inventory can be contested at once without arresting scientific practice. Should some aspect of the inventory become problematic, the rest will be held provisionally constant until the problem is resolved. This process can be iterative if resolution of one problem generates new problems to work on, and the historical record of subspecializations and the professional contributions of biologists show this to be the case.

Theory change in community ecology has involved disputes about the conceptual issues, the methods for building models, and the phenomena themselves. Those possessed with a conception of scientific progress as the steady accumulation of knowledge will draw the conclusion that community ecology's last 50 years represents a science in trouble. Scientific subspecializations need agree upon subject matter and some core methods; otherwise, it is hard to see what they are subspecializations of, and in any event it is hard to form departments, run societies, and train students without a common conceptual core. Apologists will argue that so much conceptual upheaval is just the growing pains of a relatively young field, a line often taken to explain away conceptual controversy in ecology. Apologists are also visionaries. They anticipate that oscillations in conceptual disputes will damp down to a happy equilibrium after which ecologists will have more in common behind them than simmering controversies to resolve.

The position taken here is that the significant amount of theory change witnessed in community ecology is not a sign of disorderly science or science in its infancy. Recognizing that there have been changes to theories is, rather, an indicator of the periodic need to do some conceptual housecleaning. The result is greater internal consistency in the science, which is another way of saying rigor, and the articulation of a more robust science as theories and the models they engender account more systematically for a broader range of phenomena. Implied in this statement is that the development of the science depends on the theory changes described by Ives and McCann. Without theory changes, theory, models, and data would not be separately reconsidered and their relations would not be reevaluated. Accordingly, what follows is an attempt to say why the particular developments in community ecology are significant and interesting from the standpoint of understanding how theories, models, and data can change.

10.3 | KNOWLEDGE IN ECOLOGY

In the 1990s a controversy erupted about the nature of explanation in ecology. In a book that came to be the lightning rod for the debate, Peters argued that pretty much everything that ecologists do is wrong (Peters 1991). Peters claimed that ecologists waste their time working on theories, grandiose conceptions of how nature really works, and fruitless discussions of the mechanisms underlying ecological phenomena. Many responses to Peters focused on rebutting his conception of a correlation-based and potentially predictive science driven by Popperian falsificationism. It became increasingly apparent to those reflecting on the debate that more was at issue than the cogency of Peters' arguments. What was at stake was whether an account of scientific explanation in ecology was possible, one that would restore confidence in intuitions that ecology really is about knowing the true and general causes of a collection of natural phenomena (Castle 2001b).

The prospects for a generalized knowledge in ecology have been addressed by several authors. Greg Cooper, for one, argues that the scientific status of ecology has been wrongfully denigrated by invoking an inappropriate standard of what would count as a law, or law-like explanation in ecology (Cooper 1998). By adopting a philosophically coarse conception of what generalized or law-like explanations in ecology would look like, it is hardly surprising that those seeking positive environmental outcomes, ecologists wanting to defend their science, and philosophers thinking about explanation in ecology are frustrated by ecology's lack of laws. Cooper retorts, and rightly so, that in cases like this, where the science marches on but the complaints continue, the science is probably less to blame than is the yardstick by which it is evaluated. He therefore proposes an alternate conception of law or, perhaps better, a conception of law-like generalisations, which would incorporate probabilistic notions articulated in the philosophy of science by Brian Skyrms (1990). This conception of law-like generalisations is beyond the scope of this chapter, but what is salient is Cooper's taxonomy of the kinds of general knowledge available in ecology.

Within biology in general there has been a long-standing tradition of contrasting explanations that originate in generalised theory, produce abstract models and seek data, with those that start with observation studies leading to abstract models and generalised theory. *Model-*

driven and *data-driven* ecology comprise two different camps, each with their own methodological approaches and pretheoretical commitments about what constitutes good ecological science. Model-driven ecologists tend to rely on high-level theoretical commitments (e.g., what niches must be in theory) from which they derive mathematical models to account for observable phenomena. Modellers, so characterised, extract generalisations from theoretical commitments to relate mathematical to isomorphic empirical systems, but they will maintain a distinction between their theoretical commitments and the mathematical model used to generate them. Mathematical models may not fit data, which is problematic if they are intended as causal explanations. Failure to fit a data set, however, does not constitute a direct challenge to theories that can exist somewhat independently. This is antithetical to the motivations behind data-driven ecology in which the model success is equal parts internal cogency and fitting the facts (Cooper 1998). For ecologists of this ilk, the purpose of generalizing in ecology is to round up phenomena and provide causal mechanisms, which are often represented in mathematical models. Whereas model-driven ecology suffers from well-known pitfalls associated with showing how models map onto phenomena, data-driven approaches to elucidating ecological generalizations from modelling efforts suffer from the problem of showing that the inductive derivation of models from data reflects what is really going on in nature. These are the typical tensions between deductive and inductive modes of explanation as they arise in the natural sciences.

Cooper's view is that model-driven and data-driven ecologists might benefit equally from a conception of scientific law that is rooted in probabilistic expectations. He contends that laws in ecology ought to be thought of as generalizations that prove to be resilient to epistemic challenge and are by and large invariant. This approach is advanced by Cooper to suggest that although no assurances are forthcoming about the presence of laws *qua* universal truth, ecology will not devolve into a science of particulars incapable of reaching beyond descriptions of contingent events. Instead, the kinds of generalizations generated by ecology fall into three categories: theoretical principles, causal generalizations, and empirical patterns (Cooper 1998).

At first blush this conclusion might appear to be a commonplace observation, but it challenges either model-driven ecologists or their data-driven counterparts to see that different kinds of comparably law-like statements are possible in ecology. This is an important prelude to interpreting theory change as described by Ives and McCann as the dynamics of how one of the three kinds of generalization can become preferred for a time. In other words, theory need not provide abstract truths before empirical patterns can be regarded as meaningful and reliable, and correspondingly it is not the case that a coherent but untested theory without data cannot be fruitful in other ways and tested later.

This conclusion is acceptable as a rough approximation of a more fully articulated philosophical position, and it clearly has implications for theory change in ecology. For one thing, it suggests that progress from mere data to lawlike statements is not the unspoken objective, or, perhaps better, not the *only* objective of revisiting theoretical commitments and the concepts that underlie them. Second, it suggests that where theoretical and empirical modellers clash, disagreements are likely to be epistemological rather than ontological. That is, the bone of contention is apt to be about how modellers arrived at their conclusions before turning from methodological considerations to judgments about whether the methods deployed generate knowledge about reality. If the mode of reasoning used to arrive at theoretical models differs from empirical models, surely there is less than ideal common ground to compare what is said in the models. The third implication is that disagreements about the relationships among theoretical, model, and data generalizations will be a constant source of dispute, particularly if one type is assigned logical priority or given pride of place in a scientific explanation.

In the next section the foregoing considerations will be applied as an interpretive framework for understanding where Ives and McCann are in agreement and disagreement and how they arrived at their respective positions. Before doing so, it should be noted that Ives and McCann are both theoretically oriented, so the foregoing is not meant to say that one is principally concerned with one form of generalization in ecology and the other is not.

10.4 | THEORY CHANGE IN COMMUNITY ECOLOGY

McCann argues that there are three main periods of conceptual change in the diversity-stability debate in community ecology. The first he describes is an outgrowth of intuitions in which recognition of the diversity of nature is taken to be a necessary precondition of stability. Odum and MacArthur are the prime exponents of this view in their consideration of the structure and stability of food webs being based on the number of energetic pathways. Elton, too, associated stability with diversity in his comparison of causes of mainland versus island stability. In the second period, dynamic models of communities were made possible by advances in mathematical techniques and early computer modelling in the 1970s. May's important work in this vein reversed the intuition that diversity generates stability and spawned much work devoted to modelling the dynamics of food webs. The third period, with which McCann associates himself, incorporates spatial elements of community structure suggested by empirical studies, which can be incorporated into already temporally dynamic models to give more robust accounts of community structure and stability.

Ives thinks that McCann has the history basically correct but that he draws the wrong conclusion from it—at least partially. Ives claims that the main change in theory does not lie in the changes in the conception of the relation between diversity and stability, which is McCann's focus, but in the definition of stability itself. Ives thinks that as the definition of stability changed, corresponding change in the relationship between stability and diversity was necessary. There may, however, be a sense in which the different definitions of stability represent options for exploring its connection to diversity. Ives supports this thesis by generating a model that would, were data provided, test equally all of the candidate conceptions of stability he describes. Yet, in the end, Ives and McCann share ideas about the need to take cogent but empirically empty models further by providing content for them. This is reflected in Ives' suggestion for specific ecological tests.

The theoretical framework previously developed is useful in understanding McCann's and Ives' positions. For McCann, the account of the diversity-stability relation motivated by intuitions about nature is a theoretical generalization. The conceptions of diversity and stability are linked closely, but the intuitionist account also attempts to ratify their pairing by associating the concepts with a surface appreciation of natural diversity. In this respect it is a model-driven account, which is why McCann can trace a neat progression towards May and followers' subsequent model-building efforts. McCann's account of conceptual change in the diversity-stability debate finishes with efforts to build models that are more spatially realistic. In this respect, data generalizations are driving model-building efforts, which is where the drive to have more rigorous and robust models leads McCann.

Ives' approach is quite different. His approach is to take a historical account of theory change raised by McCann and, in the first instance, turn it into a historical account by showing than an abstract model could provide a generalized platform for understanding several conceptions of stability. This is an interesting move because it couples theory-level definitions with an abstract model to make a point about the range and kind of empirical systems that could be reconciled with the model. In this respect, Ives is rather a classic model builder: one who is building conceptually plastic models and seeing if the phenomena will come. But like McCann, he makes recommendations about how to bring more generalizations about data into ecological models. So Ives and McCann are reconciled about what ecology needs—reliable generalizations about data.

Why worry about patterns of data fitting models? As Cooper has suggested, this is the principle way generalized causal mechanisms in ecology can be articulated. By having a rigorous model with robust data supporting it, generalizations at the level of data and models converge and mutually support the causal hypothesis. When data patterns and general models are consistent with one another, a weight of evidence begins to favour a candidate mechanistic explanation. This is one of the objectives of model building in the biological sciences because it provides empirically tractable explanations that not only are useful in community ecology but also might have broader use in environmental decision making, for example.

Theory-driven research aims to provide cogent theoretical explanations. Indeed, that is where Ives and McCann each begin to describe theory change in community ecology. Here, however, one might ask how it is that despite their differences about the meaning of terms such as *stability* and about the historical development, both authors essentially focus on the possible fruitful data generalizations and model generalizations.

10.5 | THEORY CHANGE, ABATED

Perhaps it is the case that the core concepts that inform community ecology models, having been the focus of inquiry for half a century, are now so firmly established that they can be held constant while other problematic aspects of community ecology are investigated. This is not to suggest that theory development is over in community ecology, for presumably community ecologists have yet to think of new theories of community ecology. It is, however, to suggest that if Ives and McCann each review the history of theory change in community ecology and, differences notwithstanding, conclude by emphasizing empirical generalizations in support of generalized models, the pendulum has swung from model-driven to data-driven explanations in ecology.

So if Ives is correct and the basic, alternative conceptions of stability have been worked through to some extent, and related concepts of diversity and complexity are comparably well characterized, then perhaps a resting point in theory change has been reached. With a satisfactory set of theoretical generalizations, ecologists can get on with the business of looking for more empirical generalizations and causal mechanisms characterized in fruitful models. In this case, it would mean that a period of theory change is drawing to a close. But this is not to say that the intellectual work is over; far from it. As McCann's chapter suggests quite clearly, new models that assimilate new parameters, such as spatial compartmentalization, are available. These models will need to prove their mettle as rigorous mathematical systems and will have to find robust applications as explanations of causal mechanisms in ecological communities.

In the near future, community ecology can be expected to generate more empirically robust generalizations about causal mechanisms, and the focus on foundational concepts and theory development may take a backseat to this development. This would be a timely development from another perspective, one that has sought, without much success, to find environmental policy applications of community ecology. These demands have been central in the criticism that ecology has yet to generalize lawlike explanations. In the foregoing description of theory change in community ecology, the focus has been on dispelling the idea that community ecology is not a source of generalized explanations. Quite the opposite is true. Generalizations can be found at levels of explanation in community ecology other than theory. The place to look is data-driven model-building exercises that will have recognizable pragmatic value.

REFERENCES

Brandon, R. (1994). Theory and experiment in evolutionary biology. *Synthese* **99**, 59–73.

Castle, D. (2001a). A gradualist theory of discovery in ecology. *Biol. Phil.* **16**, 547–571.

Castle, D. (2001b). A semantic view of ecological theories. *Dialectica* **55**, 51–65.

Cooper, G. (1998). Generalizations in ecology: A philosophical taxonomy. *Biol. Phil.* **13**, 555–586.

Ideker, T., Galitski, T., & Hood, L. (2001). A new approach to decoding life: Systems biology. *Annu. Rev. Genom. Hum. Genet.* **2**, 343–72.

Peters, R.H. (1991). "A Critique for Ecology." Cambridge University Press, Cambridge.

Ravetz, J. (1989). "Scientific Knowledge and Its Social Problems." Oxford University Press, Oxford.

Skyrms, B. (1980). "Causal Necessity." Yale University Press, New Haven, CT.

IV | HISTORICAL REFLECTION

11 | ECOLOGY'S LEGACY FROM ROBERT MACARTHUR

Eric R. Pianka and Henry S. Horn

11.1 | INTRODUCTION

Robert MacArthur brought evolutionary insights into community ecology, and he explored natural history with mathematical analyses in ways that were strikingly novel in their time and that are still important today. In this chapter we review MacArthur's work, highlighting some of his novel insights, and try to give some feeling for how pervasive his influence was during his brief life. In a sense, this is a historical viewpoint that complements the historical analysis of theoretical models in this volume. But our view is idiosyncratic and intensely personal. We use this personal stance as an excuse to review a past in which we take great joy, but we have little to say about the present and less to say about the future. There is plenty to say about MacArthur's influence in the present and the future, but that is a project for a whole book. So we content ourselves with reminding colleagues of our generation what we owe MacArthur and with bringing some of his timelessly refreshing work to the attention of the next generation of evolutionary and community ecologists.

In the less than 20 years of MacArthur's academic life, he published 52 papers and three books and he laid the cornerstones of theoretical ecology. MacArthur's academic career, from 1955 to 1972, will seem to many practicing ecology today to belong to an age of long ago, the middle of the last century of the previous millennium. Academic "generation time" is short, only a few years between the establishment of a professor and the establishment of his or her academic progeny, but overlap among generations is great. Most of us are connected by just two or three handshakes to MacArthur. It is a sad and costly consequence of his early death that few ecologists worked with and knew well this extraordinary scientist. We are two of those fortunate few.

MacArthur's legacy is so vast that we cannot possibly review it all, but having been privileged to know and interact with him, and running out of time ourselves, we feel obligated to make a beginning. We shall outline some of his contributions that continue to affect ecology as it is currently practiced. Each of us also reflects personally on some of our own experiences with MacArthur to illustrate what a truly great man he was and how his personality affected our science (as illustrated in Fretwell 1975 and Brown 1999). Although his approach was a truly novel development in ecology, its repercussions still pulse along an expanding network of ideas and personalities. In this sense, his influence has led to both a new paradigm and a subsequent evolutionary change in theoretical and practical ecology.

11.2 | THE LEGACY

MacArthur was a towering intellect in evolutionary ecology. His doctoral thesis on warblers (MacArthur 1958b) is cited in introductory textbooks and in general field guides. His joint work with E.O. Wilson codified the field of island biogeography (MacArthur & Wilson 1963, 1967), the foundation for the parts of conservation biology that deal with nature reserves and habitat fragmentation. He helped to found optimal foraging theory and developed the theory of limiting similarity, which underlies current discussions of biodiversity. He contributed to the twentieth century resurgence in interest in tropical biology at North American universities. He shared major responsibility for the current prominence and respectability of theoretical approaches to ecology. In particular, not only did he develop equilibrial theories at levels from genetic to biogeographical, but he also helped both theoreticians and empiricists to realize that variation and patchiness were worthy of study in their own rights, rather than being merely nuisances that stood in the way of accurate measurement of averages. Furthermore, when deriving equilibrial conditions, he invariably described the dynamics about that equilibrium and explicitly examined the degree to which it was "invadable" by a species with novel characteristics. So his insights were never static but were always implicitly if not explicitly evolutionary. Like his intellectual father, Hutchinson, he trained or heavily influenced an entire generation of community and evolutionary ecologists. Impressive as he was, we think that MacArthur was just "hitting his stride" when he died in midlife at only 42.

MacArthur was a genius, but as Fretwell (1975, p. 2) put it, "He wore his genius lightly, and shared it easily." He combined a love of nature with an analytical mind, an attitude that is now so commonplace that his novelty and leadership at the time is often underappreciated. For him, ideas always trumped forum and format. He was an astute observer with considerable mathematical aptitude. He had an uncanny ability to identify ecologically crucial contrasts (such as coarse versus fine-grained utilization, search versus pursuit time, generalists versus specialists, *r*- versus *K*-selection, immigration versus extinction, and strong versus weak interactions), which he used to construct simple but elegant graphical models of trade-offs. MacArthur liked to make predictions, and he revitalized parts of ecology that were stagnating. He sought generalizations in his theory and patterns in his facts, and he was convinced that a general theory of ecology would eventually emerge. MacArthur occupied centre stage during the 1960s, arousing both admiration and jealousy, and even resentment, among and from other ecologists. His interests were broad and eclectic. He introduced many new topics to ecologists, often accompanied by graphical models. Indeed, the range of his fundamental contributions is so extensive that there is not room here to describe them. (For details of a subset, see Fretwell 1975.) The list alone is so impressive that we highly recommend organizing it topically for undergraduate and graduate seminars. (See the bibliography in Cody & Diamond 1975 and remember that many of the topics are addressed in his three books: MacArthur 1972, MacArthur & Wilson 1967, and MacArthur & Connell 1966.)

MacArthur was a true visionary. His enormous effect on ecology was reviewed soon after his death in 1972 (Cody & Diamond 1975, Fretwell 1975), and his effect continues. The history of development of ideas in ecology was reviewed by Kingsland (1985, 1995), from Pearl, Lotka, Volterra, and Gause to Lack, Hutchinson, and MacArthur. In her 1995 afterword (p. 224), Kingsland states:

> An analysis of how MacArthur has been cited, how he has functioned as both the icon and villain in ecology, and how his ideas about ecology have been interpreted and used after his death would reveal a great deal about the continual process of historical revision that is an essential part of the business of doing science.

The notion of MacArthur as "villain" requires some explanation. His interest in the generality of ecological patterns was so strong that he often failed to mention the natural history that inspired his theories lest they be viewed as only applicable to a particular natural setting. (For a counter to this impression, see Kingsland 1985 or 1995, p. 180; Fretwell 1975, pp. 2 and 12; Wilson 1994, p. 294; Bonner 2002, p. 157; and Cox in Marks 1996, pp. 174–175.) When he tested those theories, he mathematically massaged his data to be appropriate to the test. This irritated many empiricists who saw mathematical transformation not as a logical necessity but as a tainting of the original data recorded in a field notebook. As ecologists have become more mathematically literate, partly because of MacArthur's legacy, they have come to view transformations as righteous when they are appropriate. But the most regrettable vilification of MacArthur comes from people who overzealously and mechanically apply the ideas of Kuhn's *The Structure of Scientific Revolutions* (1962), attempting to define and then to attack a "MacArthurian paradigm". One of us has taken this enterprise to task somewhat flippantly (Horn & Farnsworth 1986), but MacArthur (1973, p. 259) did so seriously:

> It is a pity that several promising young ecologists have been wasting their lives in philosophical nonsense about there being only one way—their own way, of course—to do science. Anyone familiar with the history of science knows that it is done in the most astonishing ways by the most improbable people and that its only real rules are honesty and validity of logic, and that even these are open to public scrutiny and correction.

MacArthur's legacy to ecologists is overdue for a comprehensive review. Major ideas in his papers, and how he developed them from his intuition, need to be identified. Many questions need to be answered. What biological assumptions did he make, and how critical were they? Which of his "conclusions" are more accurately described as insightful conjectures? Which parts of his arguments are borne by data, and which are borne by logic, faith, courage, or mathematical intimidation? What are the most novel ideas in MacArthur's work? Have those novel ideas been expanded recently? If so, how? Where did MacArthur get his ideas—from natural history, theoretical insight, previous literature, or other sources? Where was he original, convincing, right, wrong, or prescient? A wealth of relevant information is available in works by Cody and Diamond (1975), Fretwell (1975), Keddy (1994), Kingsland (1985, 1995), Brown (1999), and Swanson (2000). But MacArthur has yet to receive the detailed attention and the coherent and eloquent appreciation that his intellectual father, Hutchinson, has (Kohn 1971, Slobodkin 1993, Slobodkin & Slack 1999).

11.3 | "POPULATION BIOLOGY" OF MACARTHUR CITATIONS

A major and innovative piece of work needs to be done on the "epidemiology" of MacArthur's ideas, rather than just citations of his papers (drawing inspiration also from Kohn's 1971 analysis of the intellectual demography of Hutchinson's students). But even trying to disentangle which ideas were durable and which not, which were innovative but sparked their own early obsolescence, and so on, would be a daunting task. Therefore, we attempt something much easier as a first-order response to Kingsland's call for an analysis of his citations. We plot cumulative pages of MacArthur publications against calendar year (Fig. 11.1) and logarithmically scaled citations against the age of a particular paper or book (Fig. 11.2).

It is clear that MacArthur died as his personal productivity was accelerating (Fig. 11.1). We could observe that the obvious "spurts" are books, two of them coauthored, but we think that the number of original, valu-

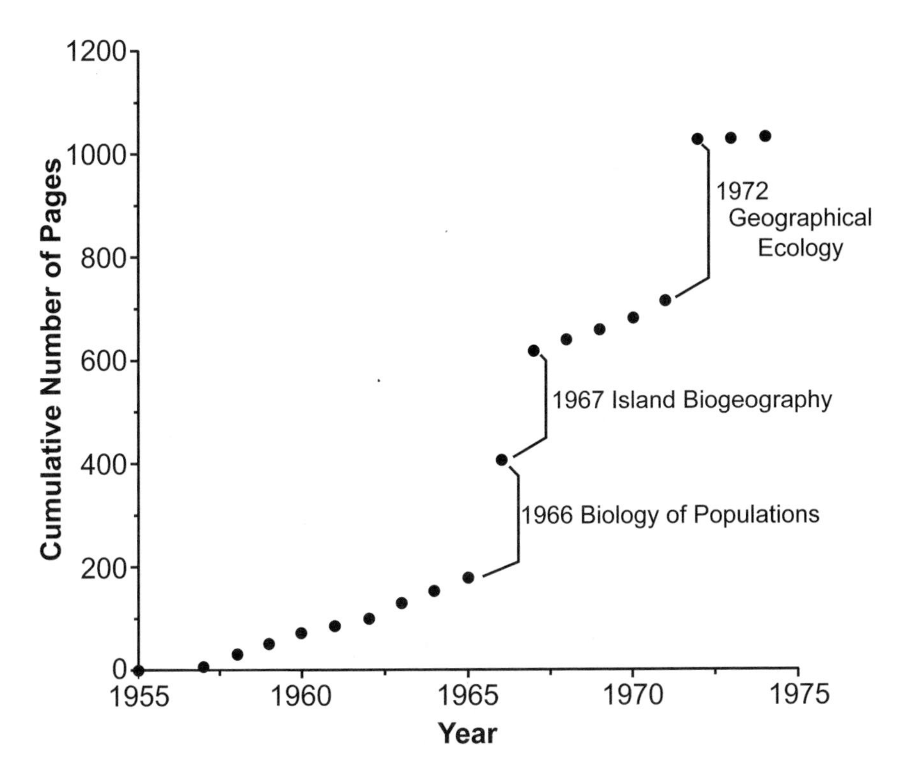

FIGURE 11.1 | Cumulative number of pages published by Robert MacArthur, suggesting that he died just as he was hitting full stride.

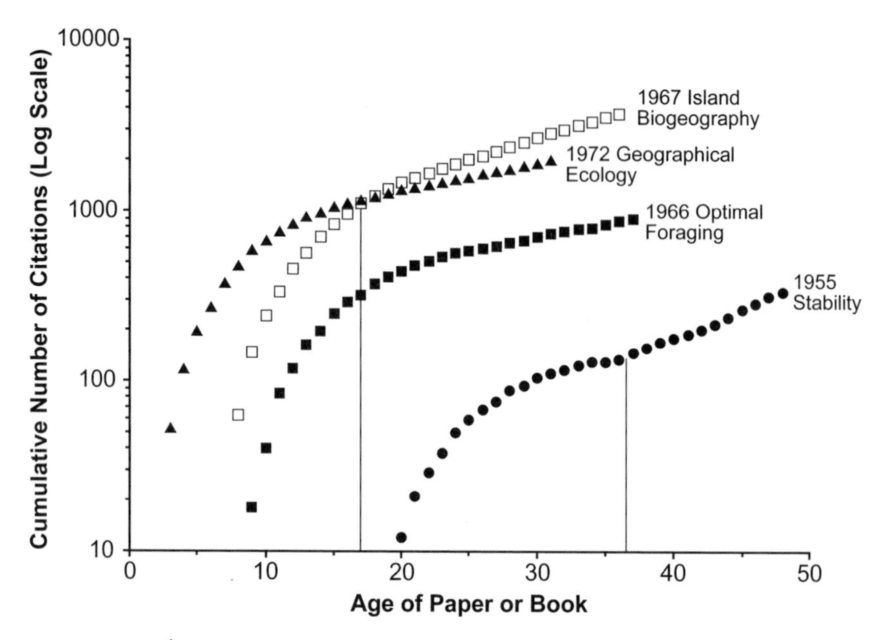

FIGURE 11.2 | Cumulative number of citations since 1975 versus age of two books and two major papers published by Robert MacArthur. Note the upturn around 37 years of age for citations to MacArthur's 1955 *Ecology* paper on the relationships among diversity, food web complexity, and stability (Data from ISI Web of Science).

able, and durable insights per page is about the same in his books as in his papers.

Figure 11.2 reports cumulative citations for two of MacArthur's books (MacArthur & Wilson 1967, MacArthur 1972) and two of his heavily cited papers (MacArthur 1955, MacArthur & Pianka 1966). Each is plotted against the age of the paper or book since publication. Regrettably, these data are limited because they report only citations since 1975, post-dating the cited references by varying numbers of years. Major trends and differences are apparent. In particular, all four references settle to a log-linear relation of citations to age, implying exponential growth. Furthermore, the paper on stability (MacArthur 1955), after settling into relative obscurity by age 37, has entered a long period of renewed growth. We interpret this to mean that in some respects MacArthur was so far ahead of his time that only now, three decades later, are the rest of us beginning to catch up.

It is often overlooked that MacArthur's 1955 classic but much misunderstood paper on diversity and stability in food webs—cited 435 times since 1975—was also his first. It was published in *Ecology* while he was serving two years in Arizona in the US Army after being drafted out of graduate school at Yale University. In this paper he proposed a method, using an information theoretical index, for quantifying the diversity of energy pathways within a food web and suggested reasons the stability of food webs might be positively associated with this measure of diversity. May (1973) challenged this traditional view, claiming that more diverse systems should be less stable and igniting a controversy that has continued. (See the Community Ecology section of this volume and Watt 1964, 1965; Horn 1974; McNaughton 1977; Lawlor 1978; Tilman, 1996; Naeem & Li 1997; Tilman, Lehman, & Bristow 1998; and Haydon 1994, 2000.) Citations to MacArthur's first paper remained low from 1975 to the mid-1990s but have shot markedly upwards during the past few years (Fig. 11.2).

MacArthur's most cited publication (more than 4200 citations since 1975) was his second book, coauthored with Wilson, on the equilibrium theory of island biogeography. In it, they suggested that immigration rate was balanced by extinction rates on islands (MacArthur & Wilson 1967). To describe the difference between what natural selection favoured in an uncrowded environment and what it favoured in a crowded one, they coined the terms *r-selection* and *K-selection*. These terms have been reviled and vilified, most recently by Reznick *et al.* (2002). The equilibrium theory of island biogeography itself was criticized by Gilbert (1980). The putative but premature "rise and fall" of this theory was chronicled by Hanski and Simberloff (1997); however, citations to it have increased markedly since 1995. MacArthur and Wilson (1967, p. v) anticipated and indeed invited criticism, saying, "We do not seriously believe that the particular formulations advanced will fit . . . the exacting results of future empirical investigation. We hope instead that they will contribute to the stimulation of new forms of theoretical and empirical studies, which will lead in turn to a stronger general theory." They got their wish, from questions of priority (Brown & Lomolino 1989), to alternative formulations (e.g., see Lomolino 2000 and Hubbell 2001), to reviews from philosophical and social perspectives (e.g., see Sismondo 2000 and Powledge 2003).

MacArthur's last book, *Geographical Ecology* (cited about 2100 times since 1975), was written as he was dying of cancer (MacArthur 1972). As Wilson (1973, p. 12) movingly describes it, "*Geographical Ecology* is both the reflective memoir of a senior scientist and the prospectus of a young man whose creative effort ended, to our immeasurable loss, at the point of its steepest trajectory." Because the book was written in seclusion as MacArthur's cancer was advancing, it explores places MacArthur wished he had time to go. It still repays reading and rereading for new ideas and for new insights into old ones. See, as one of many examples, pages 173 to 177, where he suggests ways to disentangle the roles of competitive equilibrium from historical accident in determining the number of species in a given place. It was what was on the top of his mind when he died. Annual citations to *Geographical Ecology* peaked about 1980 but have remained stable, from about 50 to 60 per year, since then.

MacArthur helped both of us launch our ideas and careers, and he generously salted, and even nucleated, some of our ideas with his own. Next, each of us briefly recounts some of our interactions with him.

11.4 | ERIC'S REFLECTIONS

I began graduate studies at the University of Washington in 1960. Evolutionary ecology did not even exist then. The only textbooks were by Elton (1927) and Odum (1959), neither of which had an evolutionary perspective. G.H. Orians was hired as a new assistant professor in 1960, and R.T. Paine was hired a year later. A cohort of bright new ecology graduate students arrived about the same time. These included J. Verner, M.F. Willson, C.C. Smith, J.M. Emlen, H.S. Horn, T.H. Frazzetta, and C.E. King. Together, we participated in the birth of evolutionary ecology and helped to put the program at the University of Washington on the map. Our group discovered and devoured Fisher's (1930) *The Genetical Theory of Natural Selection*, Lack's (1954) *The Natural Regulation of Animal Numbers*, and Slobodkin's (1962) *Growth and Regulation of Animal Populations*. We also read, argued, and discussed new developments and many now classic papers, including Cole's and Deevey's *Quarterly Review of Biology* papers (Cole 1954, Deevey 1947). Our group also relished each and every new paper by MacArthur and Levins (Levins 1966; MacArthur & Levins 1964, 1967; MacArthur 1955, 1957, 1958a, 1958b, 1959, 1960a, 1960b, 1961b, 1962, 1964, 1965; MacArthur & MacArthur 1961; MacArthur, MacArthur, & Preer 1962; MacArthur, Recher, & Cody 1966). Every time a new MacArthur paper came out, we could not wait to read and discuss it. It was an exciting time. We had the critical mass needed to bounce exciting new ideas back and forth rapidly, and everybody matured intellectually as a result.

For my PhD research project, I chose to study the latitudinal gradient in species diversity of desert lizards in western North America (Pianka 1966a, 1966b, 1967). Early in my stint in graduate school, I wrote to Robert and asked him if he would review my proposal. He wrote back (his letters were all handwritten in the holographic epistolary tradition in blue ink) and agreed. As it turned out, he commented on virtually everything

I wrote during graduate school. Personal computers did not exist in those days (everything had to be typed on typewriters), and it is hard now to imagine how extremely valuable a "clean" new draft of each manuscript was. (Each time I sent him something to read I felt I had to apologize for "my abominable typing".) Making figures was also time consuming, requiring rate-limiting LeRoy or "press-on" lettering. From the other side of the continent, Robert showered me with an incredible amount of encouragement and support for several years during my graduate career. He probably gave me at least as much useful input as any member of my UW research committee. Robert responded to every letter I wrote, always with fruitful ideas to pursue. I treasured his letters (my folder of MacArthur correspondence is half an inch thick!). He read my review of hypotheses to explain latitudinal gradients in species diversity and shared his own ideas about factors that might cause differences in diversity and how they might act, helping me to improve my own papers. I always began my letters with a respectful "Dear Dr. MacArthur" and he responded with "Dear Mr. Pianka" until we had been exchanging ideas for a couple of years, when he began with "Dear Eric". I was thrilled to be on a first-name basis with such a great man and wrote back with "Dear Robert". Robert read the first draft of my dissertation, paid painstaking attention to all details, and sent me six precious handwritten pages of insightful comments. He suggested that I could do a time series autocorrelation analysis of weather data. At the time, I did not even know what an autocorrelation coefficient was, but I read up on the subject and wrote a Fortran program (I still think of data in the 80-column format of keypunched cards!) to compute them for various monthly time lags from 1 month to 12 months—this proved to be an exceedingly powerful way to compare the climates at different sites, and I incorporated it into my thesis and subsequent publications.

In college, I had a hand-me-down World War II army-grey wool blanket with a sewn-on label that said "Made in Australia". As long as I could remember, I had been intrigued by certain Australian lizards such as the thorny devil *Moloch horridus*, clearly an ecological equivalent of North American horned lizards (*Phrynosoma*). For a postdoctoral, I decided to compare North American desert lizard assemblages with independently evolved ones "Down Under". I wrote up a brief preproposal and sent it to Robert, asking him if he would be willing to sponsor me as his postdoc. He liked my proposal and encouraged me to apply to several granting agencies for funding. It proved to be quite a battle to obtain funding, as it required 3 years of support, half of which was to be spent in Australia doing fieldwork. When National Science Foundation funding failed, Robert offered to support me on some of his own "meagre" funds shortly after he left the University of Pennsylvania (Princeton University lured him away with the offer of a large house within easy walking distance of his new office). Fortunately, soon afterwards National Institutes of Health awarded me a 3-year postdoctoral fellowship to study at Princeton with Robert and in Australia from 1965 to 1968.

When I finished defending my dissertation in Seattle in late 1965, I wrote to Robert and announced that I planned to drive across the continent and would arrive in Princeton in November. He wrote back inviting me to stay with him until I could find a place. I will never forget the warm welcome he gave me when he found me on his doorstep at our first face to face meeting. He generously put me up in the guest room in his house while I looked for an apartment. For about a week, the two of us walked back and forth together daily between the university and his home. Robert often took a mathematical book (such as Feller 1954) home with him for bedtime reading. He told me that when he was with mathematicians he claimed to be a biologist but when with biologists he was a mathematician. He combed mathematical books for ideas that could be imported into ecology. During the broken stick controversy, Robert once told me that the better people moved on to new ideas whereas lesser ones hung on to old, worn-out ideas. He once suggested that ecology should be taught in the wilderness, accessible only by canoe and hiking, so that biologists could learn in the field. (He later published this idea in *Ecology*; see MacArthur 1969.) Robert's senior graduate students Martin Cody and Michael Rosenzweig stayed behind in Pennsylvania; only one of his new graduate students, Ed Maly, accompanied him to Princeton. Thus, I found myself in an almost ideal situation for a young academic. This brilliant scientist was virtually without any colleagues, extremely approachable, and eager for interaction and intellectual stimulation! Immediately we began to discuss his newest ideas, then just a germ, on costs and benefits of various foraging activities. The speed with which Robert's mind worked, as well as its clarity, was simply dazzling. Never before had I encountered true genius. Never did he look down on those of us with normal intellectual abilities. It was exhilarating but also humbling to be part of the two-man brainstorming effort that ensued during the fall and winter of 1965 and 1966. Each evening I went home determined to think of something really neat, but precious little came. Other than acting as a sounding board for Robert's fine mind, my major contribution to *optimal use* was to propose and to outline the table summarizing its results! Robert's generosity in making me a coauthor (MacArthur & Pianka 1966) was typical of his dealings with lesser scientists. Quite simply, I was exceedingly fortunate to be in the right place at the right time.

We developed a graphical model of animal feeding activities based on costs versus profits (MacArthur & Pianka 1966). A forager's optimal diet could be specified, and some interesting predictions emerged. Prey abundance influences the degree to which a consumer can afford to be selective because it affects search time per item eaten. Diets should be broad when prey are scarce (long search time) but narrow if food is abundant (short search time) because a consumer can afford to bypass inferior prey only when there is a reasonably high probability of encountering a superior item in the time it would have taken to capture and handle the previous one. Also, larger patches should be used in a more specialized way than smaller patches because travel time between patches (per item eaten) is lower. In *Geographical Ecology*, under the rubric of *Economics of Consumer Choice*, Robert offered a more generalized, more powerful, improved, and much clearer presentation of the basic ideas of optimal foraging theory (MacArthur 1972).

Although it is difficult to believe now, foraging theory did not exist in 1965. Our paper (later recognized as a citation classic) and Emlen's 1966 paper, published back to back, ushered in the concept of optimal foraging, which has blossomed greatly. Behavioural ecologists embrace foraging theory because it confers rigor and generates testable predictions in an otherwise subjective field. The theorem that diets contract when food is abundant but expand when food is scarce has proved exceedingly robust and now constitutes a basic tenet of evolutionary ecology. Although optimal foraging theory has been savagely attacked (Pierce & Ollasen 1987), it has become an entire subdiscipline of evolutionary ecology (Kamil & Sargent 1981, Stephens & Krebs 1986).

When Robert was elected to the National Academy of Science in 1969, I sent him a congratulatory note saying that it was "about time" and that "now I didn't have to worry about G.E. Hutchinson dying off before I write my *PNAS* paper" because he could communicate it for me. How much I took for granted—Robert died before Hutchinson and my *Proceedings of the National Academy of Sciences* paper (Pianka 1974b) was communicated by E.O. Wilson. I am fortunate to have witnessed and recorded the early stages of Robert's new discipline (Pianka 1974a), and I owe it to his personal and intellectual generosity.

11.5 | HENRY'S REVERIE

I first encountered the work of MacArthur when Wilson assigned Robert's warbler paper (MacArthur 1958b) as the centrepiece of a week of lectures on ecology in a course on evolution and systematics. I was immediately seduced into ecology by Robert's elegant combination of conceptual analysis, inventive data gathering, and ability to revel in the details of outright natural history.

That had been my only formal academic exposure to ecology, when, in 1962, I went to graduate school at the University of Washington. There I fell under the spell of Gordon Orians, abetted by Robert Paine, Alan Kohn, W.T. Edmondson, and an extraordinary group of fellow graduate students, among whom my closest friend was Eric Pianka. I endorse Eric's description of those heady times. We had been brought up applying the concepts of evolution and adaptive significance as ad hoc explanations of patterns of morphology and behaviour in a taxonomic context, but now, enticed by the faculty and reinforced by our own discussions, we were using evolutionary arguments to pose ecological and behavioural hypotheses and designing field measurements to test them. Accordingly, in addition to reading Fisher's (1930) *The Genetical Theory of Natural Selection*, we devoured his *Statistical Methods for Research Workers* (1925) and *The Design of Experiments* (1935), and Eric and I were among the first of our generation to learn computer programming (bi-daily batch-processed Fortran on an IBM 7094 mainframe). We did not know it at the time, but we were taking part in the births of both *evolutionary ecology* and *behavioural ecology* as named disciplines.

My thesis was on the adaptive significance of colonial nesting in Brewer's blackbird (Horn 1968a), but Gordon and I (Orians and Horn 1969) modelled an additional interest on Robert's warblers, namely, how six species split the job of being a blackbird in the near-desert steppes of eastern Washington. For this work, I developed an index of foraging overlap, based on current indices of diversity, and I sent a manuscript describing it to *The American Naturalist* (Horn 1966; interesting side-light: The original draft paper in 1963 was a respectful parody of Robert's 1957 paper, titled " 'On the relative identity of communities,' by H. Arthur MacRoberts"). Robert was a reviewer for the manuscript, and he had just developed a similar index of overlap (MacArthur 1965). He questioned the efficiency of publishing two papers on the same topic. I withdrew my paper, and he immediately wrote to encourage me to resubmit with a simple defence of my approach vis-à-vis his. This was for me the first of many demonstrations of his charity in dealing with youngsters and his lack of pretension about his own ideas. With the correspondence about this paper, "Prof. MacArthur" became Robert, and he continued to inquire about the status of my work, just as he did about Eric's. When Eric joined Robert as a postdoctoral fellow and learned that Princeton was looking for an ecologist, he suggested "that kid, Henry, back at UW", and Robert asked me to come and give a seminar. I was overjoyed when I was offered the job, which I took with alacrity in the fall of 1966.

It was an extraordinary privilege to have my idol as a mentor, a colleague, and a guardian angel. Robert's office was next to mine and there was an interior door between them, so we would consult whenever either of us had an idea or a problem. In our discussions the ratio of ideas to problems was <1 for me and $\gg1$ for Robert. Indeed, most of my best ideas of the time started out as his. My struggle to present to my classes the contemporary controversy about *density dependence* in population regulation led to a chalkboard consultation and then a paper (Horn 1968b) that crystallized Robert's insight. My whole program of studying forest succession arose from Robert and me trying to refine the ways that he used to measure canopy coverage for his studies of the role of foliage profiles in fostering bird species diversity (Horn 1971, MacArthur & Horn 1969). The crucial insight of representing forest succession as a Markov process (Horn 1975) was Robert's idea (MacArthur 1958a, 1961a).

Robert thought that it was important to introduce students to unfamiliar environments. So he regularly led interterm field courses to the tropics (Costa Rica and Panama) and the desert (Tucson and vicinity). Students on these trips were encouraged to develop their own programs of research and observation, but Robert also invited anyone who was interested to bird-watch with him and then to map territories from repeated observations of countersinging defenders and to measure the structure of the vegetation. He was a brown-belt birder, both visually and aurally, and he continually posed questions about why individuals of one species or another were where they were and why they were not where they were not. Robert's field trips were my own introduction to truly unfamiliar environments and communities, and his search for adaptive patterns has infected everything that I have done since then.

One of our interactions illustrates Robert's pragmatic attitude towards theory. We had developed an idea of Levins and Culver (1971) into a primitive version of what is now known as the *competition-colonization trade-off* (Horn & MacArthur 1972; see also the review in Tilman 1994). Slatkin (1974) pointed to a flaw in the mathematical argument. Robert and I had distributed two mutually excluding competitive species randomly among islands of appropriate habitat in a sea of inappropriate habitat. Then we had estimated the proportion of islands with cooccurrence as the product of the proportions with each of the two species. Of course, if the species were truly mutually exclusive, there would be *no* islands of cooccurrence. Corrections of this flaw were notationally messy, but the qualitative picture of the phase plane analysis and its interpretation were unchanged from our original argument. Robert immediately lost interest in the correction because it was a technical refinement with no substantive effect on our conclusions.

Robert had many other strong effects on my personal and intellectual development. As a graduate student I had read and wrestled with the work of R.A. Fisher, leading Robert to overestimate my mathematical ability, so he set out a program of mathematical learning that has enlightened and taxed me to this day. He helped me to design a cabin for a piece of abandoned farmland in Whitehall, New York, and through his example of adding to his summer home in Marlboro, Vermont, he gave me the courage to build a 12- by 24-foot utility-free cabin, where my family lived in primitive splendour for many wonderful summers. He was an exemplar as a family man, and his treatment of students, post-docs, and junior faculty was paternal without being paternalistic. In his last months and days, his attitude towards his own illness and death was far more comforting to me and to his other friends than we were to him.

Robert left little evidence of unfinished business when he died. This is partly because when notes and correspondence concerned anything other than an active project, he would read, respond if appropriate, and add to the top of a pile on a table next to his desk. When the pile reached "theta, the critical angle of repose", he would carefully position a trash bin to catch what sloughed off. Accordingly, the files that survived him are not many. Swanson organized and catalogued them in 1998. Among the gems that she uncovered was an unpublished draft manuscript, "A note on history", in which Robert writes (Swanson 2000, p. 52):

> Joseph Grinnell was the greatest American naturalist and his outlook toward biogeography, ecology, and evolution was astonishingly modern . . . R.A. Fisher more than any other man showed how mathematics could be applied to population biology . . . To make the perfect biogeographer one would combine the wisdom of Grinnell with the talent of Fisher.

Robert came close to being his ideal of the perfect biogeographer.

11.6 | CONCLUDING REMARKS

In concluding, we take our inspiration from MacArthur's mentor, Hutchinson, whose "Concluding remarks" (1958) set the stage over which MacArthur helped to realize one of Hutchinson's great metaphors for our field, *The Ecological Theater and the Evolutionary Play* (Hutchinson 1965). Hutchinson was among the dying breed of "renaissance" folk who could be both interested and expert in subjects currently split among several academic disciplines. Such rare folk will be of increasing importance if ecology is to continue to be, as we tell our students when defining it, an *integrative* science. But in some ways, all science is counterproductive; as human knowledge expands, no one can keep up with everything.

In this context, MacArthur's legacy presents something of a paradox. In combining a naturalist's enthusiasm for organisms, their behaviour, and ecology; an evolutionist's interest in their history, geography, and potentiality; and a logician's and mathematician's analytical ability, he helped to define a new field of *evolutionary ecology*. Yet, the definition itself—and worse, the arguing about the definition—sets boundaries that threaten to divide ecology into disparate disciplines. Thus, as the training of ecologists becomes more specialized, perspectives become narrower. In particular, theory and natural history are seldom practiced by the same individual, and attempting to make realistic generalizations has fallen out of fashion. Even more troubling are the gaps in communication among parochial subdisciplines of ecology so that those few people who strive for generalization find themselves intellectually isolated.

We can only speculate about how ecology might have developed if MacArthur had lived longer. How would he have viewed the revolutions in computer technology that have facilitated the study of nonequilibrium population dynamics and hastened the marriage of ecological models and data? Or what would he think of the developing field of molecular ecology as it reveals the role of historical processes in community ecology? Would MacArthur have been able to arbitrate and expedite the sometimes contentious and lengthy disputes that have impeded progress in some areas of population ecology, such as the relation between species diversity and "stability" of community composition or function? He would have been an eloquent and effective spokesman for the importance of both evolutionary and theoretical ecology in modern science education and research. Ecology would surely enjoy a higher priority in the collective political mind if we all showed more evidence of MacArthur's integrative spirit in the practice of our discipline.

People have a tendency to stop citing, and to quit reading, someone who has passed on in favour of stroking living egos who judge grant proposals and publications. As a result, most books and journal articles cite predominantly recent papers. The older literature is seriously undervalued, and well-meaning scientists reinvent the wheel time and again. When John Lawton assigned the reading of MacArthur's warbler paper to his graduate students (Lawton 1991), they did not even know who MacArthur was. One of them, obviously impressed, asked whether "this guy MacArthur has written anything else that is similar?" We should all read the warbler paper again, and *Geographical Ecology*, and MacArthur in general, and Hutchinson, and Fisher, and Grinnell, and Darwin, and . . .

We would probably be joined by all of MacArthur's collaborators in echoing Wilson's (1994, p. 259) sentiment, "I can imagine no more inspiriting intellect or steeper creative trajectory cut so short with such a loss to others . . . I owe him an incalculable debt, that for at least once in my life I was permitted to participate in science of the first rank." By being so far ahead of his time, Robert MacArthur defined our time. His new paradigm continues to evolve.

ACKNOWLEDGEMENTS

We heartily thank Dan Haydon for bringing us together for this enterprise, for encouraging us and cajoling our editors when our energy ran low, and most of all for revising several crucial paragraphs to temper the cockiness of our youthful memories and the stodginess of our grey-haired present. We thank our editors, Bea Beisner and Kim Cuddington, for wise suggestions and for extraordinarily constructive forbearance with our responses. We thank Simon Levin for encouragement at crucial times. We thank Sharon Kingsland and Rebecca Swanson for their extensive and insightful written histories of MacArthur's role in episodes in the development of evolutionary ecology. Henry also thanks Rebecca and his recent EEB 429 undergraduate class for priming his thoughts about MacArthur's work with theirs. We both miss Robert terribly, and we are deeply grateful that the time we had with him continues to enrich our lives.

REFERENCES

Bonner, J.T. (2002). "Lives of a Biologist." Harvard University Press, Cambridge, MA, pp. 155–160.

Brown, J.H. (1999). The legacy of Robert MacArthur: From geographical ecology to macroecology. *J. Mammal.* **80**, 333–344.

Brown, J.H., & Lomolino, M.V. (1989). Independent discovery of the equilibrium theory of island biogeography. *Ecology* **70**, 1954–1957.

Cody, M.L., & Diamond, J.M. (1975). Preface. *In* "Ecology and Evolution of Communities" (M.L. Cody & J.M. Diamond, Eds.). Belknap, Harvard University Press, Cambridge, MA.

Cole, L.C. (1954). The population consequences of life history phenomena. *Q. Rev. Biol.* **29**, 103–137.

Deevey, E.S. Jr. (1947). Life tables for natural populations of animals. *Q. Rev. Biol.* **22**, 283–314.

Elton, C.S. (1927). "Animal Ecology." Sidgwick & Jackson, London.

Emlen, J.M. (1966). The role of time and energy in food preference. *Am. Nat.* **100**, 611–617.

Feller, W. (1954). "Probability and Statistics." Academic Press, New York.

Fisher, R.A. (1925). "Statistical Methods for Research Workers." Oliver & Boyd, Edinburgh.

Fisher, R.A. (1930). "The Genetical Theory of Natural Selection." Clarendon Press, Oxford.

Fisher, R.A. (1935). "The Design of Experiments." Oliver & Boyd, Edinburgh.

Fretwell, S.D. (1975). The impact of Robert MacArthur on ecology. *Annu. Rev. Ecol. Syst.* **6**, 1–13.

Gilbert, F.S. (1980). The equilibrium theory of island biogeography: Fact or fiction? *J. Biogeogr.* **7**, 209–235.

Hanski, I.A., & Simberloff, D. (1997). Chapter 1. *In* "Metapopulation Biology: Ecology, Genetics, and Evolution" (I.A. Hanski & M.E. Gilpin, Eds.). Academic Press, New York, pp. 5–26.

Haydon, D.T. (1994). Pivotal assumptions determining the relationship between stability and complexity: An analytical synthesis of the stability–complexity debate. *Am. Nat.* **144**, 14–29.

Haydon, D.T. (2000). Maximally stable ecosystems can be highly connected. *Ecology* **81**, 2631–2636.

Horn, H.S. (1966). Measurement of overlap in comparative ecological studies. *Am. Nat.* **100**, 419–424.

Horn, H.S. (1968a). The adaptive significance of colonial nesting in the Brewer's blackbird *(Euphagus cyanocephalus). Ecology* **49**, 682–694.

Horn, H.S. (1968b). Regulation of animal numbers: A model counterexample. *Ecology* **49**, 776–778.

Horn, H.S. (1971). "The Adaptive Geometry of Trees." Princeton University Press, Princeton, NJ.

Horn, H.S. (1974). The ecology of secondary succession. *Annu. Rev. Ecol. Syst.* **5**, 25–37.

Horn, H.S. (1975). Markovian properties of forest succession. *In* "Ecology and Evolution of Communities" (M.L. Cody & J.M. Diamond, Eds.). Harvard University (Belknap) Press, Cambridge, MA, pp. 196–211.

Horn, H.S., & Farnsworth, J.C. (1986). Notes on empirical ecology. *Am. Sci.* **74**, 572–573.

Horn, H.S., & MacArthur, R.H. (1972). Competition among fugitive species in a harlequin environment. *Ecology* **53**, 749–752.

Hubbell, S.P. (2001). "The Unified Neutral Theory of Biodiversity and Biogeography." Princeton University Press, Princeton, NJ.

Hutchinson, G.E. (1958). Concluding remarks. *Cold Spring Harb. Symp. Quant. Biol.* **22**, 415–427.

Hutchinson, G.E. (1965). "The Ecological Theater and the Evolutionary Play." Yale University Press, New Haven, CT.

Kamil, A.C., & Sargent, T.D., Eds. (1981). "Foraging Behavior: Ecological, Ethological, and Psychological Approaches." Garland STPM Press, New York.

Keddy, P.A. (1994). Applications of the Hertzsprung–Russell star chart to ecology: Reflections on the 21st birthday of *Geographical Ecology. TREE* **9**, 231–234.

Kingsland, S.E. (1985). "Modeling Nature: Episodes in the History of Population Ecology." University of Chicago Press, Chicago.

Kingsland, S.E. (1995). "Modeling Nature: Episodes in the History of Population Ecology—Second Edition with a New Afterword." University of Chicago Press, Chicago.

Kohn, A.J. (1971). Phylogeny and biogeography of *Hutchinsonia:* G.E. Hutchinson's influence through his doctoral students. *Limnol. Oceanogr.* **16**, 173–176.

Kuhn, T.S. (1962). "The Structure of Scientific Revolutions." University of Chicago Press, Chicago.

Lack, D. (1954). "The Natural Regulation of Animal Numbers." Oxford University Press, New York.

Lawlor, L.R. (1978). A comment on randomly constructed model ecosystems. *Am. Nat.* **112**, 445–447.

Lawton, J.H. (1991). Warbling in different ways. *Oikos* **60**, 273–274.

Levins, R. (1966). The strategy of model building in population biology. *Am. Sci.* **54**, 421–431.

Levins, R., & Culver, D. (1971). Regional coexistence of species and competition between rare species. *Proc. Natl. Acad. Sci. USA* **68**, 1246–1248.

Lomolino, M.V. (2000). A species-based theory of insular biogeography. *Global Ecol. Biogeogr.* **9**, 39–58.

MacArthur, R.H. (1955). Fluctuations of animal populations, and a measure of community stability. *Ecology* **36**, 533–536.

MacArthur, R.H. (1957). On the relative abundance of bird species. *Proc. Natl. Acad. Sci. USA* **43**, 293–295.

MacArthur, R.H. (1958a). A note on stationary age distributions in single-species populations and stationary species populations in a community. *Ecology* **39**, 146–147.

MacArthur, R.H. (1958b). Population ecology of some warblers of northeastern coniferous forests. *Ecology* **39**, 599–619.

MacArthur, R.H. (1959). On the breeding distribution pattern of North American migrant birds. *Auk* **76**, 318–325.

MacArthur, R.H. (1960a). On the relation between reproductive value and optimal predation. *Proc. Natl. Acad. Sci. USA* **46**, 143–145.

MacArthur, R.H. (1960b). On the relative abundance of species. *Am. Nat.* **94**, 25–36.

MacArthur, R.H. (1961a). Community. *In* "The Encyclopedia of the Biological Sciences" (P. Gray, Ed.). Reinhold, New York, pp. 262–264.

MacArthur, R.H. (1961b). Population effects of natural selection. *Am. Nat.* **95**, 195–199.

MacArthur, R.H. (1962). Some generalized theorems of natural selection. *Proc. Natl. Acad. Sci. USA* **48**, 1893–1897.

MacArthur, R.H. (1964). Environmental factors affecting bird species diversity. *Am. Nat.* **98**, 387–397.

MacArthur, R.H. (1965). Patterns of species diversity. *Biol. Rev.* **40**, 510–533.

MacArthur, R.H. (1969). The ecologist's telescope. *Ecology* **50**, 353.

MacArthur, R.H. (1972). "Geographical Ecology." Harper & Row, New York.

MacArthur, R.H. (1973). Coexistence of species. *In* "Challenging Biological Problems" (J. Behnke, Ed.). Oxford University Press, Oxford, pp. 253–258.

MacArthur, R.H., & Connell, J.H. (1966). "The Biology of Populations." Wiley, New York.

MacArthur, R.H., & Horn, H.S. (1969). Foliage profiles by vertical measurements. *Ecology* **50**, 802–804.

MacArthur, R.H., & Levins, R. (1964). Competition, habitat selection, and character displacement in a patchy environment. *Proc. Natl. Acad. Sci. USA* **51**, 1207–1210.

MacArthur, R.H., & Levins, R. (1967). The limiting similarity, convergence, and divergence of coexisting species. *Am. Nat.* **101**, 377–385.

MacArthur, R.H., & MacArthur, J.W. (1961). On bird species diversity. *Ecology* **42**, 594–598.

MacArthur, R.H., MacArthur, J.W., & Preer, J. (1962). On bird species diversity: II—Prediction of bird census from habitat measurements. *Am. Nat.* **96**, 167–174.

MacArthur, R.H., & Pianka, E.R. (1966). On optimal use of a patchy environment. *Am. Nat.* **100**, 603–609.

MacArthur, R.H., Recher, H., & Cody, M. (1966). On the relation between habitat selection and species diversity. *Am. Nat.* **100**, 319–332.

MacArthur, R.H., & Wilson, E.O. (1963). An equilibrium theory of insular biogeography. *Evolution* **17**, 373–387.

MacArthur, R.H., & Wilson, E.O. (1967). "The Theory of Island Biogeography." Princeton University Press, Princeton, NJ.

Marks, P.H., Ed. (1996). "Luminaries: Princeton Faculty Remembered." Association of Princeton Graduate Alumni, Princeton, NJ.

May, R.M. (1973). "Stability and Complexity in Model Ecosystems." Princeton University Press, Princeton, NJ.

McNaughton, S.J. (1977). Diversity and stability of ecological communities: A comment on the role of empiricism in ecology. *Am. Nat.* **111**, 515–525.

Naeem, S., & Li, S.B. (1997). Biodiversity enhances ecosystem reliability. *Nature* **390**, 507–509.

Odum, E.P. (1959). "Fundamentals of Ecology," 2nd edition. Saunders, Philadelphia.

Orians, G.H., & Horn, H.S. (1969). Overlap in foods and foraging of four species of blackbirds in the potholes of central Washington. *Ecology* **50**, 930–938.

Pianka, E.R. (1966a). Convexity, desert lizards, and spatial heterogeneity. *Ecology* **47**, 1055–1059.

Pianka, E.R. (1966b). Latitudinal gradients in species diversity: A review of concepts. *Am. Nat.* **100**, 33–46.

Pianka, E.R. (1967). On lizard species diversity: North American flatland deserts. *Ecology* **48**, 333–351.

Pianka, E.R. (1974a). "Evolutionary Ecology." Harper & Row, New York.

Pianka, E.R. (1974b). Niche overlap and diffuse competition. *Proc. Natl. Acad. Sci. USA* **71**, 2141–2145.

Pierce, G.J., & Ollasen, J.G. (1987). Eight reasons why optimal foraging theory is a complete waste of time. *Oikos* **49**, 111–118.

Powledge, F. (2003). Island biogeography's lasting impact. *BioScience* **53**, 1032–1038.

Reznick, D., Bryant, M.J., & Bashey, F. (2002). *r*- and *K*-selection revisited: The role of population regulation in life-history evolution. *Ecology* **83**, 1509–1520.

Sismondo, S. (2000). Island biogeography and the multiple domains of models. *Biol. Phil.* **15**, 239–258.

Slatkin, M. (1974). Competition and regional coexistence. *Ecology* **55**, 128–134.

Slobodkin, L.B. (1962). "Growth and Regulation of Animal Populations." Holt, Rinehart, & Winston, New York.

Slobodkin, L.B. (1993). An appreciation: George Evelyn Hutchinson. *J. Anim. Ecol.* **62**, 390–394.

Slobodkin, L.B., & Slack, N.G. (1999). George Evelyn Hutchinson: 20th-century ecologist. *Endeavor* **23**, 24–30.

Stephens, D.W., & Krebs, J.R. (1986). "Foraging Theory." Princeton University Press, Princeton, NJ.

Swanson, R.C. (2000). "Bridging the Gap Between the Naturalist Tradition and Ecological Science: Robert H. MacArthur, Edward O. Wilson, and *The Theory of Island Biogeography*." MA Thesis (History), North Carolina State University, Raleigh, NC.

Tilman, D. (1994). Competition and biodiversity in spatially structured habitats. *Ecology* **75**, 2–16.

Tilman, D. (1996). Biodiversity: Population versus ecosystem stability. *Ecology* **77**, 350–363.

Tilman, D., Lehman, C.L., & Bristow, C.E. (1998). Diversity–stability relationships: Statistical inevitability or ecological consequence. *Am. Nat.* **151**, 277–282.

Watt, K.E.F. (1964). Comments on fluctuations of animal populations and measures of community stability. *Can. Entomol.* **96**, 1434–1442.

Watt, K.E.F. (1965). Community stability and the strategy of biological control. *Can. Entomol.* **97**, 887–895.

Wilson, E.O. (1973). Eminent ecologist, 1973, Robert Helmer MacArthur. *Bull. Ecol. Soc. Am.* **54**, 11–12.

Wilson, E.O. (1994). "Naturalist." Island Press, Washington, DC.

V | EVOLUTIONARY ECOLOGY

12 | ON THE INTEGRATION OF COMMUNITY ECOLOGY AND EVOLUTIONARY BIOLOGY: HISTORICAL PERSPECTIVES AND CURRENT PROSPECTS

Robert D. Holt

12.1 | INTRODUCTION

In this chapter, I provide a personal reflection on a topic that is both perennial and persistently unresolved—namely, the relationship between ecology and evolution—with a focus on the community level of organization. Ecological communities are surely among the most complex entities tackled by scientists, inasmuch as communities contain many species (e.g., thousands), each with unique attributes and historical origins, interacting in all sorts of idiosyncratic ways. The structure and dynamics of communities are likely to reflect the imprint of historical processes, including evolution. Conversely, the pattern and rate of evolution is likely to be influenced by community processes.

A major intellectual (not to mention educational and sociological) challenge in the early decades of the twenty-first century is to achieve the seamless integration of the disciplines of ecology and evolutionary biology. The companion chapter by Day (Chapter 13) crisply summarizes the vigorous and continuing development of theoretical evolutionary ecology. As his overview shows, evolutionary ecology has traditionally been concerned with the traits of single species or pairs of interacting species. The interaction between community ecology and evolutionary biology is by comparison still in its infancy, but I suspect that we are poised on a threshold of exciting developments at this interface. Several authors have addressed, from different perspectives, the interplay of community ecology and evolution (e.g., see Eldredge 1995, Brooks & McLennan 1991, Orians 1962, Collins 1986, Silvertown & Antonovics 2001, Price 2003, Webb *et al.* 2002, Levin 1999, and articles in Agrawal 2003), and I will not pretend to give a thorough overview of this giant issue. Instead, I start with some general reflections on the relationship between community ecology and evolution. I then give a quick tour through high points in the history of this relationship, from the early years of the twentieth century to the present, using as a point of departure key documents that defined the discipline of community ecology and emphasizing the evolutionary perspectives taken by the authors. I am not a professional historian but rather have the vantage point of a practitioner, who read and responded to the publications I mention; doubtless I have a distorted and limited view both of the historical facts and of their societal and intellectual drivers. The theme of the interplay of ecology and evolution richly deserves the attention of professional historians of science. In the last part of this chapter, I explain some key questions that seem to me to be particularly ripe for further development at the community ecology-evolution interface. I have chosen not to dwell on explicit, formal mathematical theory but instead to reflect on the conceptual grounding that underlies all such theory.

12.2 | BACKGROUND REFLECTIONS

All biologists know Dobzhansky's famous quip, "Nothing makes sense in biology except in the light of evolution" (Dobzhansky 1973; see Box 12-A). Presumably this includes the core questions of community ecology—such as elucidating the causal underpinnings of patterns of species composition and richness of entire communities. Yet, as Webb and his colleagues (2002) note, the "integration of evolutionary biology and community ecology remains elusive". Hamilton (1995) likewise criticizes the discipline of community ecology for its "relative neglect of . . . the genetical and evolutionary change occurring within species" because "populations and species cannot be treated as entities resembling molecules any more than individuals can." I think it is accurate to state that many (most?) community ecologists do not directly consider evolution in their work. This is not a damning criticism, far from it. Much valuable work can be carried out in community ecology with little attention paid to evolution. An ecologist colleague of mine once likened his role to that of an analyst of card games; his mission is to interpret what happens when the cards are dealt, not to explain where the cards come from or how they acquired their face values. This is often a reasonable stance. A typical example is a book on marine communities (Bertness, Gaines, & Hay 2001) that synthesizes major patterns and processes in marine systems. A few chapters deal with genetics and evolution, but most do not, and they do not really need to do so to provide substantial insights into these systems.

Box 12-A

What about turning Dobzhansky's quip on its head? What would evolution be like without ecology? Would evolution make sense? Do these questions even make sense? I do not know the answer, but it is worth considering this issue briefly. Certainly a great deal of evolutionary biology can be (and is) carried out with scant or no direct reference to ecology. In particular, phylogenetic reconstruction using cladistic methods depends on the assumption of ancestor-descendent relationships and the identification of shared, derived characters using appropriate outgroup comparisons. Phylogenetic techniques do not in any obvious way depend on understanding how organisms interact with the environment or how species interact with one another.

However, I must wonder if the seemingly ecology-free phylogenetic mission simply takes for granted (viz. sweeps under the rug) key ecological forces implicitly involved in the origin, maintenance, and diversification of species and ultimately in the generation of phylogenies. For instance, would species even exist in the absence of ecological forces? The cohesive forces that are believed to keep species persistent as evolutionary units (e.g., sexual reproduction, leading to reproductive isolation; gene flow; and stabilizing selection) are hard to understand in the absence of ecology. A prime theory for the evolution of sex is that it reflects the antagonistic evolution of parasites and their hosts providing the right kind of fluctuating selection to favour recombination (Seger & Hamilton 1988). Without these interspecific interactions, sex might not exist, in which case species would not be bounded by reproductive isolation. In like manner, the evolution of dispersal (permitting gene flow) often seems to reflect either local competition for scarce resources or spatiotemporal variability in the external environment (see chapters in Clobert *et al.* 2001). Without these ecological drivers, a researcher might see little dispersal. Many forms of stabilizing selection (e.g., on body size) clearly involve interactions between organisms and their environments. Organisms do not inherit just their genes from their ancestors; they also inherit their environments, which can lead to a commonality in the selective regime experienced within a lineage (Holt 1987, Holt & Gaines 1992, Harvey & Pagel 1991, Odling-Smee, Laland, & Feldman 2003). Without cohesive forces based in ecology, it is thus not clear that a researcher would expect to see well-defined lineages—species—at all.

There thus may be a sense in which phylogenies of discrete species only exist (or at least are discernible against the background noise of mutational variation) because of the unfolding of ancestor-descendent relationships in an ecological context providing cohesion to lineages. Moreover, much of the history of life at the grandest scale potentially reflects the imprint of interactions between organisms and a variable physical environment (see Rothschild & Lister 2003 for an excellent overview). Even in a constant environment, biotic interactions within and among species can generate a ceaseless evolutionary dynamic. Without such "Red Queen" forces, abstract theoretical models suggest that evolution long ago would have ground to a halt (Stenseth & Maynard Smith 1984).

A community is a collection of species found in a particular place and time that can potentially interact (Morin 1999). Among the concerns of community ecologists are to understand how interactions among species within a community affect each species' mean abundance and pattern of fluctuations, to gauge how interactions do (or do not) influence which species co-occur in local communities (Tokeshi 1999, Holt 2001), and to ascertain how these local processes combine with forces at larger scales to determine patterns of species richness (Ricklefs & Schluter 1993, Holt 1993, Rosenzweig 1995). Traditionally, community ecology has been dominated by a concern with local mechanisms (e.g., keystone predation). Important developments in recent years have been a broadening of the spatial and temporal scales of inquiry and a growing recognition that regional and historical processes may be essential determinants of local community structure (Ricklefs 2004, Webb *et al.* 2002; see also later sections of this chapter). Ricklefs (2004) has argued that this increasing scale requires integrating evolution into community theory, because the timescale of ecological processes such as competitive exclusion becomes comparable to that of macroevolutionary processes (speciation and diversification). Whether evolution should be explicitly considered in an ecological investigation depends on questions of timescale separation.

To address the timescale issue, it is useful to step back from all the details and ask at the most basic level what we do as scientists. I think much of what we do can be boiled down into two questions. First, for a given system, what is an appropriate joint specification of the states of that system, and the forces acting upon it, that describes how it changes with time? Second, how does a scientist combine an understanding of multiple systems (all analyzed in terms of a joint specification of states and forces) to arrive at a more synthetic understanding of them all?

Theoretical community ecology is largely devoted to analyzing systems of coupled differential or difference equations that depict how communities change over time and across space. The *state* in this case is typically a vector of species abundances, defined at a particular spatial scale; the *forces* are the functional forms of the population growth equations, with parameters describing interspecific interactions such as predator functional responses and intrinsic growth rates, that allow a scientist to predict species abundances through time. But if species' properties can change because of evolutionary forces, the parameters and functional forms of the *force* equations can themselves change (see Chapter 13 for examples) and so, in some situations, may be variables over ecological timescales rather than fixed quantities. There are an increasing number of examples of rapid evolution in traits related to interspecific interactions across a range of taxa in both natural (Reznick & Ghalambor 2001) and laboratory (Mueller & Joshi 2000) settings. Most species are very small (e.g., microbes, nematodes, or collembola) and correspondingly have short generation lengths and large population sizes. Evolutionary dynamics (including speciation) of these community members may easily occur at timescales that are small, scaled against the timescale of changes in abundance of the longest-lived community members (e.g., trees or vertebrates). For instance, in studies of plant competition in terrestrial plant communities, the effect of long-lived plant species on one another may be mediated partly by how they influence both the ecological and the evolutionary dynamics of the soil microflora.

Speciation or extinction in effect changes the state vector. A consideration of macroevolutionary dynamics is essential when considering broad comparative or biogeographical questions about species richness or community composition (e.g., what explains the latitudinal diversity gradient?). In comparative studies, if the systems in question have been separated long enough, observed patterns will typically reflect both ecological and evolutionary processes.

The very definition of a *community* brings out an important difference in the frames of reference used in evolution and community ecology, respectively, which one could refer to as *Eulerian* versus *Lagrangian*. In classical fluid mechanics, a Eulerian frame of reference takes as its basic unit a bit of space (e.g., volume) and tracks the fluid fluxes coupling this unit with other such units and the local interactions. By contrast, in a Lagrangian frame of reference, the basic unit is a bit of fluid, which is followed as it moves across space (Symon 1960, p. 313). Neither frame of reference is the "correct" perspective, as both illuminate reality in usefully different ways. A community ecologist focused on dynamics in a particular spatially defined habitat, for instance, the interactions among tit species in Wytham Woods, is a Eulerian. By contrast, an evolutionary biologist concerned with history, say, the phylogenetic relationships of those same tits or the degree to which their life history variables reflect selective pressures, would properly be concerned with a spatiotemporally bounded lineage of past populations of tits in whatever local environment they happened to inhabit; this evolutionary biologist implicitly takes a Lagrangian stance. Elsewhere (Holt & Gaines 1992) I suggested that one way to link these two approaches is to develop a *phylogenetic envelope* for each species. In effect, for the individuals in each population in a local community in a given time step, the scientist identifies in the previous time step the environments (in both real space and abstract environmental space) inhabited by the parents of those individuals. Recursively, going back in time, the scientist can in principle reconstruct the actual environments and spatial locations experienced by the lineages that generated the local community (see Fig. 1 in Holt & Gaines 1992 for a graphical example). If the phylogenetic envelopes of two species overlap only in the current generation, and not before, then it is obvious that coevolution between them is irrelevant to their current interaction. If for a given species in each generation the phylogenetic envelope in prior generations was located elsewhere, the local population may be demographically a *sink* population, which is likely to be maladapted to the local abiotic and biotic environments (Holt & Gaines 1992).

The seminal ecologist G.E. Hutchinson (1965) once wrote an elegant book memorably titled *The Ecological Theatre and the Evolutionary Play.* This title (although not the author) seems to me to miss part of the fundamental dialectic that governs living systems. Namely, the basic evolutionary processes of adaptation and speciation largely (although by no means exclusively) reflect ecological forces, constraints, and opportunities. Conversely, ecological processes are mediated by individuals, species, and interacting suites of species, whose behavioural, morphological, and physiological traits have arisen as products of evolutionary processes; moreover, even abiotic forces can reflect the imprint of evolution. It is as if the props and stage in the "ecological theatre" are themselves constantly being constructed, destroyed, and moved around as the "evolutionary play" unfolds.

It can be difficult to even define the "environment" of an organism independent of the organism itself (Lewontin 1985, Brandon 1990). This is particularly the case when the organism modifies its environment so that it directly affects the selective regime it or its offspring experience (e.g., *ecological engineering* in Jones, Lawton, & Schachak 1994 or *niche construction* in Odling-Smee, Laland, & Feldman 2003). One of the most important conceptual developments in evolutionary theory in the last half-century has been the recognition that although in the short-run the environment selects adaptations, in the long run those adaptations shape and determine the selective environment, particularly when the environment includes species that are themselves evolving (Nowak & Sigmund 2004). Thus, the height of trees reflects as much competition among trees as selective influence of the physical environment on plant form and function (Falster & Westoby 2003). This dialectic between organism and environment makes evolutionary dynamics more like the unfolding of the surprising twists and turns of a game than like climbing the fixed peaks of a rigid adaptive landscape.

This is clear when considering interactions among individuals, within and among species (see Chapter 13). But even the abiotic environment can be strongly influenced by evolution. The recent literature suggests that feedbacks between organisms and ecosystem processes can take surprising and even outrageous forms. For instance, many ecosystems are dominated by fire. Yet, wildfires depend on the availability of highly flammable fuels. At first glance, it is hard to understand why some plant species generate highly flammable dead tissues, which make it more likely a fire will kill the plant that produces such tissues. Schwilk and Kerr (2002) demonstrate that if interactions and dispersal are localized, flammability can evolve because parents that burn can leave open areas to be monopolized by their offspring. As another example, consider the quintessential abiotic factor—the weather. Hamilton and Lenton (1998) provocatively proposed that microbes and small phytoplankton "have evolved to seed cloud formation to create local dispersal vehicles for themselves, winds and clouds" (the mechanism involves the production of dimethylsulphide in localized clones of algae, which can potentially influence both wind and cloud production and thereby facilitate dispersal of algal spores concentrated near the surface of water bodies). This fascinating hypothesis is not yet proven, but it points out that almost any feedback loop, from organismal activity to ecosystem process back to components of organismal fitness, could in principle have substantial evolutionary consequences. (For other examples, see Odling-Smee, Laland, & Feldman 2003.)

These ecosystem effects of evolution surely have many effects at the community level (e.g., flammability can determine the intensity and temporal pattern of fire disturbance regimes, which can influence species richness and composition), but the knock-on effects of evolved ecosystem feedbacks for community dynamics are mostly unexplored (although see, e.g., Loreau 1998 and Mazancourt, Loreau, & Abbadie 1998). Among the many challenges we face as community ecologists, understanding the coevolution of organisms and their environments may be "the most central one in understanding natural communities and ecosystems" (Lewontin, quoted in Levin 1983).

12.3 | A CAPSULE HISTORY OF THE RELATIONSHIP BETWEEN EVOLUTION AND COMMUNITY ECOLOGY

The disciplines of ecology and evolutionary biology have had intertwined histories since their inception (although the relationship has often been a nodding acquaintance rather than an intimate marriage). Charles Darwin was as great an ecologist as he was an evolutionist, and his corpus of work is imbued with a keen awareness of the concrete environmental context of evolutionary processes. The very term *struggle for existence* is an ecological concept, involving the interplay of demography (births and deaths) and constraints on population growth (e.g., resource limitation), arising from interactions among individuals and between individuals and the external environment.

A seminal work in the founding of community ecology was the monograph *Animal Ecology*, written by Charles Elton (1927) when he was only 27 years old (reprinted in 2001 with useful commentaries by the University of Chicago Press). This book introduced many core concepts that even today permeate community ecology, such as food webs, the niche, succession, and the interplay of community and ecosystem dynamics. Elton paid particular attention to population processes such as fluctuations in abundance and dispersal, linking community ecology with population dynamics. This book preceded the modern synthesis in evolutionary biology, yet it is sprinkled with comments about evolution, many prescient of current themes and concerns. A founding father of the discipline, Elton was certainly aware of the interplay of ecology and evolution (for examples, see Box 12-B). The first lines in the book (Preface, p. vii) are "Ecological methods may be employed in many different branches of biology. For instance . . . in the study of evolution and adaptation". Yet Elton immediately states that "structural and other adaptations [are] . . . the final results of a number of processes in the lives of animals, summed up over thousands or millions of years . . . [And] although of great intellectual interest and value, a knowledge of these throws curiously little light on the sort of problems which are encountered in field studies of living animals." This assumption of a rigid decoupling of timescales is still quite common in ecology, as in the anecdote about playing cards noted previously (and indeed is the practice of much theoretical evolutionary ecology even today, as described in Chapter 13).

Box 12-B

Elton's text contains many comments that suggest the germ of potential linkages between evolutionary and ecological processes, linkages that did not emerge until much later in intellectual history. Here I provide a few examples. For instance, in Chapter 4 of *Animal Ecology*, Elton describes the sorting of species in response to limiting factors along environmental gradients. He evokes the importance of adaptation and natural selection when he notes that a scientist can "use the psychological reactions of animals as an indication of their physiological 'abilities'" and can assume that conditions are unsuitable simply because animals avoid places where they occur, a pattern "presumably brought about by the process of natural selection acting over very long periods, since animals which chose a habitat which turned out to be unsuitable would inevitably die or fail to breed successfully . . . By making the assumption that animals are fairly well adapted to their surroundings we certainly run a risk of making serious mistakes in a few cases, because owing to the lag in the operation of natural selection animals are not by any means always perfectly adapted to their surroundings. But the rule is useful in a general way" (p. 41). This is an interesting passage because it suggests that a scientist should view the habitat distributions of organisms as a reflection of adaptive responses to the environment, including behaviour. Elton's thought here presages the study of habitat selection as a key force by vertebrate community ecologists (e.g. Rosenzweig 1987, Morris 2003). The passage also shows that Elton is not a Panglossian, because he recognizes the potential for imperfect adaptation, particularly in changing environments, presaging current concerns with *ecological traps* and other ecological expressions of maladaptation.

In Chapter 5, Elton introduces the core concepts of a food web and the ecological niche (as the *role* of a species in its community). Evolutionary considerations are rather oblique, but he hints that the linkage pattern of food webs should reflect adaptive behavioural choices by consumers in that there is an "optimum size of food which is the one usually eaten, and the [upper and lower prey size] limits actually possible are not usually realized in practice" (p. 60). Chapters 8 and 9 focus on determinants of population abundance. Much of the description seems implicitly to involve group selection (viz., the notion of an "optimum" density), but there is a clear appreciation of how evolution in species' traits can influence population dynamics, as in the following passage (pp. 114–115): "the habits and other characteristics of the species . . . are continually changing during the course of evolution, and any such change is likely to cause a corresponding alteration in the optimum density of numbers. For instance, if the cats on Tristan da Cunha had possessed poison fangs like a cobra they might have been able to maintain themselves with a small population." (The cats were introduced to eliminate rats, but the latter fought back and were sufficiently abundant to overwhelm the cats—an instance of what we now call *intraguild predation*; see Holt 1997).

The only chapter in the text squarely concerned with evolution is Chapter 12. Despite the preceding passages, it is clear that Elton felt that ecology existed as a discipline largely apart from evolutionary biology. "It may at first sight seem out of place to devote one chapter [Chapter 12] of a book on ecology to evolution." Although I find the thinking in this chapter rather unclear, it ends with two provocative assertions about the potential importance of ecology for evolutionary biology: "Ecological studies upon animal numbers from a dynamic standpoint are a necessary basis for evolution theories[, and an] important result of the periodic fluctuations which occur in the numbers of animals is that the nature and degree of severity of natural selection are periodic and constantly varying." These are important insights that are still being actively developed.

Scientists must be careful not to over-interpret texts from the past, viewed through the lenses of today; nonetheless, in rereading Elton, I am struck with how, despite his initial dismissal of evolution, a conscious concern with using insights from evolution pervades his text (see Box 12-B). He points out the importance of considering traits of organisms as adaptations to their environments and how these adaptations have ecological consequences, ranging from the determination of food web structure, to influencing average abundance, to underpinning patterns of dispersal. His notion of *optimal* population size requires that selection operate on levels of organization greater than that of the individual organism (for recent thoughts along these controversial lines, see, e.g., Wilson 1997). He notes that differences among species perceived by taxonomists may not translate into differences in how those species are regulated in their distribution and abundance, a key assumption of contemporary *neutral* models of community organization (Hubbell 2001, Bell 2001).

A later important textbook by Robert Whittaker (1975) in many ways echoes the evolutionary stance taken 50 years earlier by Elton. The book is sprinkled with interesting thoughts about the adaptive underpinnings of community and ecosystem patterns and about how "diversity begets diversity" over long timescales. Much of Whittaker's thinking strikes me as either explicitly or implicitly *group-selectionist* (e.g., aspects of community organization are viewed as *super-organismal* adaptations), and there is little sense of the potential importance of rapid evolutionary change. Nonetheless, a concern with the evolutionary dimension of ecology is clear.

By contrast to Elton and Whittaker, in perusing the text by Eugene Odum (1971), for many years *the* canonical textbook of ecology, it is striking the degree to which evolution is decoupled from ecology. I suspect that this neglect of evolutionary perspectives arose because Odum and the school of ecosystem ecology he helped found emphasized holistic approaches to ecological systems (e.g., aggregate variables such as energy flows and nutrient stocks). Odum defined ecology as "the totality or pattern of relations between organisms and their environment" (p. 3) and stated "ecology is concerned largely with system levels beyond that of the organism" (p. 4). Because most adaptations are of organisms struggling with their environments, this holistic stance automatically leads to a diminished concern with evolutionary processes.

It would not be fair to state that Odum's text ignores evolution. Odum writes that "organisms are not just slaves to the physical environment; they adapt themselves and modify the physical environment so as to reduce the limiting effects of temperature [etc.]. Such . . . compensation is particularly effective at the community level . . . but also occurs within the species. Species with wide geographical ranges almost always develop locally adapted populations . . . [with] optima and limits of tolerances adjusted to local conditions" (p. 109). The chapter on the individual level of organization considers evolution, but compared with the authoritative feel of most of the text, the material here is somewhat disjointed and perfunctory (my feeling is that this material was included more out of a sense of obligation to the student to be comprehensive rather than out of a heartfelt belief that these themes were essential to the conceptual foundations of ecology). Overall, I think it is fair to say that the role of evolution in the conceptual framework of ecology provided in Odum's text is rather muted.

The main evolutionary chapter in Odum's text (Chapter 9), titled "Development and Evolution of the Ecosystem", focuses mostly on succession. Odum says "the 'strategy' of succession as a short-term process is basically the same as the 'strategy' of long-term evolutionary development of the biosphere, namely, increased control of, or homeostasis with, the physical environment." He briefly describes how selection pressures shift during succession (from *r*- to *K*-selection). A consideration of evolution comes up explicitly in a treatment of biosphere changes over geological timescales (pp. 270–273), echoing Elton's sense that evolution mainly deals with processes over vast geological epochs. Odum concludes by noting that ecologists are intrigued by the possibility of natural selection at higher levels of organization than those of conventional neo-Darwinism (he mentions coevolution and group selection). The notion that there can be "adaptations" at these higher levels has for some decades been anathema to many biologists (following the critique of Williams 1966). However, the degree to which features of these higher levels of organization can be viewed as adaptive is still a matter of active debate (e.g., see Agrawal 2003; Leigh & Vermeij 2002; Leibold & Norberg, in press; and Leibold, Holt, & Holyoak, in press).

Much of ecosystem and community ecology over the last 50 years developed in Odum's footsteps without explicit linkage to evolutionary biology. Allen and Hoekstra (1992), for instance, remark that although "ecosystems depend on evolved entities . . . evolution is only tenuously connected to ecosystems." In a 500-plus page monograph reviewing the literature on competition (a central theme in community ecology), Keddy (2000) touches only glancingly on evolutionary themes. A reviewer of a paper of mine (Holt 1994) once argued that knowledge of evolution gave no greater insight into community and ecosystem processes than provided by just knowing the set of species present and their demographic and ecosystem parameters. In his fine community ecology text, Morin (1999) only slightly refers to evolutionary issues. Many otherwise excellent undergraduate textbooks hardly consider evolution (although some do—e.g., those by Ricklefs and by Pianka). In 1987 the British Ecological Society carried out a survey of its members to compile a list of the 50 most important concepts in ecology (Cherrett 1989). Of these 50, 7 have an evolutionary flavour (#9, life history strategies; #12, ecological adaptation; #24, coevolution; #33, *r*- and *K*-selection; #34, plant–animal coevolution; #37, optimal foraging; and #40, ecotype). Yet less than a third of the respondents chose "ecological adaptation" as a key ecological concept. Phylogenetics, speciation, and other concepts related to macroevolution and macroecology are not even mentioned in the list. It would be an interesting exercise to repeat this survey today to see if this lack of connection with evolution is still the norm.

Ricklefs (2004) has provided a crisp overview of the development of community ecology, emphasizing the development of a worldview in which population processes and competitive exclusion in local communities determined patterns of species diversity. A central player in the articulation of this body of ideas was Hutchinson (e.g., see Hutchinson 1959), who formalized ideas about constraints on community membership and niche theory. These ideas stimulated work on topics such as limiting similarity and community assembly in the 1970s and 1980s. The papers collected by Hazen in a volume of readings (Hazen 1970) for students of that era encapsulated the core concerns of community ecology. In many of these papers, a consideration of evolution and genetics is absent. A notable exception is the paper by Harper (1967); although his main focus is on single-species demography, he notes the potential for genetic shifts in competition to facilitate the coexistence of competing species (as suggested by Pimentel *et al.* 1965). Reference to adaptation is also made in papers by Hairston, Smith, & Slobodkin (1960) and their opponents dealing with trophic level regulation; for instance, Hairston, Smith, & Slobodkin note that the absence of obvious evolved mechanisms for interference competition among many herbivores argues for their limitation by factors other than food availability (e.g., predation), whereas Ehrlich and Birch (1967) counter that herbivores could be limited by food because many plants are well adapted to escape herbivore effects. (It should be noted that evolutionary ecologists now recognize that there can be a decoupling of the importance of a given environmental force as a limiting factor in population dynamics and as a selective factor in evolutionary dynamics, particularly in trophic interactions; see Abrams 1986 and Holt, Hochberg, & Barfield 1999). The famous "Santa Rosalia" article by Hutchinson (1959) on the regulation of diversity explicitly invokes natural selection leading to divergence as a determinant of species coexistence (p. 146).

Hazen's collection also included several influential papers by a student of Hutchinson, Robert MacArthur, who throughout his career incorporated explicit evolutionary thinking into his ecological theories (see Box 12-C). Elsewhere (Holt 2003) I have outlined how MacArthur's perspectives on the dynamics of species' ranges included important evolutionary issues, such as the evolution of specialization and dispersal, that are still the focus of active research. The famous monograph *The Theory of Island Biogeography* (MacArthur & Wilson 1967) has a chapter devoted to evolutionary changes following colonization, including both microevolutionary shifts in traits and increases in species richness during speciation. However, it is fair to say that MacArthur tended to underemphasize both the role of history in determining present-day patterns in ecology and the potential for rapid evolution occurring at timescales commensurate with ecological dynamics.

Box 12-C

A central figure in the field of community ecology was MacArthur, who in contrast to Odum embedded evolutionary perspectives into his thinking. Even in his 1958 graduate thesis (which appeared in the Hazen (1970) collection), he used an argument based on natural selection on foraging behaviour to discount one possible mode of niche differentiation among closely related species. In his final book-length testament, *Geographical Ecology* (1972), MacArthur proposed that biogeographical patterns (including those in community structure) reflected four essential ingredients: the structure of the environment, the morphology of the species, the economics of species behaviour, and the dynamics of population change. The term *economics* indicates the importance of natural selection and adaptation. He was writing this book the last years of his life, by which point I had come to know him.

His thinking, as expressed in conversation and class, moved readily between ecological and evolutionary perspectives. I still recall the first essay question given in his biogeography class the fall of my sophomore year (which I took despite being a physics major because I was [and still am] a keen birder, and MacArthur was the closest thing to an ornithologist on the faculty at Princeton University). We were to choose one of the following two thought experiments: what would life be like on the Earth if the day lasted 48 hours, rather than 24? or, what would life be like on the Earth if it took 10 times as long to rotate around the sun? (This is from memory, so these numbers may be approximate.) What he expected were essays outlining how biomes would shift, life histories would be altered, and so on, if evolution played out on these "new worlds"; in effect, we were expected to work through the ideas in Chapter 1 of *Geographical Ecology*, "Climates on a Rotating Earth," with these new conditions and think through the biological consequences. For me, as I recall I picked the second one, which was a piece of cake. I was an amateur astronomer, so I knew Kepler's laws, which describe how there is a mathematical relationship between how long it takes a planet to go around the sun and how far the planet is from the sun. A short calculation shows that with the longer year, the earth would be far enough away from the sun that water would doubtless be frozen all year. Voila, no life. End of essay. A very short essay.

This was not the answer MacArthur was expecting, so he had me stop by his office, where we had the first of several long conversations about ecology, evolution, and science and life in general. He ended up becoming my undergraduate advisor at Princeton in a special programme and invited me to tag along on his last lengthy field trip with other faculty and graduate students (a spring month in the splendid landscapes of southeastern Arizona). This was a wonderful experience. I recall conversations around the campfire, moving seamlessly between ecological questions (e.g., determinants of altitudinal range limits in these desert mountains) and evolutionary questions (e.g., implications of temporal variation in the desert for life history evolution). I had the good fortune of being clueless about how famous he and various of these other faculty (e.g., John Terborgh and Jared Diamond) were, and in any case I had no idea that I would ever be making a living as an ecologist, so I was not abashed about entering into this talk. I will always be grateful for the kindness and generosity shown to me by Dr. MacArthur and these other ecologists, who perturbed me onto a new career path and led me to appreciate throughout my research and teaching career the desirability of viewing ecological systems through evolutionary lenses.

What has happened in the 30-plus years since MacArthur's death? There have certainly been several attempts made to nurture links between community ecology and evolution. For instance, a volume commemorating MacArthur's brief life brought together the thinking of MacArthur's students and friends (Cody & Diamond 1975). Several contributors dealt with explicitly evolutionary themes. Levins, for example, dealt with the effect of selection on population parameters on the stability and population abundances of interacting species in complex communities, and Rosenzweig developed the implications of viewing species diversity as emerging from the interplay of speciation and extinction rates mediated through the effect of diversity on species' range sizes. A later volume edited by Diamond and Case (1986) provided a survey of community ecology, and again several authors explicitly focused on evolutionary themes. Grant, for instance, dealt with the interplay of selection and competition in fluctuating environments; other authors considered topics such as sexual selection (Colwell), major adaptive syndromes in plant competition (Tilman and Cody), and the adaptive implications of indirect effects in multispecies communities (Wilson). The publication of a text by Roughgarden (1979), considering in detail topics at the interface of evolution and ecology (e.g., character displacement, niche evolution, and density-dependent selection) seemed to beckon the final fusion of ecology and evolution by making the needed formalism accessible to graduate students (Case 2000 updates much of this material). An outside observer might conclude that these books in the Hutchinsonian lineage marked the intellectual fusion of ecology and evolution more broadly throughout the discipline.

12.4 | WHAT DERAILED THE FUSION OF EVOLUTION AND COMMUNITY ECOLOGY?

But as noted in the previously cited quote from Webb *et al.* 2002, this fusion did not happen. In 2000 there was a joint meeting of the British

Ecological Society and the Ecological Society of America. The symposium volume produced from this meeting deals with evolution, but with the exception of brief comments by Richard Lenski (2001), the evolutionary issues considered are almost entirely at the level of individual properties (e.g., life history traits).

It is useful to reflect briefly on why the seeming inevitability of a broader unification of community ecology and evolution may have been derailed by issues both within and outside ecology. On the evolutionary side, I feel that this largely had to do with several broad intellectual and sociological currents.

First, there were numerous attacks on *adaptationism*, stimulated (irritated?) by the emergence of sociobiology in the mid-1970s (the history of critiques of adaptationist thinking is crisply summarized in Rose & Lauder 1996). If evolutionists were squabbling among themselves about whether they could even identify adaptations at the level of individual species without enormous effort (despite stout defences, particularly by behavioural ecologists, such as in Reeve & Sherman 1993 and Mitchell & Valone 1990), outsiders such as community ecologists would likely have been disinclined to view their systems in a way that highlighted the importance of adaptive mechanisms and processes.

Second, there was a reinvigoration of the field of systematics. Part of this involved a vigorous attack on the practice of evolutionary taxonomy, which, roughly speaking, combined historical information about ancestor-descendent relationships with notions about adaptive zones and niches to characterize higher taxa (Wiley 1981). This debate was decisively settled in favour of cladistic approaches, which in practice deliberately and necessarily dealt with organismal characters stripped of their environmental context and adaptive significance. Again, however, this meant that an important group of evolutionary biologists (those focused on reconstructing the history of life) to an ecologist seemed largely focused on characters that did not directly pertain to the *struggle for existence.*

Third, the explosion of molecular genetic techniques and information led to an inexorable pull of young evolutionists towards the use of molecular data and mathematical models for analyses of molecular variation, evolution, and phylogenetic reconstruction, issues on the whole well removed from phenotypic evolution and the broader ecological context of life. In the 1950s and 1960s, the field of ecological genetics focused on conspicuous polymorphisms, which could easily be related to patterns of selection from environmental causes (e.g., see Ford 1975). However, starting with protein electrophoresis in the late 1960s and continuing with the development of deoxyribonucleic acid techniques, a staggering amount of genetic variation was revealed, so much so that it was difficult to imagine classical balancing selection arising from the environment as the primary cause (Lewontin 1974). The neutral theory of molecular evolution (Kimura 1983) proposed that patterns of genetic variation could be explained by the interplay of mutation, gene flow, and drift, with a minor role for selection. This proposal evoked a great controversy (e.g., see Gillespie 1991) and involved many issues far removed from aspects of the phenotype of interest to ecologists.

On the ecological side, I think several factors influenced the relative neglect of evolutionary perspectives in the last three decades of the twentieth century. The part of community ecology that shades into ecosystem ecology was largely dominated by the nonevolutionary stance represented by Odum's text. Beyond this, many of the associates of MacArthur who championed the evolutionary dimension of ecology also emphasized a single ecological interaction—competition—and moreover often assumed that the world could be viewed as at or near equilibrium. Part of the internal intellectual dynamic of community ecology in this period was an increasing appreciation of the importance of alternative interactions (e.g., predation and mutualism) and of the crucial importance of disturbance and temporal variability. There was also a sharpened concern with the need for experimental manipulations to test ecological hypotheses concurrent with a general scepticism about inferring process from pattern (Strong *et al.* 1984), leading to an increasing (and at times myopic) focus on examining processes at local scales. I wonder if with the proliferation of local mechanisms being considered, and with the emphasis on experimental tests of theory, there was not an incidental submerging of a nascent evolutionary dimension in community theory essentially because of the sociology of science; those ecologists who tended to be thinking about the evolutionary dimension of their systems (in the Elton-Hutchinson-MacArthur lineage) were being sharply criticized for other reasons (e.g., because they were proponents of competition, equilibrial reasoning, etc.), leading to a kind of submergence of evolutionary themes, which were caught in the cross fire of these academic disputes.

Be that as it may, I find it intriguing that the potential avenues suggested by Lewontin (1974) to explain the *paradox of variation* in population genetics all have strong parallels in the intellectual history of community ecology over the last several decades. He suggested three basic changes needed in evolutionary theory: an explicit focus with the concrete history of populations, a concern with non-equilibrial dynamics, and a focus on linkage and interaction among loci. In like manner, with the growth of phylogenetics, community ecologists have become much more aware of the importance of explicitly weaving history into the study of community assembly (e.g., see Webb *et al.* 2002); disturbance and non-equilibrial dynamics are now considered the norm, rather than the exception, in natural communities, and the pattern of spatial associations of interacting species, interacting locally but dispersing in *metacommunities*, is widely recognized as an essential driver of community dynamics (Holyoak *et al.*, in press).

It is fascinating that, comparable to the neutral theory of genetic variation, a similar theory has emerged in community ecology, in which patterns of local and regional species diversity are explained in terms of speciation, immigration, and extinction, in the absence of niche differences among species—the *neutral* theory first proposed by Hal Caswell then greatly developed, in particular, by Steve Hubbell (Hubbell 2001; see also Bell 2001). This body of ideas explicitly drew inspiration from the neutral theory of molecular evolution. In a literal interpretation of neutral community theory, neither intraspecific evolution nor macroevolution matter, because all species are assumed to be competitively equivalent, regardless of their specific traits. If the neutral theory of community ecology proves to be even approximately true, this makes it more likely researchers would expect neutral evolution of within-species genetic variation (inasmuch as phenotypic differences among individuals in the same species are usually minor compared with differences among individuals of different species). The general utility of neutral theories in community ecology is a topic of considerable debate (e.g., see Leibold *et al.* 2004 and chapters in Holyoak, Leibold, & Holt, in press).

12.5 | POINTERS TO THE FUTURE

One seemingly dispiriting message that seems to emerge from the vast experimental and observational literature of community ecology is that there are relatively few commonalities among communities, viewed at very local scales. As John Lawton (2000, pp. 56–57) notes, "the local rules of engagement, both the details and many of the key drivers, appear to be different from system to system in virtually every published study in community ecology . . . we have no means of predicting which processes will be important in which types of system. The Devil is in the contingent detail. Almost every place, time and species assemblage is sufficiently different to make general patterns and rules about local community membership and population abundances impossible to find."

One interpretation of Lawton's remark is that community ecology has now devolved into a disorganized collection of special cases. My own sense is more optimistic. Instead of general theories (in the sense of familiar laws), we seem to have a trend towards developing structured suites of theories tailored to particular situations. Community ecology is in the position of chemistry in the nineteenth century, seeking a small number of organizing principles that can lead to a kind of "periodic table" of communities. There are several fruitful directions being developed. One promising direction is to draw on fundamental models of how organisms work as metabolic machines, extracting resources from their environments, and then to use metabolic principles to generate scaling relationships (e.g., of life history variables versus body size) (Brown & West, 2000; Allen, Brown, & Gillooly 2002). Understanding the adaptive significance of key organismal metabolic traits (e.g., of resource acquisition, stoichiometry, retention, and allocation strategies), as governed by constraints and played out in phylogenies, provides an obvious bridge between community and ecosystem ecology and evolutionary theory (Loehle & Pechmann 1988, Holt 1994).

In my own work, I have championed the utility of what I call *community modules,* which are small number of species (e.g., 3 to 10 species) linked in a specified structure of interactions (Holt 1997). Familiar modules include unbranched food chains, exploitative competition among consumers sharing a single resource, and keystone predation on competing prey. Sometimes, ecological systems closely resemble a particular module (e.g., agricultural pests with their natural enemies). Species in multispecies communities often interact strongly with just a few other species. The operational hope is that modules provide bite-size conceptual units that permit a researcher to discern key aspects of processes operating in full communities (e.g., indirect interactions or the interplay of top-down and bottom-up forces). These are rather modest goals (relative, say, to predicting diversity as a function of *niche packing* rules). Which module pertains to a particular system depends on many contingent details of that system (Holt & Lawton 1994).

Thus, instead of identifying universal laws (as in physics), maybe we community ecologists will end up with clusters of laws, tailored for particular settings, and a metalaw that allows us to know which local laws apply to which situation. Several community ecologists have championed such pluralist approaches to the development of community theory. Schoener in particular, in the Case-Diamond volume noted previously, sketched a kind of periodic table of communities and pointed out that MacArthur had also been clearly aware of the likely importance of pluralistic perspectives: "The future principles of the ecology of coexistence will be of the form 'for organisms of type A, in environments of structure B, such and such relations will hold' . . . With different initial conditions, different things will happen . . . Initial conditions and their classification in ecology will prove to have vastly more effect on outcomes than they do in physics." (cited in Schoene 1986, p. 468) A concern with initial conditions entails an interest in history, and so leads naturally to an appreciation of the evolutionary dimension of community ecology.

A powerful set of tools provided by evolutionary biology that can be used to craft the "periodic table" of ecological communities is the explicit use of phylogenies (Webb *et al.* 2002). Phylogenetic reconstruction by its very nature permits a community ecologist in principle to identify key organismal traits shared by many species using their shared evolutionary histories. This can lead to novel insights into community assembly and organization. For instance, Webb (2000) showed that in Bornean rainforest there was a strong phylogenetic signal in the distribution of trees across space, with related species more likely to co-occur in environmental space than to be far apart. This in general accords with the growing theoretical appreciation that for species to coexist in communities, it is not enough that they differ along some niche axis; they also must be equalized in their responses to the environment among other niche axes (Holt 2001, Chesson 2000, Chase & Leibold 2003). If species are roughly equal, then even if competitive exclusion occurs, it may do so very slowly and on a timescale commensurate with within-species evolutionary dynamics and even with speciation (Hubbell 2001). A very active area in community ecology is linking the insights of neutral theory with more classical niche-based approaches (McPeek & Gomulkiewicz, in press). Phylogenetic perspectives are essential to this effort.

In the past 20 years there has been a growing appreciation of the need to understand biodiversity dynamics at large spatial and temporal scales, ranging from metapopulation and metacommunity dynamics at landscape and regional scales to macroevolutionary and palaeobiological timescales (Ricklefs & Schluter 1993; Ricklefs 2004; Rosenzweig 1995; Brown 1995; Maurer 1999; McKinney & Drake 1998; Polis, Power, & Huxel 2004; Holyoak, Leibold, & Holt, in press). Tokeshi (1999) for instance starts his monograph on species coexistence by emphasizing the evolutionary drivers of species origination and differentiation.

The field of community ecology is in flux. As the spatial and temporal scales considered by community ecologists grow, there are many important arenas in which theoretical and empirical community ecology will be enriched when more explicitly linked to evolutionary analyses and theory. Moreover, many questions in evolutionary biology may only be fully resolved when evolutionary analyses are linked with community perspectives. Carroll and her colleagues (2004) noted that "one of the great genetic surprises of the molecular era is the formidable collection of genes for which no obvious phenotypes have been identified" and go on to suggest that the phenotypic effects of these genes on fitness may only be revealed in natural conditions in which individuals can interact and experience ecological competition. A full understanding of the factors governing the maintenance and expression of genetic variation may thus require analysis of the community context within which species have evolved.

12.5.1 | Evolution and Ecology at Commensurate Timescales

The timescales of evolutionary change can often be commensurate with those relevant to population and community dynamics. Ignoring evolution can lead to a serious misunderstanding of the processes governing the community, even at local spatial and temporal scales. A key question is thus: what is the role of ongoing evolutionary dynamics in population and community dynamics over short ecological timescales? Theoretical studies (e.g., Abrams 2000) suggest that permitting short-term parameter evolution (caused by evolution by natural selection) can at times qualitatively alter the nature of population dynamics and constraints on species coexistence. This recent theory formalizes David Pimentel's notion of *genetic feedback* (Pimentel *et al.* 1965, Pimentel 1968) as contributing to population stability in predator-prey systems and coexistence in competing guilds (described later).

The issue of how the maintenance of genetic variation influences stability and persistence of ecological systems is important in both single species and community dynamics. There has been a great deal of interest in how (or if) community diversity influences ecosystem stability (e.g., see Loreau, Naeem, & Inchausti 2002). A parallel question at the level of single-species population dynamics is whether the presence of genetic variation promotes population stability. In community ecology the focus is on whether genetic variation and short-term evolution influence the likelihood of stability, given potentially unstable interspecific interactions.

Similar forces may maintain diversity at different levels of biological organization (Antonovics 1976). At the community-ecosystem interface, among-species variation is maintained by mechanisms known to maintain the diversity of interacting species within communities, such as niche differentiation, food web interactions, disturbance, and spatial dynamics. At the level of individual species and small sets of interacting species, comparable mechanisms can selectively maintain adaptive genetic variation within species. Diversifying spatial and temporal heterogeneity in selection, with gene flow, provides a potent set of mechanisms for the maintenance of genetic variation, broadly comparable to the forces that maintain diversity in communities (e.g., see Levin & Muller-Landau 2000). Broadly speaking, the stabilizing factors through which biodiversity is believed to stabilize entire ecosystems (e.g., the *insurance hypothesis* and *niche complementarity*; Yachi & Loreau 1999, Loreau 2000) should also pertain to the stabilizing influence of genetic variation on population dynamics.

Species diversity may be particularly important in maintaining ecosystem functioning in changing environments (Norberg *et al.* 2001). Similar processes can operate within species, through shifts in genetic composition, because genetic variation is required for adaptation to facilitate persistence in changed environments (Lande & Shannon 1996).

Microevolutionary processes may play a role in determining some of the most basic processes considered by community ecologists. Realized patterns of community richness reflect controls on both colonization and extinction rates. The process of community assembly involves "testing" local communities with propagules of invading species drawn from a regional species pool; community structure arises from sequences of successful invasions, failures, and extinctions of prior residents. Most ecological theory of assembly dynamics assumes that species' properties remain fixed. In the initial stages of invasion, a species will typically be rare locally and prone to extinction for that reason alone. Yet if genetic variation is present, natural selection can increase the growth rate of the population in the local environment and thus facilitate invasion. Among species with otherwise similar responses to the local environment (e.g., in resource requirements), those with more genetic variation (e.g., because of their mating system) should be more responsive to local selection and thus should be more likely to increase when rare.

The relationship between population size and extinction risk is a core concern of island biogeography (MacArthur & Wilson 1967) and related areas such as metapopulation ecology and conservation biology. What is the relative importance of purely ecological factors versus genetic and evolutionary factors in determining the extinction risks of small populations? There are numerous genetic factors that potentially enhance extinction risks in small populations, including inbreeding depression, accumulation of deleterious mutations, and the loss of adaptive potential. Newman and Pilson (1997) described an experiment with the annual plant *Clarkia pulchella* in which they manipulated the genetic effective population size but maintained a fixed census population size. They found that lower genetic effective sizes substantially increased extinction risk. If extinction dynamics is an important determinant of community structure, then microevolutionary processes that alter extinction risk could have effects at the community level.

For instance, consider the determinants of species-area relationships. Such relationships are almost universally observed, whether the researcher is considering nested mainland samples or samples drawn from islands differing in size. However, the relationship is usually weaker for the former (e.g., as measured by the slope of a log(species) versus log(area) regression). A general explanation for this is that colonization rates are lower on islands than in similar-sized areas of the mainland (MacArthur & Wilson 1967, Rosenzweig 1995). Another explanation is that local extinction rates are similarly lower for the latter. The greater immigration rate expected within a continuous continental area both enhances recolonization of sites that have experienced extinction and makes extinction less likely in the first place. There are purely ecological reasons that enhanced immigration might reduce extinction risk (as in the *rescue effect* of Brown & Kodric-Brown 1977), but genetic mechanisms may be important as well. Immigrants can reduce the effect of local inbreeding and replenish stocks of genetic variation that can be acted on by natural selection (e.g., to deal with local edaphic conditions or temporal variability; Lande & Shannon 1996, Gomulkiewicz & Holt 1995). With recurrent immigration, evolution can draw on genetic variation available at much larger scales than just the local population, so the phenotypes of local populations can be sculpted to match local environments; this may eventually permit a species to occupy a wider range of habitats than would otherwise be possible. When these evolutionary effects are iterated across many species, researchers might expect a shallower species-area relationship to emerge (because species can be more widespread, having adapted to a wider range of environments). An open question in community ecology is thus determining the potential relative contributions of genetic and evolutionary versus ecological explanations for explaining the shallower species-area relationships observed in continental communities compared with sets of oceanic islands.

Evolution can influence interspecific interactions at short timescales. Thompson (1999), for instance, asserts that "dozens of species interactions are known to have evolved during the past 100 years." He suggests "the stability of communities may rest on the ability of species to make short-term evolutionary changes to each other." The classic example of a rapid evolutionary shift in an interspecific interaction may be the interaction between the myxoma virus and the introduced rabbit in Australia. The virus initially had a strong effect on rabbit abundance, but because of selection towards both increased resistance in the rabbit and decreased virulence in the virus, the strength of the interaction gradually waned. In agricultural systems, genetically homogeneous monocultures can suffer devastating outbreaks of pests. This suggests that in natural systems, genetic diversity in prey and hosts may be essential in reducing the effect of any given natural enemy and may contribute to community stability.

For each model of a community module that I have described in Holt (1997), researchers can revisit issues such as coexistence, stability, and alternative states and ask how the model predictions would be modified by incorporating ongoing evolution into the model parameters, occurring at roughly the same timescale as ecological dynamics in numbers. Theoretical studies by Levin, Abrams, Hochberg, and others suggest that evolutionary dynamics can either stabilize or destabilize predator-prey or host-parasite dynamics. A microcosm experiment by Hendry and his colleagues (2000) showed that ongoing evolutionary change could contribute to the generation of predator-prey cycles. The interplay of evolutionary and population dynamics is of immense practical importance, for instance, in the control of pest species. Abrams (e.g., see Abrams 2000) in particular has championed how incorporating evolution into dynamic models of predator-prey interactions can have a major effect on the expected behaviour of the system. The specific effects of such coevolution depend on the details of how a focal predator-prey interaction is coupled with the remainder of the food web (see also Levin, Segel, & Adler 1990).

Rapid evolutionary responses within species may at times foster the maintenance of diversity within trophic levels of competing species. This is not a new idea. Pimentel (Pimentel 1968, Pimentel *et al.* 1965) explored the idea that genetic feedbacks could permit species declining because of competition nonetheless to persist in the community. The basic idea is that if a dominant species is pushing a subordinate species towards extinction, selection in the former will be dominated by intraspecific interactions, whereas selection in the latter will be increasingly dominated by the need to cope with interspecific competition as its numbers decline. This asymmetry in the target of selection could potentially permit the maintenance of species in communities. Wardle (2002, p. 205) argues that the ability of soil organisms (e.g., microbes) to adapt rapidly to changing conditions could be an important contributor to observed high levels of diversity in soil communities. It strikes me that elucidating the influence of ongoing evolution to community stability and the maintenance of diversity is a question worth a great deal more empirical and theoretical scrutiny.

12.5.2 | Final Thoughts on the Interplay of Ecology and Evolution

Even if species' traits are relatively fixed, evolutionary perspectives can help enrich comparisons among communities. This is not a new point and, indeed, is implicit in several of the historical references noted previously. We need a new, tempered, phylogenetically sensible, adaptationist perspective to help organize the numerous contingent facts noted by Lawton (2000). Webb and his colleagues (2002) and Ricklefs (2004) point out several important questions that can be addressed using phylogenetic techniques.

Over ecological time-scales, community assembly requires sampling species already present in regional species pools (MacArthur & Wilson 1967, Holt 1993, Rosenzweig 1995). Over evolutionary timescales, community assembly reflects the processes of speciation and adaptive radiation. Gillespie's (2004) study of Hawaiian spider communities reveals that the adaptive radiation of habitat associations is not random, that similar sets of *ecomorphs* arise through both dispersal and evolution, and that species diversity is maximal for communities of intermediate age. She suggests that "the similar patterns of species accumulation through evolutionary and ecological processes suggest universal principles underlie community assembly." Sorting of species along environmental gradients can permit communities to exhibit the properties of 'complex adaptive systems', with adaptive matches between the distributions of phenotypes and local environmental conditions (Leigh & Vermeij 2002, Leibold, Holt, & Holyoak, in press).

The last few years have seen a resurgence of interest in the structure and dynamics of food webs (e.g., see Polis, Power, & Huxel 2004), but most of this work has to date not directly dealt with evolutionary questions. As one example of a food web attribute that may require an evolutionary explanation, consider food chain length. Pimm (1982) distinguished four basic hypotheses for why food chain length may be limited: energy flow, size and other design constraints, optimal foraging, and local dynamic constraints. More recently, other authors have proposed that ecosystem size (Post *et al.* 2000) and metacommunity dynamics (Holt 1997) could also constrain food chain length. Of this set of alternative hypotheses, two (body size and optimal foraging) are essentially evolutionary in nature. As another example, consider patterns of connectance. There are purely dynamic forces that can influence connectance in food webs (Pimm 1982), but it must also be true that connectance reflects the evolutionary dynamics of trophic specialization and generalization, and coevolutionary arms races, across various trophic levels. The relative importance of evolutionary dynamics, versus ecological dynamics and constraints such as ecosystem productivity and size in determining food web attributes is an open and important question that needs much more focused work.

Moreover, whether a species' traits are fixed or not is not an absolute but depends on the structure of the environment in which species live and the traits in question. Evolutionary analyses can help sharpen our understanding of when we can reasonably assume that a species' traits are fixed for the purpose of a particular ecological analysis. A second important question is thus: when does the *absence* of evolution reflect ecological constraints?

It is clear that, at times, population dynamics may be decoupled from evolutionary dynamics. Moreover, sometimes the ecological structure of a system can preclude an important role for evolution, even though genetic variation is present. I, with Hochberg (1997), argue that in heterogeneous landscapes, a factor may be strongly limiting (in terms of its effect on distribution and abundance) without it necessarily being a strong selective factor. For instance, consider a predator-prey system with an effective predator that is nonetheless stabilized because a spatial refuge is present. Prey individuals in the refuge can breed without fear of predation and tend to stay there. If they nonetheless generate emigrants who leave the refuge, this "surplus" can stably sustain the predator. Yet because most successful reproduction occurs in the refuge, if this movement rate is low there is an automatic bias in adaptive evolution towards maintaining adaptation to this habitat at the expense of improvement in adaptation outside. Hence, a predator may effectively limit prey abundance and distribution (largely to the refuge) without simultaneously generating strong selection on the prey species to withstand predation when it is exposed outside the refuge. The effect is particularly strong when the predator-prey system is dynamically unstable (Holt, Hochberg, & Barfield 1999).

A striking phenomenon that suggests the existence of constraints on the basic ecological properties of species is *niche conservatism* (Peterson, Soberon, & Sanchez-Cordero 1999; Holt & Gaines 1992). I should note that here I am using the word *niche* in a sense that goes back to Grinnell (via Hutchinson), as a summary of the requirements a species must meet to persist in a given environment (i.e., be able to increase when rare), rather than in Elton's sense, which has to do with the *role* of species in communities. The niche is an abstract character of a species, in effect a mapping of expected growth rate, when rare, onto an abstract environmental space (with axes of temperature, soil moisture, food levels, etc.). Niche conservatism is to be expected if species experience environmental change sharply enough (in time or across space) such that new environments are associated with high probabilities of extinction (as in the experiments on thermal tolerance in *Escherichia coli* describe by Lenski 2001). I have argued that niche conservatism can be fostered by demographic asymmetries among populations (Holt & Gaines 1992, Holt 1996, Holt & Gomulkiewicz 2004; see also Kawecki 1995, 2004). Broadly speaking, this is because niche evolution is self-referential in a way that most other character evolution is not. Consider the familiar evolutionary catechism: evolution by natural selection occurs if there is (1) variation that is (2) heritable and that (3) leads to different expected fitnesses by individuals bearing different phenotypes. But fitness is always relative to a particular environment, because it is always a joint function of the phenotype and of the environment; if we change the environment, we can often change fitness. For instance, optimal beak size may vary with fluctuations in the food supply (as occurs in Darwin's finches; Grant & Grant 1989). What defines the environment in which a researcher evaluates niche evolution? To a first approximation, it is the niche itself! Environmental states well outside the niche are those in which a species is expected to go rapidly extinct—and there is little scope for evolution in extinct populations.

There are many other issues that could be considered when contemplating the interplay of evolution and ecological interactions. Evolution depends on heritable variation, but the heritable component of genetic variation depends on the environments in which individuals live (Holt 1990, Hoffmann & Merila 1999). Natural selection is not the only force that can potentially influence community dynamics. Evolutionary forces such as mutation, gene flow, and drift can perturb species from adaptive optima or evolutionarily stable states. This may lead to predictable expectations about when to observe species with trait values far from an optimum (e.g., when local density is low, increasing the importance of drift). Gene flow can also at times facilitate adaptive evolution. Rosemary and Peter Grant (1989) have suggested that low levels of hybridization between closely related species can provide significant sources of variation for selection to respond to environmental shifts; species that are competitors over ecological timescales might be mutualists over evolutionary timescales.

A final important consideration is that the effectiveness of selection on a single character can be hampered by genetic correlations with other characters also undergoing selection. Consider a prey species in a complex food web. It may experience selection to avoid predation by a specific predator, but the evolutionary response to such selection will likely reflect the costs of evolved responses, as measured, for instance, in resource uptake rates or exposure to other predators. An important task for future work is to take the web of interactions impinging on a species, affecting its numerical dynamics (e.g., resource limitation, predation, direct interference, and tolerance to disturbance), and then to translate the interaction web into the interlinked suite of selective forces simultaneously acting upon that species. The absence of evolution may reflect the action of a multiplicity of selective forces, tugging in contradictory directions. Ackerly (2003) has suggested that interspecific interactions, in particular multiple interactions arising in a multispecies context, may provide part of the explanation for observed patterns of niche conservatism. Following a change in the environment, species sorting can permit a locally superior species to supplant a resident species much more rapidly than microevolutionary dynamics can occur in the resident. Conversely, if a species can persist across a range of environments with different adaptive optima, given enough time a researcher might expect speciation to be driven by spatial variation in selection (Doebeli & Diekmann 2003, Gavrilets 2003). Thus, the ultimate explanation for a central conundrum in evolutionary biology—the observation of *stasis* in many species over substantial spans of evolutionary history and bursts of adaptive radiation in other species (Eldredge 1989)—may in the end only be resolved by an explicit treatment of the combined effects of interspecific interactions and community assembly and sorting processes, played out across spatially and temporally heterogeneous landscapes.

12.6 | CONCLUSIONS

Thompson and his colleagues (2001) reported the musings of a National Science Foundation "white paper" committee formed to delineate major frontiers in ecology. These scientists concluded that there are four frontiers in understanding the earth's biological diversity. Two of these frontiers are as much frontiers in evolutionary biology as they are in ecology. It was argued that understanding the "dynamics of coalescence in complex communities" was a key desideratum. The term *community coalescence* denotes the development of complex ecological communities from a regional species pool, which in turn is generated by evolutionary processes creating species and constraining their phenotypes (e.g., because of adaptive trade-offs) so as to comprise functional groups. Another important frontier is to understand "evolutionary and historical determinants of ecological processes". This led to a consideration of six distinct crucial issues that need to be addressed to develop a theory of ecology that takes into account the genetics and evolution of organisms: the phylogenetic structure of ecological processes, rapid evolution and ecological dynamics, coevolution and ecological dynamics, the spatial scales of evolutionary dynamics relative to ecological processes, genetic diversity and ecological dynamics, and the genomics of ecological dynamics.

As I have argued in this chapter, aspects of an evolutionary perspective can be found even from the earliest days of the discipline of ecology. Only recently, however, with the concordant refinement of phylogenetics and molecular technologies, has there been an appreciation of the critical importance of taking an explicit evolutionary stance when considering the core questions of community ecology. There has also been an underappreciation of the potential for rapid evolutionary change, of the need to incorporate organismal-driven change in the environment into descriptions of evolution, and finally, of the potential for ecological forces themselves to determine whether a scientist observes stasis, versus rapid evolution, in species' ecological traits. Our field is ripe for major advances at the ecology-evolution interface, and in my opinion undergraduate and graduate training, and the directives of funding agencies, should be modified to foster the long overdue intellectual coalescence of these two disciplines.

ACKNOWLEDGEMENTS

I thank the editors, Mark Vellend, and an anonymous reviewer for thoughtful comments, Erin Taylor for assistance in manuscript preparation, and the University of Florida Foundation for support.

REFERENCES

Abrams, P.A. (1986). Adaptive responses of predators to prey and prey to predators: The failure of the arms race analogy. *Evolution* **40**, 1229–1247.

Abrams, P.A. (2000). The evolution of predator–prey interactions: Theory and evidence. *Annu. Rev. Ecol. Syst.* **31**, 79–105.

Ackerly, D.D. (2003). Community assembly, niche conservatism, and adaptive evolution in changing environments. *Int. J. Plant Sci.* **164** (Suppl. 3), S165–S184.

Agrawal, A.G., Ed. (2003). Special feature: Community genetics. *Ecology* **84**, 543–601.

Allen, A.P., Brown, J.H., & Gillooly, J.F. (2002). Global biodiversity, biochemical kinetics, and the energetic–equivalence rule. *Science* **297**, 1545–1551.

Allen, T.F.H., & Hoekstra, T.W. (1992). "Toward a Unified Ecology." Columbia University Press, New York.

Antonovics, J. (1976). The input from population genetics: "The new ecological genetics." *Syst. Bot.* **1**, 233–245.

Bell, G. (2001). Neutral macroecology. *Science* **293**, 2413–2418.

Bertness, M.D., Gaines, S.D., & Hay, M.E. (2001). "Marine Community Ecology." Sinauer Associates, Sunderland, MA.

Brandon, R. (1990). "Adaptation and Environment." MIT Press, Cambridge, MA.

Brooks, D., & McLennan, D. (1991). "Phylogeny, Ecology and Behavior: A Research Program in Comparative Biology." University of Chicago Press, Chicago.

Brown, J.H. (1995). "Macroecology." University of Chicago Press, Chicago.

Brown, J.H., & Kodric-Brown, A. (1977). Turnover rates in insular biogeography: Effect of immigration on extinction. *Ecology* **58**, 445–449.

Brown, J.H., & West, G.B., Eds. (2000). "Scaling in Biology." Oxford University Press, Oxford.

Carroll, L.S., Meagher, S., Morrison, L., Penn, D.J., & Potts, W.K. (2004). Fitness effects of a selfish gene (the Mus T complex) are revealed in an ecological context. *Evolution* **58**, 1318–1328.

Case, T.J. (2000). "An Illustrated Guide to Theoretical Ecology." Oxford University Press, Oxford.

Chase, J.M., & Leibold, M.A. (2003). "Ecological Niches: Linking Classical and Contemporary Approaches." University of Chicago Press, Chicago.

Cherrett, J.M. (1989). "Ecological Concepts: The Contribution of Ecology to an understanding of the Natural World." Blackwell, Oxford.

Chesson, P. (2000). Mechanisms of maintenance of species diversity. *Annu. Rev. Ecol. Syst.* **31**, 343–366.

Clobert, J., Danchin, E., Dhondt, A.A., & Nichols, J.D., Eds. (2001). "Dispersal." Oxford University Press, Oxford.

Cody, M.L., & Diamond, J.M., Eds. (1975). "Ecology and Evolution of Communities." Harvard University Press, Cambridge, MA.

Collins, J.P. (1986). Evolutionary ecology and the use of natural selection in ecological theory. *J. Hist. Biol.* **19**, 257–288.

de Mazancourt, C., Loreau, M., & Abbadie, L. (1998). Grazing optimization and nutrient cycling: When do herbivores enhance plant production? *Ecology* **79**, 2242–2252.

Diamond, J., & Case, T.J. (1986). "Community Ecology." Harper & Row, New York.

Dobzhansky, T. (1973). Nothing in biology makes sense except in the light of evolution. *Am. Biol. Teach.* **35**, 125–129.

Doebeli, M., & Dickemann, U. (2003). Speciation along environmental gradients. *Nature* **421**, 259–264.

Ehrlich, P.R., & Birch, L.C. (1967). The "balance of nature" and "population control." *Am. Nat.* **101**, 97–107.

Eldredge, N. (1989). "Macroevolutionary Dynamics: Species, Niches, and Adaptive Peaks." McGraw-Hill, New York.

Eldredge, N. (1995). "Reinventing Darwin." John Wiley & Sons, New York.

Elton, C.S. (1927). "Animal Ecology." Sidgwick & Jackson, London. Reprinted by University of Chicago Press, Chicago (2001), with commentary by M. Leibold & T. Wootton.

Falster, D.S., & Westoby, M. (2003). Plant height and evolutionary games. *Trends Ecol. Evol.* **18**, 337–343.

Ford, E.B. (1975). "Ecological Genetics," 3rd edition. Chapman & Hall, London.

Gavrilets, S. (2003). Perspective: Models of speciation—What have we learned in 40 years? *Evolution* **57**, 2197–2215.

Gillespie, J.H. (1991). "The Causes of Molecular Evolution." Oxford University Press, Oxford.

Gillespie, R. (2004). Community assembly through adaptive radiation in Hawaiian spiders. *Science* **303**, 356–359.

Gomulkiewicz, R., & Holt, R.D. (1995). When does evolution by natural selection prevent extinction? *Evolution* **49**, 201–207.

Grant, B.R., & Grant, P.R. (1989). "Evolutionary Dynamics of a Natural Population." University of Chicago Press, Chicago.

Hairston, N.G., Smith, F.E., & Slobodkin, L.B. (1960). Community structure, population control, and competition. *Am. Nat.* **94**, 421–425.

Hamilton, W.D. (1995). Ecology in the large: Gaia and Genghis Khan. *J. Appl. Ecol.* **32**, 451–453.

Hamilton, W.D., & Lenton, T.M. (1998). Spora and Gaia: How microbes fly with their clouds. *Ethol. Ecol. Evol.* **10**, 1–16.

Harper, J.L. (1967). A Darwinian approach to plant ecology. *J. Ecol.* **55**, 242–270.

Harvey, P.H., & Pagel, M.D. (1991). "The Comparative Method in Evolutionary Biology." Oxford University Press, Oxford.

Hazen, W.E. (1970). "Readings in Population and Community Ecology." Saunders, Philadelphia.

Hendry, A.P., Wenburg, J.K., Bentzen, P., Volk, E.C., & Quinn, T.P. (2000). Rapid evolution of reproductive isolation in the wild: Evidence from introduced salmon. *Science* **290**, 516–518.

Hoffmann, A.A., & Merila, J. (1999). Heritable variation and evolution under favorable and unfavorable conditions. *Trends Ecol. Evol.* **14**, 96–101.

Holt, R.D. (1987). Population dynamics and evolutionary processes: The manifold roles of habitat selection. *Evol. Ecol.* **1**, 331–347.

Holt, R.D. (1990). The microevolutionary consequence of climate changes. *Trends Ecol. Evol.* **5**, 311–315.

Holt, R.D. (1993). Ecology at the mesoscale: The influence of regional processes on local communities. In "Species Diversity in Ecological Communities: Historical and

Geographical Perspectives" (R.E. Ricklefs & D. Schluter, Eds.). University of Chicago Press, Chicago, pp. 77–88.

Holt, R.D. (1994). Linking species and ecosystems: Where's Darwin? In "Linking Species and Ecosystems" (C. Jones & J. Lawton, Eds.). Chapman & Hall, New York, pp. 273–279.

Holt, R.D. (1996). Demographic constraints in evolution: Towards unifying the evolutionary theories of senescence and niche conservatism. *Evol. Ecol.* **10**, 1–11.

Holt, R.D. (1997). Community modules. In "Multitrophic Interactions in Terrestrial Systems," 36th Symposium of the British Ecological Society. (A.C. Gance & V.K. Brown, Eds.). Blackwell, Oxford, pp. 333–350.

Holt, R.D. (2001). Species coexistence. In "Encyclopedia of Biodiversity," Vol. 5. Academic Press, San Diego, pp. 413–426.

Holt, R.D. (2003). On the evolutionary ecology of species ranges. *Evol. Ecol. Res.* **5**, 159–178.

Holt, R.D., & Gaines, M.S. (1992). Analysis of adaptation in heterogeneous landscapes: Implications for the evolution of fundamental niches. *Evol. Ecol.* **6**, 433–447.

Holt, R.D., & Gomulkiewicz, R. (2004). Conservation implications of niche conservatism and evolution in heterogeneous environments. In "Evolutionary Conservation Biology" (U. Diekmann, D. Couvet, & R. Ferriere, Eds.). Cambridge University Press, Cambridge.

Holt, R.D., & Hochberg, M.E. (1997). When is biological control evolutionarily stable (or is it)? *Ecology* **78**, 1673–1683.

Holt, R.D., Hochberg, M.E., & Barfield, M. (1999). Population dynamics and the evolutionary stability of biological control. In "Theoretical Approaches to Biological Control" (B.A. Hawkins & H.V. Cornell, Eds.). Cambridge University Press, Cambridge, pp. 219–320.

Holt, R.D., & Lawton, J.H. (1994). The ecological consequences of shared natural enemies. *Annu. Rev. Ecol. Syst.* **25**, 495–520.

Holyoak, M., Leibold, M., & Holt, R.D., Eds. (in press). "Metacommunities: Spatial Dynamics and Ecological Communities." University of Chicago Press, Chicago.

Hubbell, S.P. (2001). "The Unified Neutral Theory of Biodiversity and Biogeography." Princeton University Press, Princeton, NJ.

Hutchinson, G.E. (1959). Homage to Santa Rosalia, or why are there so many kinds of animals? *Am. Nat.* **93**, 145–159.

Hutchinson, G.E. (1965). "The Ecological Theatre and the Evolutionary Play." Yale University Press, New Haven, CT.

Jones, C.G., Lawton, J.H., & Schachak, M. (1994). Organisms as ecosystem engineers. *Oikos* **69**, 373–386.

Kawecki, T.J. (1995). Demography of source-sink populations and the evolution of ecological niches. *Evol. Ecol.* **9**, 38–44.

Kawecki, T.J. (2004). Ecological and evolutionary consequences of source-sink population dynamics. In "Ecology, Genetics, and Evolution of Metapopulations" (I. Hanski & O.E. Gaggiotti, Eds.). Academic/Elsevier, Burlington, MA, pp. 387–414.

Keddy, P.A. (2000). "Competition," 2nd edition. Kluwer, New York.

Kimura, M. (1983). "The Neutral Theory of Molecular Evolution." Cambridge University Press, Cambridge.

Lande, R., & Shannon, S. (1996). The role of genetic variation in adaptation and population persistence in a changing environment. *Evolution* **50**, 434–437.

Lawton, J.H. (2000). Community ecology in a changing world. In "Excellence in Ecology Series," Vol. 11 (O. Kinne, Ed.). International Ecology Institute, Oldendorf/Luhe, Germany.

Leibold, M.A., Holt, R.D., & Holyoak, M. (in press). Adaptive and coadaptive dynamics in metacommunities: Tracking environmental change at different spatial scales. In "Metacommunities: Spatial Dynamics and Ecological Communities" (M. Holyoak, M.A. Leibold, & R.D. Holt, Eds.). University of Chicago Press, Chicago.

Leibold, M.A., Holyoak, M., Mouquet, N., Amarasekare, P., Chase, J.M., Hoopes, M.F., Holt, R.D., Shurin, J.B., Law, R., Tilman, D., Loreau, M., & Gonzalez, A. (2004). The metacommunity concept: A framework for multiscale community ecology. *Ecol. Lett.* **7**, 601–613.

Leibold, M.A., & Norberg, J. (in press). Plankton metacommunities as self-organized adaptive systems. *Limnol. Oceanogr.*

Leigh, E.G. Jr., & Vermeij, G.J. (2002). Does natural selection organize ecosystems for the maintenance of high productivity and diversity? *Phil. Trans. Roy. Soc. Lond. B* **357**, 709–718.

Lenski, R.E. (2001). Testing Antonovics' five tenets of ecological genetics: Experiments with bacteria at the interface of ecology and genetics. In "Ecology: Achievement and Challenge" (M.C. Press, N.J. Huntly, & S. Levin, Eds.). Blackwell, Oxford, pp. 25–26.

Levin, S.A. (1983). Coevolution. In "Population Biology" (H.I. Freedman & C. Strobeck, Eds.), Lecture Notes in Biomathematics, Vol. 52. Springer-Verlag, Berlin, pp. 328–334.

Levin, S.A. (1999). "Fragile Dominion." Perseus Publishing, Cambridge, MA.

Levin, S.A., & Muller-Landau, H.C. (2000). The emergence of diversity in plant communities. *Life Sci.* **323**, 129–139.

Levin, S.A., Segel, L.A., & Adler, F.R. (1990). Diffuse coevolution in plant–herbivore communities. *Theor. Popul. Biol.* **37**, 171–191.

Lewontin, R.C. (1974). "The Genetic Basis of Evolutionary Change." Columbia University Press, New York.

Lewontin, R.C. (1985). The organism as subject and object of evolution. In "The Dialectical Biologist" (R. Levins & R.C. Lewontin, Eds.). Harvard University Press, Cambridge, MA.

Loehle, C., & Pechmann, J.H.K. (1988). Evolution: The missing ingredient in systems ecology. *Am. Nat.* **132**, 884–899.

Loreau, M. (1998). Ecosystem development explained by competition within and between material cycles. *Proc. Roy. Soc. Lond. B* **265**, 33–38.

Loreau, M. (2000). Biodiversity and ecosystem function: Recent theoretical advances. *Oikos* **91**, 3–17.

Loreau, M., Naeem, S., & Inchausti, P., Eds. (2002). "Biodiversity and Ecosystem Functioning: Synthesis and Perspectives." Oxford University Press, Oxford.

MacArthur, R.H. (1972). "Geographical Ecology." Harper & Row, New York.

MacArthur, R.H., & Wilson, E.O. (1967). "The Theory of Island Biogeography." Princeton University Press, Princeton, NJ.

Maurer, B.A. (1999). "Untangling Ecological Complexity: The Macroscopic Perspective." University of Chicago Press, Chicago.

McKinney, M.L., & Drake, J.A., Eds. (1998). "Biodiversity Dynamics: Turnover of Populations, Taxa, and Communities." Columbia University Press, New York.

McPeek, M.A., & Gomulkiewicz, R. (in press). Assembling and depleting species richness in metacommunities: Insights from ecology, population genetics, and macroevolution. In "Metacommunities: Spatial Dynamics and Ecological Communities." (M. Holyoak, M. Leibold, & R.D. Holt, Eds.). University of Chicago Press, Chicago.

Mitchell, W.A., & Valone, T.J. (1990). The optimization research program: Studying adaptations by their function. *Q. Rev. Biol.* **65**, 43–52.

Morin, P.J. (1999). "Community Ecology." Blackwell, Oxford.

Morris, D.W. (2003). Toward an ecological synthesis: A case for habitat selection. *Oecologia* **136**, 1–13.

Mueller, L.D., & Joshi, A. (2000). "Stability in Model Populations." Princeton University Press, Princeton, NJ.

Newman, D., & Pilson, D. (1997). Increased probability of extinction due to decreased genetic effective population size: Experimental populations of *Clarkia pulchella. Evolution* **51**, 354–362.

Norberg, J., Swaney, D.P., Dushoff, J., Lin, J., Casagrandi, R., & Levin, S.A. (2001). Phenotypic diversity and ecosystem functioning in changing environments: A theoretical framework. *Proc. Natl. Acad. Sci. USA* **98**, 11,376–11,381.

Nowak, M.A., & Sigmund, K. (2004). Evolutionary dynamics of biological games. *Science* **303**, 793–799.

Odling-Smee, F.J., Laland, K., & Feldman, M.W. (2003). "Niche Construction: The Neglected Process in Evolution." Princeton University Press, Princeton, NJ.

Odum, E.P. (1971). "Fundamentals of Ecology," 3rd edition. Saunders, Philadelphia.

Orians, G.H. (1962). Natural selection and ecological theory. *Am. Nat.* **96**, 257–263.

Peterson, A.T., Soberon, J., & Sanchez-Cordero, V. (1999). Conservatism of ecological niches in evolutionary time. *Science* **285**, 1265–1267.

Pimentel, D. (1968). Population regulation and genetic feedback. *Science* **159**, 1432–1437.

Pimentel, D., Feinberg, E.G., Wood, P.W., & Hayes, J.T. (1965). Selection, spatial distribution, and the coexistence of competing fly species. *Am. Nat.* **94**, 97–109.

Pimm, S.L. (1982). "Food Webs." Chapmen & Hall, London.

Polis, G.A., Power, M.E., & Huxel, G.R., Eds. (2004). "Food Webs at the Landscape Level." University of Chicago Press, Chicago.

Post, D.M., Pace, M.L., & Hairston, N.G. (2000). Ecosystem size determines food-chain length in lakes. *Nature* **405**, 1047–1049.

Price, P.W. (2003). "Macroevolutionary Theory on Macroecological Patterns." Cambridge University Press, Cambridge.

Reeve, H.K., & Sherman, P.W. (1993). Adaptation and the goals of evolutionary research. *Q. Rev. Biol.* **68**, 1–32.

Reznick, D., & Ghalambor, C.K. (2001). The population ecology of contemporary adaptations: What empirical studies reveal about the conditions that promote adaptive evolution. *Genetica* **112–113**, 183–198.

Ricklefs, R.E. (2004). A comprehensive framework for global pattern in biodiversity. *Ecol. Lett.* **7**, 1–15.

Ricklefs, R.E., & Schluter, D. (1993). "Species Diversity in Ecological Communities: Historical and Geographical Perspectives." University of Chicago Press, Chicago.

Rose, M.R., & Lauder, G.V. (1996). Postspandrel adaptationism. In "Adaptation" (M. Rose & G.V. Lauder, Eds.). Academic Press, San Diego, pp. 1–6.

Rosenzweig, M.L. (1987). Habitat selection as a source of biological diversity. *Evol. Ecol.* **1**, 315–330.

Rosenzweig, M.L. (1995). "Species Diversity in Space and Time." Cambridge University Press, Cambridge.

Rothschild, L.J., & Lister, A.M. (2003). "Evolution on Plant Earth: The Impact of the Physical Environment." Academic/Elsevier, London.

Roughgarden, J. (1979). "Theory of Population Genetics and Evolutionary Ecology: An Introduction." Macmillan, New York.

Schoener, T.W. (1986). "Community Ecology." Harper & Row, New York.

Schwilk, D.W., & Kerr, B. (2002). Genetic niche-hiking: An alternative explanation for the evolution of flammability. *Oikos* **99**, 431–442.

Seger, J., & Hamilton, W.D. (1988). Parasites and sex. In "The Evolution of Sex" (R.E. Michod & B.R. Levin, Eds.). Sinauer Associates, Sunderland, pp. 176–198.

Silvertown, J., & Antonovics, J., Eds. (2001). "Integrating Ecology and Evolution in a Spatial Context." Blackwell, London.

Stenseth, N.C., & Maynard Smith, J. (1984). Coevolution in ecosystems: Red Queen evolution or stasis. *Evolution* **38**, 870–880.

Strong, D.R., Simberloff, D., Abele, L.G., & Thistle, A.B. (1984). "Ecological Communities: Conceptual Issues and the Evidence." Princeton University Press, Princeton, NJ.

Symon, K.R. (1960). "Mechanics." Addison-Wesley, Reading, MA.

Thompson, J.N. (1999). The evolution of species interactions. *Science* **284**, 2116–2118.

Thompson, J.N., Reichman, O.J., Morin, P.J., Polis, G.A., Power, M.E., Sterner, R.W., Couch, C.A., Gough, L., Holt, R., Hooper, D.U., Keesing, F., Lovell, C.R., Milne, B.T., Molles, M.C., Roberts, D.W., & Strauss, S.Y. (2001). Frontiers of ecology. *BioScience* **51**, 15–24.

Tokeshi, M. (1999). "Species Coexistence: Ecological and Evolutionary Perspectives." Blackwell, Oxford.

Wardle, D.A. (2002). "Communities and Ecosystems: Linking the Aboveground and Belowground Components." Princeton University Press, Princeton, NJ.

Webb, C.O. (2000). Exploring the phylogenetic structure of ecological communities: An example for rain forest trees. *Am. Nat.* **156**, 145–155.

Webb, C.O., Ackerly, D.D., McPeek, M.A., & Donoghue, M.J. (2002). Phylogenies and community ecology. *Annu. Rev. Ecol. Syst.* **33**, 475–505.

Whittaker, R.H. (1975). "Communities and Ecosystems," 2nd edition. Macmillan, New York.

Wiley, E.O. (1981). "Phylogenetics: The Theory and Practice of Phylogenetic Systematics." Wiley, New York.

Williams, G.C. (1966). "Adaptation and Natural Selection." Princeton University Press, Princeton, NJ.

Wilson, D.S. (1997). Biological communities as functionally organized units. *Ecology* **78**, 2018–2024.

Yachi, S., & Loreau, M. (1999). Biodiversity and ecosystem productivity in a fluctuating environment: The insurance hypothesis. *Proc. Natl. Acad. Sci. USA* **96**, 1463–1468.

13 | MODELLING THE ECOLOGICAL CONTEXT OF EVOLUTIONARY CHANGE: DÉJÀ VU OR SOMETHING NEW?

Troy Day

13.1 | INTRODUCTION

A principle interest in evolutionary ecology is to understand how ecological interactions within and between species generate natural selection and, in turn, how such natural selection shapes these ecological interactions through evolutionary change (MacArthur 1972; Pianka 1974; Roughgarden 1996; Cockburn 1991; Bulmer 1994; Real 1994; Fox, Roff, & Fairbairn 2001). This feedback between ecological and evolutionary processes lies at the heart of this area of research. Creating a theory that adequately represents this mutual dependence, and that makes testable predictions about ecological and evolutionary processes, has presented a considerable challenge. It is difficult enough to obtain a faithful theoretical description of either ecological or evolutionary processes, let alone a coherent melding of the two. The development of such a theory is desirable because presumably the resultant bridge between ecology and evolutionary biology will inject new ideas into both fields and lead to an important consolidation and extension of our understanding of the earth's biota.

Given the complexity of both ecological and evolutionary processes, it is necessary to make several simplifying assumptions in the development of theory. In this chapter I will describe various theoretical developments in this area, organizing and presenting them in a fashion that is chronological and is meant to illustrate the connections, similarities, and differences among them. My treatment is by no means an exhaustive review, and the perspective taken is necessarily biased towards areas in which I have some knowledge (and thus is biased towards my interests). Although I was alive throughout most of these developments, the earliest of them happened when I was too young to be fully cognizant (or even interested) in such arcane topics; therefore, much of my historical perspective has been gleaned from the literature and talking with more senior scientists.

I begin by briefly presenting some background on theoretical ecology and theoretical evolutionary biology as independent fields of study, and I introduce some examples that will illustrate various approaches later in this chapter. In the bulk of this chapter, I consider the various ways in which researchers have sought to merge these two areas. I conclude by asking where we stand, by asking where we go next, and by considering whether there have been quantum leaps or paradigm shifts along the way.

13.2 | THEORETICAL ECOLOGY

There are many processes and types of interactions between organisms that have been the focus of ecological theory. A few of the most important include competition for resources, predation, parasitism, mutualism, and facilitation (Begon, Harper, & Townsend 1986). Although each process is distinct from the others, the ecological theories developed for each of them share at least one common feature: they have been directed towards describing and explaining the distribution and abundance of different kinds of organisms (typically species) as a result of these processes. Individuals within a given species (or sometimes within an age, condition, or size class of that species) are treated as being effectively identical. No allowance is made for genetic variation among individuals in traits that affect these ecological interactions, thereby precluding any evolutionary change.

At first this neglect of evolutionary potential might appear surprising. When theoretical ecology was born, it was well appreciated that populations can and have evolved. The reason for this omission was a feeling (by some, at least) that evolutionary change proceeds on a timescale much longer than that of ecological change. Therefore, the inclusion of evolutionary change is not critical for understanding the implications of various ecological interactions. After all, the point of theory is to simplify reality in a way that captures only those features important for the question at hand. It is now well recognized that this separation of time-scales is artificial and that rapid evolution can (and often does) occur (see Hendry & Kinnison 1999 for a review). This recognition formed an important part of the motivation for the development of theoretical evolutionary ecology. (Ironically, the most recent theoretical developments in evolutionary ecology have returned to the assumption of a separation of time-scales, but I will explain this later.)

To better illustrate these ideas, I will consider one of the simplest examples in theoretical ecology: a discrete-time version of logistic growth (Case 2000). This model is meant to capture the population dynamics of a single species under density dependence. There are many formulations of this model, and one possibility that has been used extensively in the literature is to suppose that the number of individuals in the next generation is given by the number in the current generation plus the number of new individuals produced. Suppose that each individual in generation t gives rise to $r(1 - N(t)/K)$ new individuals, where r and K are constant parameters representing the number of individuals produced in the absence of competition and the population carrying capacity, respectively, and $N(t)$ is the population size in generation t. Then we have:

$$N(t+1) = N(t) + N(t)r\left(1 - \frac{N(t)}{K}\right). \tag{13.1}$$

The effects of within-species competition are represented by the fact that an individual's reproductive output, $r(1 - N(t)/K)$, declines (linearly) as population size increases, and it reaches zero when the population attains carrying capacity (i.e., $N = K$). Equation 13.1 can be rewritten as:

$$N(t+1) = \left\{1 + r - r\frac{N(t)}{K}\right\}N(t), \tag{13.2}$$

where the quantity in the braces of Equation 13.2 is the *total* contribution of an individual (i.e., per capita) to the next generation (i.e., it is the total per capita number of individuals produced in generation t, including an individual's own survival—which happens with probability 1 in this model). The absence of the possibility for evolutionary change is reflected by the fact that this per capita production is identical for all N individuals of the population.

The analogous model for both intra- and interspecific competition has also received a large amount of attention in the ecological literature.

I present this here because it features prominently in the merger of ecological and evolutionary theory:

$$N_1(t+1) = N_1(t) + r_1\left(1 - \frac{N_1(t) + \alpha_{12}N_2(t)}{K_1}\right) \quad (13.3A)$$

$$N_2(t+1) = N_2(t) + r_2\left(1 - \frac{N_2(t) + \alpha_{21}N_1(t)}{K_2}\right), \quad (13.3B)$$

where α_{ij} (the competition coefficients) represents the competitive effect of an individual of species j on an individual of species i, relative to a conspecific individual i. The subscript numbers refer to species 1 and 2. Equations (13.3A, B) are often referred to as the Lotka-Volterra competition equations (Case 2000), and like the logistic-growth model (13.1), this model is phenomenological because it does not treat the dynamics of resource consumption and the competition that results in mechanistic manner. Rather, this is described qualitatively because higher densities of individuals (of either species) reduce the per capita production of individuals of either species. The strength of these effects is controlled by the parameters α_{ij}. More realistically, we could construct a mechanistic model of competition for resources by modelling the resource dynamics, leading to a so-called consumer-resource model (MacArthur 1972). Interestingly, it has been shown (MacArthur 1970, 1972) that if the dynamics of the resource turnover are fast relative to those of the consumer, then a system analogous to (13.3A, B) can be obtained from consumer-resource models.

As with the logistic model, the assumption of a separation of timescales between ecological and evolutionary processes in model (13.3A, B) is reflected by the lack of within-species variation in the traits that affect competition for resources. Thus, the parameters governing the interactions between the two species (e.g., r, K, and α_{ij}) remain constant during the ecological dynamics. There are an enormous number of extensions and further developments of this sort of model, but all are dynamic systems (often in continuous time) in which there are several state variables describing the density of different organisms and in which all parameters governing the interactions are treated as constants. In other words, the parameters do not change during the dynamics. This theory is typically used to understand and predict population dynamics over time (or space or both). For example, do we expect stable equilibrium population sizes, cycling, or other more complex nonequilibrium behaviour? Moreover, how do the various parameters affect the outcome? Issues surrounding this last question are of interest to evolutionary ecologists because the community dynamics themselves will generate natural selection on these parameters, causing them to evolve and thereby altering these dynamics. Thus, we need to include evolutionary change to understand how ecological interactions shape their own evolutionary trajectories.

13.3 | THEORETICAL EVOLUTIONARY BIOLOGY

As with theoretical ecology, the field of theoretical evolutionary biology is now enormous. Here I restrict attention to two relatively self-contained and influential areas: classical population genetics (which has developed into its own subdiscipline) and optimization–game theory. To my knowledge there is not yet a comprehensive treatment of the history and development of game theory in evolutionary biology, but interested readers should consult the book by Provine (2001) for a wonderful historical account of the development of population genetics.

13.3.1 | Classical Population Genetics

Most theory in population genetics (and virtually all such theory in the classical population-genetic literature) treats population densities as being either constant or irrelevant (Hartl & Clark 1989). In addition, although ecological interactions will often be important causes of natural selection through their effects on the fitness of different individuals, most population-genetic theory ignores the particular causes of natural selection and instead treats it in a phenomenological fashion. The most frequent approach is to suppose that different alleles (or genotypes) have different fitnesses, and then to simply specify these fitnesses. Thus, the fitnesses of various alleles are specified as constant parameters in classical population genetics in much the same way that the parameters governing ecological interactions are treated as constants in ecological theory.

To illustrate this approach, consider a single-locus, diallelic model for a diploid species with nonoverlapping generations. For simplicity I focus on an autosomal locus with alleles A and a. As a result there are three genotypes: AA, Aa, and aa. In such models we then need to specify the fitness of these three genotypes: W_{AA}, W_{Aa}, and W_{aa}. If you are measuring the frequency of the A allele in each generation (denoted by $p(t)$) in the gamete pool, then you can view W_{ij} as the number of gametes produced by an individual with genotype ij. Letting $N(t)$ denote the population size in generation t, there will be $N(t)p(t)^2$ AA homozygotes in that generation, $N(t)2p(t)(1 - p(t))$ Aa heterozygotes, and $N(t)(1 - p(t))^2$ aa homozygotes. Each AA homozygote will produce W_{AA} gametes (all of which carry the A allele), each heterozygote will produce W_{Aa} gametes (only half of which carry the A allele), and each aa homozygote will produce W_{aa} gametes (none of which carry the A allele). Thus, the total number of A-carrying gametes in generation $t + 1$ will be $N(t)p(t)^2W_{AA} + N(t)p(t)(1 - p(t))W_{Aa}$, whereas the total number of gametes in generation $t + 1$ will be $N(t)p(t)^2W_{AA} + N(t)2p(t)(1 - p(t))W_{Aa} + N(t)(1 - p(t))^2W_{aa}$. Thus, the frequency of the A allele in generation $t + 1$ is:

$$p(t+1) = \frac{N(t)\left(p(t)^2 W_{AA} + p(t)(1-p(t))W_{Aa}\right)}{N(t)\overline{W}(t)}$$
$$= \frac{p(t)^2 W_{AA} + p(t)(1-p(t))W_{Aa}}{\overline{W}(t)}, \tag{13.4A}$$

where $\overline{W}(t) = p(t)^2W_{AA} + 2p(t)(1 - p(t))W_{Aa} + (1 - p(t))^2W_{aa}$ is the average fitness of the population at time t (see Hartl & Clark 1989, p. 151). Equation (13.4A) reveals that if the fitnesses W_{ij} do not depend on population density, the evolutionary dynamics of the population are unaffected by population density.

Further insight can be gained by dividing the numerator and the denominator of equation (13.4A) by W_{aa} to obtain:

$$p(t+1) = \frac{p(t)^2 w_{AA} + p(t)(1-p(t))w_{Aa}}{\overline{w}(t)}. \tag{13.4B}$$

In this case, $w_{ij} = W_{ij}/W_{aa}$. W_{ij} is referred to as the absolute fitness of genotype ij, whereas w_{ij} is the relative fitness of genotype ij (i.e., relative to genotype aa, although you can use any genotype as the *standard* in this normalization). Equation (13.4B) reveals that relative fitness, not absolute fitness, determines the evolutionary dynamics. Thus, even if the *absolute* fitnesses depend on the population density (i.e., W_{ij} is a function of N), the evolutionary dynamics will still be independent of this "ecological" variable provided that the *relative* fitnesses do not. For example, if the genotypic absolute fitnesses depend on population density and have the form $W_{ij} = F(N)c_{ij}$, where c_{ij} is a genotype-specific constant and $F(N)$ is some function of population density, then the evolutionary dynamics will still be independent of population density because the relative fitnesses are $w_{ij} = c_{aa}/c_{ij}$. This observation, that population density often cancels out the equation for allele frequency change, has lead to the widespread use of evolutionary models that ignore explicit ecological interactions involving population densities.

Another useful formulation for the evolutionary dynamics is obtained by deriving an equation for the *change* in allele frequency in one generation—that is, $\Delta p(t) = p(t + 1) - p(t)$. Equation (13.4A), after some rearrangement, yields Wright's equation (Wright 1935, 1969):

$$\Delta p = \frac{p(1-p)}{2} \frac{1}{\overline{W}} \frac{d\overline{W}}{dp}, \tag{13.5A}$$

or from (13.4B), we find the equivalent equation in terms of relative fitness:

$$\Delta p = \frac{p(1-p)}{2} \frac{1}{\overline{w}} \frac{d\overline{w}}{dp}. \tag{13.5B}$$

Equations (13.5A, B) reveal that natural selection results in a change in allele frequency such that mean absolute fitness, $\overline{W}$, and mean relative fitness, $\overline{w}$, increase (Crow & Kimura 1970, Hartl & Clark 1989, Hofbauer & Sigmund 1988). Moreover, because equation (13.4B) or (13.5B) reveals that relative fitness (rather than absolute fitness) is the determinate of evolutionary change, many researchers standardize the fitnesses such that $w_{Aa} = 1 + s$, $w_{Aa} = 1 + s/2$, and $w_{aa} = 1$, where s is the selective advantage (or cost, if it is negative) of the A allele and the $s/2$ for the heterozygote assumes that alleles act additively (Crow & Kimura 1970). In this case, Equation (13.5) reduces to the following, particularly simple form:

$$\Delta p = \frac{p(1-p)}{2} \frac{s}{\overline{w}}. \tag{13.6}$$

From equation (13.6) we can clearly see that the ecological dynamics (in terms of population density) can be safely ignored when trying to understand evolutionary change (in this simple model, at least) provided that population density has no effect on the relative selective advantage of the A allele, s.

Equation (13.6) represents a simple evolutionary model, but it has been widely used to address a variety of issues and partly forms the basis for the initial neglect of ecological details when studying evolutionary dynamics. As with ecological theory, evolutionary theory has gone far beyond this simple incarnation to explore how a range of other factors affects evolutionary change. It was the recognition that the selective advantage of any given allele, s, likely will depend on ecological context in many circumstances that lead to the first attempts to integrate the two. Treating the effects of natural selection arising from ecological interactions as a constant parameter, s, is simply not good enough for many situations.

13.3.2 | Optimization and Game Theory

An alternative approach for modelling evolution is the use of optimization and game-theoretic models. I treat them together because optimality models can be viewed as a special case of game-theoretic models. Typically, optimality models ignore the details of how the genotype of an organism gives rise to its phenotype and simply seek to characterize the phenotype that yields the highest fitness. Thus, optimality models require the specification of a fitness function, and the underlying assumption is that natural selection proceeds so as to maximize this function (Maynard Smith 1978, Parker & Maynard Smith 1990).

Optimality thinking and modelling has a long history in evolutionary biology, but the introduction of game-theoretic thinking and modelling to evolutionary biology took this approach to an entirely new level. Optimization models assume that the fitness of an individual depends only on that individual's phenotype, but it has long been appreciated that an individual's fitness is determined by the phenotypes of other individuals in the population as well. The introduction of game-theoretic ideas addressed this complexity, and it was motivated largely to model the evolution of social interactions for which optimality models were simply not tenable (Maynard Smith & Price 1973, Maynard Smith 1982). An individual's fitness as a result of some social interaction depends on the behaviours of all individuals involved; therefore, it no longer even makes sense to ask the question of what is optimal. The optimal behaviour is context specific, depending upon the behaviour of other individuals. As a result, focus moved from optimal phenotypes to evolutionarily stable phenotypes (Maynard Smith 1982). An evolutionarily stable strategy (ESS) is one such that if all individuals are using this phenotype, then no single individual can do better by unilaterally altering its phenotype (Maynard Smith 1982, Bulmer 1994). Optimality models are then a special case of such game-theoretic models in which the fitness of an individual depends only on its own phenotype.

As with optimality models, these original game-theoretic ideas were focused on the end point of evolution. The underlying idea was that new mutations periodically arise, and these either sweep to fixation or die out. Thus, the population is imagined as being monomorphic with the periodic introduction of new mutations. Eventually, after a series of new mutations and periodic allelic replacements, you might expect the population to arrive at a phenotype that is evolutionarily stable.

The game-theoretic approach has been extended to many other situations involving different roles played by individuals (e.g., male versus female) and the possibility that a single phenotype is not an ESS but, rather, that a polymorphism is maintained. In addition, although this approach was often used to model the evolution of social interactions, it was soon appreciated that its utility extended well beyond this (e.g., see Lawlor & Maynard Smith 1976 and Reed & Stenseth 1984). For this chapter's purposes, it is important to note that this approach also proved useful for modelling ecological interactions because, for example, the resources available to an individual depend not only on its phenotype but also on the phenotypes of other individuals in the population (e.g., a particular resource will be abundant if few other individuals use it; see Lawlor & Maynard Smith 1976). Analogous considerations hold for other ecological interactions, making this a powerful approach for developing theory in evolutionary ecology (Abrams 2001).

13.4 | THEORETICAL EVOLUTIONARY ECOLOGY

A primary motivation for the development of theoretical evolutionary ecology was the realization that the separation of timescales assumed in much of the ecological literature, along with the lack of explicit ecological detail in the evolutionary literature, was unrealistic. Are there new insights to be gained by creating a theory that bridges these two areas? Can evolutionary biology inform ecology by providing a new perspective on the study of the distribution and abundance of organisms? Can ecology inform evolutionary biology by providing a new perspective on the study of how natural selection guides evolutionary change? To answer these questions, a theory was built that explicitly examines the feedback between ecological and evolutionary processes.

As seen in the previous section, both ecological and evolutionary theory has centred on the development of dynamic systems models describing population dynamics and allele frequency dynamics, respectively. (Although game-theoretic models originally had no explicit dynamic, there was an implicit underlying dynamic.) As a result, from a mathematical standpoint, the mutual dependence and feedback between ecological and evolutionary processes has typically been modelled using some form of a coupled dynamic system between the two. As you will see here, this general structure underlies virtually all of the various approaches used in theoretical evolutionary ecology. In this section I highlight and explain four of these: (1) single-locus theory, (2) quantitative-genetic theory, (3) game theory, and (4) adaptive dynamics.

13.4.1 | Single-Locus Theory

Some of the earliest attempts to create a synthetic theory in evolutionary ecology simply merged models of classical population genetics with those from ecology (Roughgarden 1996). Underlying this idea was the recognition that the per capita production of an individual in models such as that of equation (13.2) (i.e., $\{1 + r - r(N/K)\}$) is the absolute fitness, W, in classical population-genetic models. Therefore, you can construct an ecological-evolutionary model by specifying different per capita productions (i.e., different fitnesses) for different potential genotypes. For example, in the logistic model of equation (13.2) we might use:

$$W_{AA}(N)=\{1+r_{AA}-r_{AA}(N/K_{AA})\} \tag{13.7A}$$

$$W_{Aa}(N)=\{1+r_{Aa}-r_{Aa}(N/K_{Aa})\} \tag{13.7B}$$

$$W_{aa}(N)=\{1+r_{aa}-r_{aa}(N/K_{aa})\}, \tag{13.7C}$$

where the parameters of the per capita production are now genotype specific. You can still define the population average fitness as:

$$\overline{W}(N,p)=p^2W_{AA}(N)+2p(1-p)W_{Aa}(N)+(1-p)^2W_{aa}(N), \tag{13.8}$$

and equation (13.5A) is still valid for the evolutionary dynamics. Now, however, we must also have an equation that governs the ecological dynamics because the population density, N, does not cancel out of the equation for allele frequency change. Adding up the production of the three different genotypes in the population yields the following equation:

$$N(t+1)=\overline{W}(N(t),p(t))N(t). \tag{13.9}$$

If we instead derive an equation for the *change* in population size, $\Delta N(t) = N(t+1) = N(t)$, we get the coupled evolutionary-ecological model:

$$\Delta N=(\overline{W}(N,p)-1)N \tag{13.10A}$$

$$\Delta p=\frac{p(1-p)}{2}\frac{1}{\overline{W}(N,p)}\frac{\partial \overline{W}}{\partial p}. \tag{13.10B}$$

The model (13.10A, B) represents one of the first attempts to construct a theory of evolutionary ecology (Roughgarden 1971, 1996; Charlesworth 1971). Notice, however, that the form of the fitness functions (13.7) is somewhat restrictive in that a genotype's reproductive success depends only on the total *density* of the population but not on its genetic composition. More generally, you might expect different genotypes to have different competitive effects on one another (e.g., perhaps similar genotypes compete more strongly with one another). In this case, the fitness of an AA homozygote would generalize to:

$$W_{AA}(p,N)=\left\{1+r_{AA}-r_{AA}\frac{(\alpha_{AA,AA}Np^2+\alpha_{AA,Aa}N2p(1-p)+\alpha_{AA,aa}Np(1-p))}{K_{AA}}\right\}$$
$$=\left\{1+r_{AA}-r_{AA}\frac{N\bar{\alpha}_{AA}(p)}{K_{AA}}\right\}, \qquad (13.11)$$

where $\alpha_{ij,kl}$ is the competitive effect of genotype ij on genotype kl, and $\bar{\alpha}_{AA}(p) = (\alpha_{AA,AA}p^2 + \alpha_{AA,Aa}2p(1-p) + \alpha_{AA,aa}p(1-p))$ is the population average competitive effect on genotype AA. Thus, in general, we have:

$$W_{ij}(p,N)\left\{1+r_{ij}-r_{ij}\frac{N\bar{\alpha}_{ij}(p)}{K_{ij}}\right\}, \qquad (13.12)$$

where $\bar{\alpha}_{ij}(p) = (\alpha_{ij,AA}p^2 + \alpha_{ij,Aa}2p(1-p) + \alpha_{ij,aa}p(1-p))$.

Notice that the fitness of each genotype is now both density and *frequency* dependent (i.e., it depends on allele frequency, p); therefore, equation (13.5A) is no longer valid because it was derived under the assumption that the genotypic fitnesses were not functions of allele frequency. One can generalize this equation for the present purposes, however, to obtain (e.g., see Taper & Case 1992):

$$\Delta N=(\bar{W}(N,p)-1)N \qquad (13.13\text{A})$$

$$\Delta p=\frac{p(1-p)}{2}\frac{1}{\bar{W}(N,p)}\left[\frac{\partial\bar{W}}{\partial p}+\overline{\frac{\partial W}{\partial p}}\right]. \qquad (13.13\text{B})$$

This is the coupled ecological-evolutionary model, where $\bar{W}(N, p) = p^2W_{AA}(p,N) + 2p(1-p)W_{Aa}(p,N) + (1-p)^2W_{aa}(p,N)$ and $\overline{\partial W/\partial p} = p^2(\partial W_{AA}/\partial p) + 2p(1-p)(\partial W_{Aa}/\partial p) + (1-p)^2(\partial W_{aa}/\partial p)$.

One limitation of the preceding approach is that the evolutionary dynamics are restricted to those alleles that start in the system. No allowance is made for the introduction of new alleles through mutation. As a result, the model makes explicit predictions about short-term evolutionary change, but it has nothing to say about the more long-term process of evolutionary change that occurs as a result of continued mutation and repeated allelic replacements. This difficulty has been alleviated to some degree by considering multiple alleles; still, mutation and longer-term evolution are neglected using this approach.

Interestingly, the single-locus approach has been used to make predictions about the ultimate end point of long-term evolution by using what amounts to a game-theoretic argument. To do so, one asks is there an allele that, if present, can exclude all other possible mutations? For example, in some models similar to equation (13.10B), it can be shown that the allele that can exclude all others is that which sustains the highest population density (Roughgarden 1971, Charlesworth 1971). Once this allele is determined, one can then use the preceding theoretical framework to predict its short-term dynamics in terms of its frequency. Of course, this theoretical approach still cannot make predictions about the long-term evolutionary dynamics of a population towards this end point as a result of recurrent mutation and selection. For that, an alternative approach is required.

13.4.2 | Quantitative-Genetic Theory

The underpinnings of quantitative-genetic theory date back to the development of classical population genetics (Provine 2001), but the specific incarnation most frequently used today was developed at roughly the same time as the preceding single-locus theory in evolutionary ecology (Kimura 1965; Lande 1976a, 1976b, 1979; Lande & Arnold 1983). The initial development of this theory was not motivated by (or even clearly suited to) modelling in evolutionary ecology. Rather, the first versions of this theory took the population geneticist's perspective of not treating ecological interactions explicitly (for reasons outlined earlier) and simply assuming a largely fixed selective regime under which a population evolves. A major advantage of this theory was that it allowed for standing genetic variation in a trait (as is commonly observed), and this variation was maintained through a balance between mutation and recombination, with selection. Also, as opposed to the single-locus theory, the quantitative-genetic approach typically supposed that there were numerous loci affecting the trait of interest, with each locus having a small effect. As a result, the distribution of genotypes in the population could often be well approximated by a Gaussian (normal) distribution with a particular mean and variance.

The central question of interest in this framework is, then, How does the distribution of genotypes (and the resulting distribution of phenotypes) evolve over time (Lande 1976a)? As selective conditions change, evolutionary change occurs, with abundant genetic variation being maintained through a balance between loss of alleles and mutational input. If a researcher is willing to assume that this distribution remains Gaussian, then its evolutionary dynamics can be tracked simply by following the evolution of the mean and the variance of this distribution (because these two parameters completely specify a Gaussian distribution). Even more simply, many researchers have further assumed that the variance of the distribution remains largely constant over the time span of interest and, therefore, that evolutionary change can be tracked simply by following the population mean.

This quantitative-genetic framework was soon generalized to allow ecological interactions. One of the primary interests in doing so was to model competition for resources and character displacement (Brown & Wilson 1956, Slatkin 1980). When two species compete for a common resource pool, do you expect evolutionary divergence in their resource use? If so, then you would expect the phenotypic characteristics related to resource extraction in such species to be divergent where their geographical ranges overlap (Brown & Wilson 1956, Grant 1972).

To gain an appreciation for how this theoretical approach is used, consider a model analogous to that of the logistic growth model used earlier. Now, rather than having a few discrete genotypes (and thus phenotypes), you have a continuous distribution of genotypes (and thus phenotypes; Roughgarden 1983; Taper & Case 1992). An individual's fitness might be density dependent, in which case you have the following fitness of an individual with quantitative trait z:

$$W(z,N)=1+r(z)-r(z)\frac{N}{K(z)}, \tag{13.14}$$

where now r and K are functions of the quantitative trait z.

Denote the average phenotypic value in the population in generation t by $\bar{z}(t)$. This will also be the average genotypic value in the population in that generation if we assume that an individual's phenotype, z, is equal to its genotype, x, plus some random environmental deviation, e; that is, $z = x + e$. Also denote the average phenotypic value in the population after selection has occurred by $\bar{z}(t)_s$. Importantly, this is no longer the average genotypic value of the population because natural selection has acted on the phenotypes, and if, for example, it favoured larger phenotypes, then some of the "selected" population will have genotypes coding for small traits because they happened to have a large positive environmental deviation. Thus, the average phenotype in the next generation, $\bar{z}(t+1)$, (which is equivalent to the average genotype after selection assuming random mating) is given by $\bar{z}(t+1) = \bar{z}(t) + h^2(\bar{z}(t)_s - \bar{z}(t))$, where h^2 is the heritability of the trait given by $h^2 = \sigma_g/\sigma_p$. Here, σ_g is the additive genetic variance of the trait, and $\sigma_p = \sigma_p + \sigma_e$ is the total phenotypic variance of the trait (assumed to equal the additive genetic variance plus the variance in the environmental deviation, σ_e) (Lande 1976a, 1976b). Therefore, the evolutionary change in one generation is:

$$\Delta\bar{z}(t) = h^2(\bar{z}(t)_s - \bar{z}(t)). \tag{13.15}$$

In the present model, population average fitness is $\overline{W} = \int_{-\infty}^{\infty} p(z)W(z,N)dz$; therefore, one can verify that, under the assumption that the phenotypic trait distribution, $p(z)$, is Gaussian with mean $\bar{z}$ and variance σ_p, system (13.15) can be rewritten in terms of the population average fitness, giving the following coupled evolutionary-ecological system (Slatkin 1980, Taper & Case 1992):

$$\Delta N = (\overline{W} - 1)N \tag{13.16A}$$

$$\Delta\bar{z}(t) = \sigma_g \frac{1}{\overline{W}} \frac{\partial \overline{W}}{\partial \bar{z}}. \tag{13.16B}$$

Note the similarity between this system (13.16A, B) and the single-locus system (13.10A, B). The ecological dynamics and evolutionary dynamics take an identical form in each. This is readily apparent for the ecological dynamics, and the evolutionary dynamics can both be viewed as the genetic variance in trait ($p(1-p)/2$ versus σ_g) multiplied by the proportional increase in mean fitness that occurs with an increase in population mean trait value.

Interestingly, we can also account for frequency-dependent selection using this approach, just as was done in the single-locus approach. In this case, the fitness of an individual with trait z is (Slatkin 1980, Taper & Case 1992):

$$\begin{aligned} W(z,N) &= 1 + r(z) - r(z)\frac{N\int_{-\infty}^{\infty} p(y)\alpha(z,y)dy}{K(z)} \\ &= 1 + r(z) - r(z)\frac{N\bar{\alpha}(z)}{K(z)}. \end{aligned} \tag{13.17}$$

Here, $\alpha(z, y)$ is a function of two variables, giving the competitive effect of an individual with phenotype y on an individual with phenotype z, and I have defined $\overline{\alpha}(z) = \int_{-\infty}^{\infty} p(y)\alpha(z, y)dy$ to be the average competitive effect on an individual with trait z. One can verify that this system (13.16A, B) now becomes (Taper & Case 1992):

$$\Delta N = (\overline{W} - 1)N \tag{13.18A}$$

$$\Delta \bar{z} = \sigma_g \frac{1}{\overline{W}} \left[\frac{\partial \overline{W}}{\partial \bar{z}} + \overline{\frac{\partial W}{\partial \bar{z}}} \right], \tag{13.18B}$$

where $\overline{\partial W / \partial \bar{z}} = \int_{-\infty}^{\infty} p(z)(\partial W / \partial \bar{z})dz$. Again, when there is both frequency and density dependence, note the correspondence between the quantitative-genetic system (13.18A, B) and the single-locus system (13.13A, B).

The preceding case of intraspecific competition (i.e., either 13.18A or B) has received considerable attention as an example of quantitative-genetic theory in evolutionary ecology; therefore, it is worth examining in more detail. To do so, we need to specify explicit functions for $r(z)$, $\alpha(z,y)$ and $K(z)$ in the fitness function (13.17). Common assumptions are that r is independent of z, that K (the carrying capacity) is maximal for some intermediate value of z, and that $\alpha(z, y)$ is a unimodal function with a value of unity when $y = z$ (the competitive effect of any individual on an individual with phenotype z is one when the two have identical phenotypes) and it decreases to zero as the difference between the two phenotypes increases (Roughgarden 1972, Taper & Case 1992). With these assumptions, the model (13.18A, B) predicts that the species will evolve towards a phenotype that maximizes the carrying capacity. Note that here I have focused solely on the evolution of the mean trait value, and I have assumed that the distribution remains Gaussian (in accord with most quantitative-genetic models).

Interestingly, some of the earliest work in this area examined the simultaneous evolution of the genetic variance in the trait as well, and these results demonstrated that some parameter values result in a stable equilibrium variance whereas others result in the variance decreasing to zero (Slatkin 1980). The latter case occurs whenever the carrying capacity function, $K(z)$, is very narrow relative to the competition coefficient function $\alpha(z,y)$. Conversely, variation is maintained in the former case when the resource base is broad enough to support the evolution of a variety of resource extraction strategies (relative to the competition coefficient function).

As mentioned earlier, a two-species version of the preceding model has received the most attention in the literature, particularly in the context of studying interspecific competition and evolutionary character displacement (Slatkin 1980; Roughgarden 1983; Brown & Vincent 1987; Taper & Case 1992; Vincent, Cohen, & Brown 1993). The details of such models are a natural extension of the equations above (13.18A, B) but for the Lotka-Volterra competition equations (13.3A, B). If you assume that the genetic variance does not evolve (or does so slowly enough that you can ignore it and simply follow the mean phenotype), such models have demonstrated that evolutionary character displacement occurs for some parameter values and not for others (Taper & Case 1992). Moreover, character displacement occurs precisely for those conditions under which the natural selection favours a stable level of genetic variance for the single-species model (Slatkin 1980). This is intuitively reasonable. If the resource base is broad enough relative to the spectrum of resource use by any given species (i.e., the carrying capacity function, K, is wide relative to the width of the competition coefficient functions, α), then the system can support two different (i.e., divergent) resource extraction phenotypes. Indeed, as you will see shortly in the section on adaptive dynamics, natural selection favours this evolutionary divergence into two phenotypes in the single-species model, but the assumptions of sexual reproduction and recombination that underlie the use of a Gaussian distribution in quantitative genetics prevent a single species from evolving such a dimorphism. Assortative mating within phenotype would also have to evolve to allow such divergence when starting with a single species (e.g., see Dieckmann & Doebeli 1999).

The quantitative-genetic framework has been extremely influential in evolutionary ecology, and it has been extended and used for a variety of questions and ecological interactions. In more complex models, however, equations (13.18A, B) (or analogous equations for the ecological situation of interest) become difficult to use because it is often not possible to explicitly calculate the expression on the right-hand side of equation (13.18B). As a result, some researchers have begun to use an approximation of this equation. This approximation is most easily derived by first noting that equation (13.18B) can be rewritten as:

$$\Delta\bar{z} = \frac{\sigma_g}{\bar{W}}\overline{\frac{\partial W}{\partial z}}, \tag{13.19}$$

where $\overline{\partial W/\partial z} = \int_{-\infty}^{\infty} p(z)(\partial W/\partial z)dz$ (Lande & Arnold 1983; Iwasa, Pomiankowski, & Nee 1991; Taylor 1996). Notice the subtle distinction between this and the second term on the right-hand side of equation (13.18B). In equation (13.18B), the expectation (over the distribution p) is of the derivative of W with respect to $\bar{z}$ (the population mean phenotype), whereas in equation (13.19), both terms on the right-hand side of equation 13.18B have reduced to the expectation (again over the distribution p) of the derivative of W with respect to z (an individual's phenotype). Importantly, equations (13.18B) and (13.19) are equivalent, but equation (13.19) is written in a much simpler form. It reveals that evolutionary change in the population mean phenotype occurs in a direction given by the sign of the change in fitness that occurs with an increase in an individual's phenotype, averaged over all individuals in the population (Lande & Arnold 1983).

An approximation to equation (13.19) is then easily obtained under the condition that the variance among phenotypes in the population is relatively small. In such cases, the expectation of any function is approximately equal to the function evaluated at the mean; therefore, you obtain the following approximation (Iwasa, Pomiankowski, & Nee 1991; Taylor 1996):

$$\Delta\bar{z} = \frac{\sigma_g}{\overline{W}} \frac{\partial W}{\partial z}\bigg|_{z=\bar{z}}. \tag{13.20}$$

Because the right-hand side of equation (13.20) is readily calculated under most circumstances, this approximation has been used extensively in recent years as a simpler means for constructing quantitative-genetic models (reviewed in Abrams 2001).

13.4.3 | Game Theory

The introduction and development of game theory in evolutionary biology essentially paralleled the development of the single-locus and the quantitative-genetic theories. As with quantitative genetics, the game-theoretic approach was not originally devised explicitly for modelling in evolutionary ecology. Rather, its initial focus was largely on the evolution of social traits (Maynard Smith & Price 1973, Maynard Smith 1982). Nevertheless, it was occasionally used to model the evolutionary consequences of ecological interactions (e.g., see Lawlor & Maynard Smith 1976; Reed & Stenseth 1984; Brown & Vincent 1987a, 1987b; Vincent, Cohen, & Brown 1993; and Abrams *et al.* 1993), and the most recent incarnation of the game-theoretic approach (which will be described shortly under the heading *adaptive dynamics*) focuses largely on such ecological interactions.

Although one of the main motivations for the original development of theoretical evolutionary ecology was to dispense with the artificial separation of timescales between ecological and evolutionary processes, it is interesting to note that such a separation is invariably used in game-theoretic models of ecological interactions (e.g., see Lawlor & Maynard Smith 1976 and Reed & Stenseth 1984). The typical approach is to suppose that the population in question is monomorphic (i.e., all individuals have identical phenotypes, and this phenotype is termed the resident) and then to assume that this population reaches a population dynamic equilibrium in terms of the underlying ecological model of interest. Then imagine introducing a rare mutant allele coding for different phenotypes and ask if this mutant can increase in numbers (i.e., if it can invade).

As a simple example, consider the logistic model of one species presented earlier (i.e., 13.1), but where the carrying capacity, K, depends on some quantitative trait, $\hat{z}$ ("hats" are often used to signify the phenotype of the resident in game theory):

$$\begin{aligned} N(t+1) &= N(t) + N(t) r\left(1 - \frac{N}{K(\hat{z})}\right) \\ &= N(t)\left\{1 + r - r\frac{N}{K(\hat{z})}\right\}. \end{aligned} \tag{13.21}$$

At ecological equilibrium, the population size will be $N = K(\hat{z})$, which reveals that the equilibrium density depends on the resident phenotype. Then one can ask what the growth rate of a rare mutant will be if it has phenotype z. When rare, it will have a negligible effect on the population size; therefore, its initial (i.e., invasion) dynamics will be governed by the equation:

$$\begin{aligned} N_{mut}(t+1) &= N_{mut}(t)\left\{1 + r - r\frac{N_{res}}{K(z)}\right\} \\ &= N_{mut}(t)\left\{1 + r - r\frac{K(\hat{z})}{K(z)}\right\}, \end{aligned} \tag{13.22}$$

where we use the equilibrium density of the resident type in the mutant's per capita growth factor because the mutant is rare and therefore the density of the resident type will be the main determinate of the mutant's growth factor. Thus, we can see that the mutant will invade if (and only if):

$$\lambda(z, \hat{z}) \equiv 1 + r - r\frac{K(\hat{z})}{K(z)} > 1, \tag{13.23}$$

where I have defined $\lambda(z,\hat{z})$ to be the growth factor of a rare mutant using strategy z in a population dominated by strategy $\hat{z}$. $\lambda(z,\hat{z})$ is sometimes called the mutant's invasion fitness (Metz, Nisbet, & Geritz 1992) or the mutant's fitness, but it (and more general extensions of it for other ecological interactions) have also been referred to as the fitness generating-function or G-function in the literature (Brown & Vincent 1987a, 1987b; Vincent, Cohen, & Brown 1993). Notice that the mutant dies out if $\lambda(z,\hat{z})$ is less than one, and it is neutral if $\lambda(z,\hat{z})$ equals one (which, of course, occurs when $z=\hat{z}$).

The primary goal of the game-theoretic approach is to characterize phenotypes that are evolutionarily stable (i.e., ESSs). An ESS has the property that, if all individuals in the population adopt this strategy, no alternative can invade (Maynard Smith 1982). Using the definition in Equation 13.23, you can see that an ESS, z^*, must satisfy:

$$\lambda\left(z,z^*\right)\leq\lambda\left(z^*,z^*\right), \tag{13.24}$$

all mutant strategies z (and equality occurs when $z=z^*$). This inequality (13.24) is referred to as the Nash equilibrium condition (Bulmer 1994), and using equation 13.23 in this produces:

$$1+r-r\frac{K\left(z^*\right)}{K(z)}\leq 1+r-r\frac{K\left(z^*\right)}{K\left(z^*\right)}, \tag{13.25A}$$

or

$$K(z)\leq K\left(z^*\right), \tag{13.25B}$$

which reveals that the ESS trait value in this model maximizes the carrying capacity.

Often it is difficult to use condition (13.24) to characterize the ESS; therefore, researchers use "local" conditions instead. In particular, because the condition (13.24) states that $\lambda(z,\hat{z})$ must be maximized in its first argument (i.e., in z) at $z=\hat{z}=z^*$, we know from calculus that the first derivative at this point must equal zero. In addition, if this point is to represent a maximum rather than a minimum, we know that the second derivative at this point must be negative. This gives the following two conditions:

$$\left.\frac{\partial \lambda}{\partial z}\right|_{z=\hat{z}=z^*} = 0 \tag{13.26A}$$

$$\left.\frac{\partial^2 \lambda}{\partial z^2}\right|_{z=\hat{z}=z^*} \leq 0. \tag{13.26B}$$

In the preceding example, one can verify that the conditions in (13.26A, B) evaluate to $dK/dz = 0$ and $d^2K/dz^2 \leq 0$.

The preceding game-theoretic approach was simple because the underlying ecological model of interest was simple. As the ecological scenario becomes more sophisticated, the underlying approach remains the same, but the expression for a mutant's fitness becomes more complex (Metz, Nisbet, & Geritz 1992). This approach has also been used to model the coevolutionary dynamics of more than one species, and such cases are simply treated as *two-player* games in which each species has an expression specifying mutant fitness that, in general, depends on the densities of both species, the resident phenotypes of both species, or the combination of these (Hofbauer & Sigmund 1988, Abrams 2001).

The game-theoretic approach initially placed most emphasis on characterizing ESSs with the underlying idea that such phenotypes would be the ultimate end points of evolutionary change. Implicit in this technique is the notion that, when a new mutation invades, the evolutionary-ecological system is perturbed and the mutant strategy then sweeps through to fixation (it is almost always assumed that a polymorphism does not result). The ecological dynamics will then have reached new equilibrium, and the invasion process occurs again. Thus evolution is viewed as a succession of mutants arising, but on a timescale much slower than the ecological dynamics, and the notion was that the system would eventually attain the uninvadable strategy (i.e., the ESS).

Although the previously mentioned evolutionary processes implicitly formed the foundation of game theory, initially little attention was paid to the evolutionary dynamics of the population as it approached this ESS. This shortcoming was recognized relatively early in the development of game theory (Eshel 1983), and attempts were made to address this issue more quantitatively. One of the most profound insights to come out of this research was the finding that populations need not evolve towards an ESS (Eshel 1983, Taylor 1989, Christiansen 1991, Abrams *et al.* 1993, Geritz *et al.* 1998). Evolutionarily stable strategies can be evolutionarily unattainable (Eshel 1983). This counterintuitive finding arises because natural selection in most game-theoretic models is frequency dependent. As a result, a phenotype can be an ESS in that, if most members of the population adopt this phenotype then no alternative can do better. Nevertheless, it can still be evolutionarily unattainable in the sense that, if most members of the population adopt a phenotype slightly different from this ESS, only those phenotypes that are even more different from the ESS can invade. Thus, natural selection can drive the evolution of a population from an ESS even though, if the population was started at the ESS, it would remain there (Taylor 1989, Christiansen 1991).

Even more interesting, it was found that the conditions for evolutionary attainability and the ESS conditions (i.e., the Nash equilibrium condition) are essentially independent. There can be phenotypes that are evolutionarily attainable but not ESSs, phenotypes that are ESSs but not evolutionarily attainable, and phenotypes that are both ESSs and evolutionarily attainable (Geritz *et al.* 1998). As we will see later, the first of these situations has come to be the primary focus of adaptive dynamics.

As an example, again consider model (13.21), but now include the assumption that selection is both density-dependent and frequency-dependent as in model (13.12), (13.13B) and in model (13.17), (13.18A, B). In this case, we have:

$$N(t+1)=N(t)\left\{1+r-r\frac{\alpha(\hat{z},\hat{z})N}{K(\hat{z})}\right\}, \tag{13.27}$$

where again, $\alpha(x, y)$ is the competitive effect of an individual with phenotype y on an individual with phenotype x (and $\alpha(x, x) = 1$). The mutant's fitness function, (13.23), then becomes:

$$\lambda(z,\hat{z})\equiv 1+r-r\frac{\alpha(z,\hat{z})K(\hat{z})}{K(z)}. \tag{13.28}$$

To be more concrete, I will use the particular functions $K(z)=\kappa e^{-z^2/2\sigma_k}$ and $\alpha(z,\hat{z})=e^{-(z-\hat{z})^2/2\sigma_\alpha}$. These functions have been used numerous times in the literature (reviewed in Day 2000) and are chosen largely for mathematical convenience. With these, the conditions (13.26A, B) become $z^* = 0$ and $\sigma_k \le \sigma_\alpha$, respectively. Thus, the phenotype "0" is an ESS if (and only if) $\sigma_k \le \sigma_\alpha$.

Now consider the question of the evolutionary attainability of $z^* = 0$. To begin, suppose that the majority of the population is using a phenotype slightly below z^*; that is, $\hat{z} < z^*$. For natural selection to drive the population towards z^*, mutants slightly above $\hat{z}$ must be able to invade (i.e., have higher fitness than the resident) and mutants slightly below $\hat{z}$ must not be able to invade (i.e., have lower fitness than the resident). Mathematically, we can express this by requiring that:

$$\left.\frac{\partial \lambda}{\partial z}\right|_{z=\hat{z}} > 0 \quad \text{when} \quad \hat{z} < z^*. \tag{13.29A}$$

Inequality (13.29A) states that the fitness gradient (i.e., the direction of increasing fitness) points towards z^* when the resident phenotype is below z^*. An analogous consideration also leads to the condition:

$$\left.\frac{\partial \lambda}{\partial z}\right|_{z=\hat{z}} < 0 \quad \text{when} \quad \hat{z} > z^*, \tag{13.29B}$$

which states that the fitness gradient points towards z^* when the resident phenotype is above z^* as well. Together, these conditions (13.29A, B) imply that the fitness gradient (which is a function of the population resident strategy $\hat{z}$ only; i.e., $\partial\lambda/\partial z|_{z=\hat{z}}$) decreases as the population resident strategy, $\hat{z}$, increases, passing from positive to negative at $\hat{z} = z^*$. Locally (i.e., near $\hat{z} = z^*$), we can express this by requiring that the derivative of $\partial\lambda/\partial z|_{z=\hat{z}}$ with respect to $\hat{z}$ be negative at $\hat{z} = z^*$:

$$\frac{d}{d\hat{z}}\left\{\left.\frac{\partial \lambda}{\partial z}\right|_{z=\hat{z}}\right\}_{\hat{z}=z^*} < 0, \tag{13.30A}$$

which can also be expressed as:

$$\left[\frac{\partial^2 \lambda}{\partial z^2} + \frac{\partial^2 \lambda}{\partial \hat{z}\partial z}\right]_{z=\hat{z}=z^*} < 0. \tag{13.30B}$$

Condition (13.30A) or equivalently (13.30B) is often referred to as the convergence stability condition (Bulmer 1994) because it implies that natural selection acts in such a way as to cause the population resident strategy to converge to z^*.

Returning to the example in equation (13.28), we can use either (13.30A) or (13.30B) to show that $z^* = 0$ is convergence stable provided that $-r/\sigma_k < 0$, which is always satisfied. Therefore, we have the following two possibilities: (1) $\sigma_k < \sigma_\alpha$, in which case $z^* = 0$ is convergence stable and an ESS, or (2) $\sigma_k > \sigma_\alpha$, in which case $z^* = 0$ is convergence stable but *not* an ESS. In case 1, we can expect the population to evolve towards $z^* = 0$ and to remain there indefinitely. In case 2, we again can expect the population to evolve towards $z^* = 0$, but once there, natural selection becomes disruptive, favouring any phenotype other than $z^* = 0$. At this point some form of evolutionary diversification will occur (Taylor 1989, Christiansen 1991).

Biologically, case 2 can be understood as follows. Competition for resources always makes it beneficial to have a phenotype that is different from other individuals. At the same time, because the carrying capacity is maximized at $z = 0$, natural selection favours evolution towards this phenotype. When the population is not at this phenotype, mutants closer to $z = 0$ gain in both ways (i.e., they have the benefit of being different and the benefit of having a higher carrying capacity). This is why $z^* = 0$ is always convergence stable. If the width of the competition function is narrow relative to the carrying capacity function, however (i.e., if $\sigma_\alpha < \sigma_k$, meaning that any given phenotype is specialized in its resource use), then once the population reaches $z^* = 0$, the strength of selection for being different is strong enough to more than outweigh the loss in carrying capacity that comes from having a phenotype $z \neq 0$ and evolutionary diversification occurs (see Werner & Sherry 1987, Bolnick *et al.* 2002, Bolnick *et al.* 2003, and Bolnick 2004 for interesting empirical examples). At this stage, the preceding model no longer provides an adequate description of the evolutionary dynamics and therefore must be extended in some way to allow for a polymorphism (Christiansen 1991, Geritz *et al.* 1998).

This distinction between stability against invasion of rare mutants (i.e., ESS) and convergence stability is biologically interesting because it illustrates the potential for a trait to evolve to a point at which natural selection becomes disruptive. This finding, that biological interactions give rise to endogenously generated disruptive selection, was implicit in the early results of Eshel (1983), and it was noted more explicitly by Taylor (1989) that this should result "in a polymorphic population which is [no longer] described by the [original fitness] function." Christiansen (1991) developed these ideas of the evolution of polymorphisms more explicitly and illustrated them with the preceding example, as did Brown and Vincent (1987a, 1987b; Vincent, Cohen, & Brown 1993; Abrams *et al.* 1993).

Researchers using the game-theoretic approach to modelling evolutionary ecology have also noted the fundamental similarities between the preceding results and the previous quantitative-genetic models (Charlesworth 1990; Iwasa, Pomiankowski, & Nee 1991; Taper & Case 1992; Abrams *et al.* 1993; Abrams 2001; Taylor 1996; Taylor & Day 1997). In particular, Eshel's (1983) idea that natural selection should drive the evolution of a population in a direction given by the sign of $\partial\lambda/\partial z|_{z=\hat{z}}$ closely parallels the quantitative-genetic equation for the evolutionary dynamics of the mean trait value. More specifically, Eshel's ideas were based on the idea that the evolutionary change in the population resident strategy is proportional to $\partial\lambda/\partial z|_{z=\hat{z}}$:

$$\Delta\hat{z} \propto \left.\frac{\partial\lambda}{\partial z}\right|_{z=\hat{z}}. \tag{13.31}$$

Note the correspondence between this equation and equation (13.19) or its approximation, equation (13.20). These results are identical if we make the identification $\lambda = W$ except that, unlike quantitative-genetic models, the game-theoretic approach assumes a separation of ecological and evolutionary time-scales and therefore does not have a coupled equation for the ecological dynamics. This correspondence also reveals that the convergence stability condition of Eshel (1983) corresponds to the dynamics stability of equilibria in quantitative-genetic models (provided that the ecological dynamics are fast relative to evolution).

It has also been shown that the ESS condition (13.26B) corresponds to stability of the genetic variance in quantitative-genetic models (Taylor & Day 1997). It should come as no surprise that the conditions under which evolutionary diversification occurs in game-theoretic models of competition for resources are essentially identical to those in quantitative genetics under which a single species reaches an equilibrium variance. The potential for evolutionary diversification into a polymorphism was not fully recognized in single-species quantitative-genetic models because sexual recombination maintained a unimodal distribution of genotypes.

Lastly, it is important to stress that the preceding game-theoretic ideas and techniques are often employed under the assumption of asexual reproduction but this need not be the case. Indeed, several studies have used this approach in the context of explicit classical genetic models involving sexual populations with various forms of inheritance, including diploidy and haplodiploidy (Taylor 1989, 1996; Christiansen 1991 and references therein; Eshel, Motro, & Sansone 1997). It is also interesting to note that this game-theoretic approach is closely aligned with more recent developments of Fisher's geometrical model of evolution in which mutations periodically arise and either sweep to fixation or die out (Orr 1998). The chief difference with these recent developments is the focus on making predictions about the distribution of sizes of allelic effects for those mutations that reach fixation (Orr 1998, 2003). These recent models do not involve frequency-dependent selection, however, and it would be interesting to extend them in this direction so that they might be more readily applicable to modelling in evolutionary ecology.

13.4.4 | Adaptive Dynamics

In recent years there has been a flood of interest in modelling ecological-evolutionary feedbacks using a technique that has come to be referred to as adaptive dynamics (Gavrilets & Waxman, in press). Different researchers have different, and often strongly held, opinions about what this approach represents and how it differs from previous theoretical developments. In line with the motivation for the symposium that spawned this volume, in this section I present these recent developments and consider the question of whether they represent a paradigm shift from previous approaches or whether they are simply a refinement and natural extension.

In short, my perspective is that adaptive dynamics as a field of study is best thought of as a natural outgrowth of previous game-theoretic ideas (as they have been applied to evolutionary ecology). The seeds of, and even some of the most fundamental developments in, adaptive dynamics were clearly present in game-theoretic modelling and in some aspects of quantitative-genetic modelling. That is not to say that the developments embodied by adaptive dynamics have not been important; instead, these contributions are better thought of as developments within game theory rather than as a new approach.

One fundamental focal point of adaptive dynamics is on situations in which a trait value is convergence stable but not an ESS (Geritz *et al.* 1998). Such trait values have been given various names within the game-theoretic literature, but adaptive dynamics refers to them explicitly as branching points under the idea that evolutionary branching (e.g., speciation) is favoured by selection at these points (Geritz *et al.* 1998). As already mentioned, it was well known that such points occur within game-theoretic models (Eshel 1983, Taylor 1989), including those for the evolution of traits involved in ecological interactions such as exploitative competition for resources (e.g., see Christiansen 1991 and Abrams *et al.* 1993). It was also appreciated in this literature that such points will tend to lead to some sort of evolutionary diversification, such as a genetic polymorphism (Christiansen 1991). Thus, the existence of such interesting phenomena, as well as their evolutionary significance and implications, is not a finding that can be attributed to developments of adaptive dynamics. There are, however, at least three important developments (in my opinion) that have grown out of this research. I consider each of these in turn.

First, research in adaptive dynamics has demonstrated that branching points might be a general feature of natural systems, because it is common for models of all sorts of ecological interactions, as well as all sorts of traits, to give rise to such phenomena (Doebeli & Dieckmann 2000). Of course, these findings might just as well have been developed within the game-theoretic approach of the previous section simply by examining various models using this approach (and indeed, from a conceptual standpoint, that is what was done). Nevertheless, these developments have been carried out largely by researchers who work under the rubric of adaptive dynamics.

Second, although it was recognized that diversification is favoured by selection at branching points by earlier game-theoretic approaches, and even though some treatments even modelled the initial stages of such diversification, this evolutionary divergence was not the focus of much modelling until the field of adaptive dynamics began to grow (Geritz *et al.* 1998; Waxman & Gavrilets, in press). Again, I would argue that there is nothing distinct in doing this that necessarily warrants giving it a name other than game theory, but these developments have also been carried out largely by researchers working in adaptive dynamics.

Third, and I would argue most significantly, researchers in this field provided a coherent and explicit mathematical underpinning for the somewhat heuristic evolutionary dynamic that Eshel implicitly used (i.e., 13.31). Eshel (1983) and subsequent authors (Taylor 1989, Christiansen 1991, Abrams *et al.* 1993) identified branching points and their evolutionary significance (using a different terminology), but it was research within adaptive dynamics that provided an explicit account of the implicit evolutionary dynamic used in game theory. I briefly review this development here because it turns out to have a simple connection to quantitative-genetic models.

The underlying notion in game-theoretic models is that evolution is a mutation-limited process. The population (or community) reaches demographic equilibrium while it contains only a single phenotype (per species), and then a new mutation arises and either replaces the former resident or dies out. If it replaces the resident, then a new population dynamic equilibrium is attained. At this stage, another mutation arises and the process repeats. As such, these models assume a separation of ecological and evolutionary timescales.

Research in adaptive dynamics provided an explicit model of this process (in continuous time), and it involves two important elements of stochasticity (Dieckmann & Law 1996, Proulx & Day 2001): (1) stochasticity in the mutations that arise and (2) stochasticity in whether or not these mutations reach fixation. The stochasticity in element 1 is probably clear, and the stochasticity in element 2 is meant to reflect that, in real biological populations, even if the new mutant that arises is selectively advantageous, it might still be lost because of chance events when it is initially present in small numbers (Dieckmann & Law 1996).

To begin the derivation, imagine a very large number of independent populations in which this mutation-limited evolutionary process occurs. Each population can be viewed as following a series of successive "jumps" to new resident phenotypes. Each population has its own series of jumps, and populations differ in these patterns solely because of chance in which mutations arise and in whether or not they reach fixation. Let $p(z,t)$ be the frequency distribution of populations with a current resident value of z at time t. The average resident value at time t is therefore $\bar{z}=\int_{-\infty}^{\infty} zp(z,t)dz$, and the rate of change in $\bar{z}$ is given by:

$$\frac{d\bar{z}}{dt}=\int_{-\infty}^{\infty} z\frac{dp(z,t)}{dt}dz. \tag{13.32}$$

We now need to obtain a more explicit expression for the right-hand side of (13.32).

The frequency of the collection of populations that has resident trait value z after a small amount of time, Δt, has passed will be given by what this frequency was initially, plus the frequency of all other types of populations that have moved into that state during this time interval, minus the frequency all populations in that state that have moved to other states in this time interval:

$$p(z,t+\Delta t)=p(z,t)+\int\Omega(z,\tilde{z})\Delta t p(\tilde{z},t)d\tilde{z}-\int\Omega(\tilde{z},z)\Delta t p(z,t)d\tilde{z}, \tag{13.33}$$

where $\Omega(z,\tilde{z})\Delta t$ is the probability that a population in state $\tilde{z}$ moves to state z in the time interval Δt. Rearranging, dividing by Δt, and taking the limit $\Delta t \to 0$ gives:

$$\lim_{\Delta t\to 0}\frac{p(z,t+\Delta t)-p(z,t)}{\Delta t}=\frac{\partial p}{\partial t}=\int\Omega(z,\tilde{z})p(\tilde{z},t)d\tilde{z}-\int\Omega(\tilde{z},z)p(z,t)d\tilde{z}. \tag{13.34}$$

Therefore, equation (13.32) becomes:

$$\begin{aligned}\frac{d\bar{z}}{dt}&=\int_z\int_{\tilde{z}} z\Omega(z,\tilde{z})p(\tilde{z},t)d\tilde{z}dz-\int_z\int_{\tilde{z}} z\Omega(\tilde{z},z)p(z,t)d\tilde{z}dz\\&=\int_z\int_{\tilde{z}}(z-\tilde{z})\Omega(z,\tilde{z})p(\tilde{z},t)d\tilde{z}dz.\end{aligned} \tag{13.35}$$

Now, if you assume that the frequency distribution, $p(z, t)$, is tightly centred around its mean, $\bar{z}$ (analogous to the assumption in going from equation 13.19 to equation 13.20 in quantitative-genetic models), then equation 13.35 can be approximated as follows:

$$\frac{d\bar{z}}{dt} = \int_z (z - \bar{z})\Omega(z, \bar{z})dz. \tag{13.36}$$

To complete the derivation, we then need to be more explicit about the function $\Omega(z, \tilde{z})$. In particular, the probability that a population moves to state z from state $\tilde{z}$ in the time interval Δt is the product of the probability that a mutation occurs in that time interval (denoted by $\rho(\tilde{z})\Delta t$—this might depend on the current trait value, $\tilde{z}$), with the probability that this new mutation has trait value z (denoted by $M(z, \tilde{z})$—this might depend on the current trait value, $\tilde{z}$) and the probability that this new mutation ultimately reaches fixation (denoted by $U(z, \tilde{z})$—this might depend on the resident trait value, $\tilde{z}$):

$$\Omega(z, \tilde{z})\Delta t = \rho(\tilde{z})\Delta t M(z, \tilde{z})U(z, \tilde{z}). \tag{13.37}$$

Now, because we are supposing that all population states are clustered tightly around the mean, $\bar{z}$, you must at least also assume that the allowable mutational jumps are not very large. In this case, the probability density $M(z, \tilde{z})$ must be narrowly clustered around its mean; therefore, you can approximate (13.37) using the first two terms of a Taylor series in z near $\bar{z}$:

$$\Omega(z, \tilde{z})\Delta t \approx \rho(\bar{z})\Delta t M(z, \bar{z})U(\bar{z}, \bar{z}) + \rho(\bar{z})\Delta t M(z, \bar{z})\left.\frac{\partial U}{\partial z}\right|_{z=\tilde{z}=\bar{z}} (z - \bar{z}). \tag{13.38}$$

Substituting this into equation (13.36) gives the final result (Proulx & Day 2001):

$$\frac{d\bar{z}}{dt} = \rho(\bar{z})\mu(\bar{z})U(\bar{z}, \bar{z}) + \rho(\bar{z})\sigma^2(\bar{z})\left.\frac{\partial U}{\partial z}\right|_{z=\tilde{z}=\bar{z}}. \tag{13.39}$$

Here, $\mu(\bar{z})$and $\sigma^2(\bar{z})$ are the mean and variance in the mutational distribution, $M(z, \bar{z})$. Equation (13.39) reveals that the evolutionary change in $\bar{z}$ is the result of two forces: any mutational bias (the first term) and selection (the second term). In the absence of mutational bias we have $\mu(\bar{z}) = 0$, and equation (13.39) simplifies to (Dieckmann & Law 1996, Proulx & Day 2001):

$$\frac{d\bar{z}}{dt} = \rho(\bar{z})\sigma^2(\bar{z})\left.\frac{\partial U}{\partial z}\right|_{z=\hat{z}=\bar{z}}. \qquad (13.40)$$

Notice that equation (13.40) is analogous to Eshel's equation (13.31) and to the approximated quantitative-genetic equation (13.20), where the probability of fixation, $U(z, \bar{z})$, plays the role of the fitness function.

We can obtain even closer correspondence between these modelling approaches if you assume a particular model for the way in which stochasticity affects the probability of fixation. It can be shown that, under a stochastic model based on branching processes (and therefore one in which selectively disadvantageous mutants *never* reach fixation; see Proulx and Day 2001), you have the relationship $U(z, \bar{z}) = (b(z, \bar{z}) - d(z, \bar{z}))/b(z, \bar{z})$, where $b(z, \bar{z})$ and $d(z, \bar{z})$ are the expected birth and death rates of the mutant with trait z in a population with resident trait, $\bar{z}$ (Dieckmann & Law 1996). Using this relationship in equation 13.40 gives:

$$\frac{d\bar{z}}{dt} = V(\bar{z})\left.\frac{\partial \lambda}{\partial z}\right|_{z=\hat{z}=\bar{z}}, \qquad (13.41)$$

where $V(\bar{z}) = \rho(\bar{z})\sigma^2(\bar{z})/b(\bar{z}, \bar{z})$ is a measure of the rate at which genetic variation is introduced into the population through mutation, and $\lambda(z, \bar{z}) = b(z, \bar{z}) - d(z, \bar{z})$ is the per capita growth rate of the mutant (i.e., its fitness) (Dieckmann & Law 1996). This is identical in form to the quantitative-genetic equation (13.20), as well as to Eshel's (1983) equation (13.31), and thus provides an explicit mathematical justification for the evolutionary dynamic used by Eshel in distinguishing evolutionary stability from convergence stability. Notice, however, that unlike the quantitative-genetic models, there is no corresponding equation for the population dynamics because these are assumed to occur on a timescale much faster than evolutionary change. Therefore, the population is assumed to always be in population-dynamic equilibrium.

There have been other developments and elaborations on the adaptive-dynamic approach that take into account multiple species, finite-population sizes, environmental stochasticity, and non-equilibrium attractors (e.g., limit cycles and chaos) for the ecological dynamics, to name just a few (Metz, Nisbet, & Geritz 1992; Ferriere & Fox 1995).

These results have broadened the scope of applicability of this approach (but see Proulx & Day 2001 for a description of some limitations), but I believe it is fair to say that all of these developments are well within the normal scientific development of game theory and do not constitute a fundamentally different approach to theoretical evolutionary ecology. I would even question the need for using a separate (and potentially confusing) new label for these developments. Nevertheless, the adaptive-dynamic approach represents the latest development in theoretical evolutionary ecology, and its results have contributed important and interesting insights to this field.

13.5 | WHERE DO WE STAND? WHERE DO WE GO? IS ANYTHING NEW?

13.5.1 | Future Empirical Directions

The existence of phenotypic values that are evolutionary attractors yet give rise to disruptive selection is one of the most interesting findings to come out of theoretical evolutionary ecology. There are still few explicit tests of such predictions, but numerous opportunities exist for exploring these issues empirically. Some steps have been taken in this direction, with perhaps the most direct attempt being a study involving artificial selection in *Drosophila* (Bolnick 2001; reviewed in Day & Young 2004). This study (2001) did not, however, address the critical prediction of the occurrence of evolutionary diversification (Day & Young 2004).

Interestingly, there have been several experiments carried out for reasons unrelated to this theory that nevertheless provide some of the most relevant data for testing the predictions about such branching points (Travisano & Rainey 2000, Rainey *et al.* 2000, Kassen 2002). Most of these have been conducted using microbial cultures such as *Pseudomonas* or *Escherichia coli.* Such model organisms are ideal for testing this theory because their rapid generation times and well-defined genetic stocks make appropriate evolutionary experiments feasible. Also, because such organisms are asexual, they represent the most conducive systems for finding evolutionary branching; unlike as occurs in quantitative-genetic models, there is no sexual recombination to hinder evolutionary diversification.

Several experiments have been conducted in which a single microbial clone is propagated for several generations in some well-defined resource medium. Although these experiments were not designed to look for branching points, the theory based on exploitative competition outlined earlier predicts that researchers should initially observe an evolutionary adaptation to the highest carrying capacity. At this stage, under some conditions, evolutionary diversification should occur.

Several experiments display this sort of evolutionary diversification. For example, in single strains of *E. coli* propagated in a glucose medium, evolutionary diversification eventually took place, resulting in the stable maintenance of two distinct physiological types (Turner, Souza, & Lenski 1996; Travisano & Rainey 2000). Similarly, in colonies of a single strain of *Pseudomonas* propagated in a complex liquid medium, evolutionary diversification eventually took place, resulting in three well-defined types that appear to coexist indefinitely (the *fuzzy spreader*, *wrinkly spreader*, and *smooth* types; see Rainey & Travisano 1998). These morphs appear to exploit different spatial niches in the liquid medium. Perhaps even more remarkably, these patterns of evolutionary diversification appear to be highly repeatable between experiments.

These results are extremely exciting, and it has been noted that these experiments have inadvertently provided empirical data consistent with these recent theoretical predictions (Travisano & Rainey 2000). Diversification occurred as expected in accord with theory. It still remains, however, to determine if this sort of phenomenon is relevant in organisms other than microbes. There is clearly increasing interest in this issue (Bolnick *et al.* 2003), and there are several documented examples of disruptive natural selection in the wild, but whether these examples are best explained by the sort of endogenously generated selective pressures predicted by the theory remains an open question deserving further study. A preliminary survey of these examples indicates that most do not fit well within this theoretical explanation (Day and Abrams, unpublished results), but more rigorous examinations (and experiments) are required to reach any definitive conclusion.

13.5.2 | Future Theoretical Directions

The preceding microbial examples clearly show that evolutionary diversification in experimental systems occurs; however, it is important to ask whether the ecological interactions embodied in the theory are likely to be the cause of this diversification. Most theory has focused on evolutionary diversification as a result of competition for resources, but is this the primary reason for the diversification seen in microbial systems?

Importantly, facilitation has been well documented in many of the aforementioned microbial experiments. Facilitation is an ecological interaction in which the presence of one species enhances the fitness of another (Whittaker 1977; Bruno, Stachowicz, & Bertness 2003). For example, in the experiments in which *E. coli* diversified during propagation in glucose, it has been demonstrated that the new variants that arise are specialized on acetate, a metabolite produced by the consumption of glucose by the original strain. This is often referred to as cross-feeding in the literature, and it demonstrates that evolutionary diversification in this case occurred primarily as a result of the first species having a facilitative effect on the second through its introduction of additional resources into the environment (Turner, Souza, & Lenski 1996). Similar facilitative interactions are likely important in the *Pseudomonas* system. For example, it has been demonstrated that the *fuzzy spreader* type cannot invade a population of the *wrinkly spreader* type without the third, *smooth* type also being present (Figure 4 in Travisano & Rainey 2000).

These findings contrast with competitive diversification in which the different consumer types do not introduce new resources; rather, they affect the relative value of specializing on the various resources already present. Under facilitation, the addition of new species creates new niches (Levins & Lewontin 1985) and thereby represents a fundamentally different type of ecological interaction that likely plays an important role in evolutionary diversification. To better understand the relative roles of competition and facilitation in diversification, researchers require a clear approach for distinguishing between the two. Currently, there is little theory addressing facilitative diversification in evolutionary ecology (e.g., see Doebeli 2002), but it is likely that the two interactions can be distinguished using relatively simple ecological experiments (Day & Young 2004). Nevertheless, further theoretical results in this area would be invaluable for better dissecting the causes of evolutionary diversification.

One other area requiring further theoretical development is a better characterization of the potential evolutionary outcomes at so-called branching points. Evolutionary splitting is one possibility (even for sexual populations; Dieckmann & Doebeli 1999), but it is not the only outcome or even necessarily a likely one. Other possibilities include a simple increase in genetic variance, the evolution of a within-species polymorphism among age or size classes (Taylor & Day 1997; Day, unpublished results), and the evolution of within-species sexual dimorphism (Bolnick & Doebeli 2003). All of these outcomes effectively fill the available niche space, but the most likely end point of evolution will depend on the specifics of within-species interactions, relative to between-species interactions, coupled with the particular genetic constraints imposed by the system of inheritance for the organism in question.

As seen in the description of the various approaches presented earlier, one chief difference between quantitative-genetic and single-locus models versus game-theoretic and adaptive-dynamic models is that the latter assume a separation of ecological and evolutionary timescales. Ironically, these recent approaches in theoretical evolutionary ecology have returned to using an assumption whose dubious validity was part of the motivation for the development of theoretical evolutionary ecology. An important question remains: what does this separation of timescales do in terms of predictions? For example, are the evolutionary consequences of branching points different if a researcher allows evolutionary change to proceed on a time-scale comparable with ecological change? A powerful way to explore this question is through the use of the quantitative-genetic approximation in equation (13.20) because no restriction on relative timescales is made in its derivation, and it is directly comparable to the evolutionary dynamics for game-theoretic and adaptive-dynamic models (i.e., 13.31 and 13.41, respectively). This is an interesting area deserving further attention.

Finally, recent studies based in adaptive dynamics have demonstrated that sympatric speciation can occur seemingly easily in models of competitive diversification (like that presented earlier) by allowing assortative mating to evolve simultaneously (Dieckmann & Doebeli 1999). To some extent these results contradict earlier suggestions that sympatric speciation is unlikely as a result of tension caused by the buildup of linkage disequilibrium between alleles coding for ecological traits and alleles at other loci coding for mate preferences (Felsenstein 1981). It seems as though these recent models of sympatric speciation should suffer from the same tension because the chief difference from earlier theory on speciation is the inclusion of a mechanism by which the population is intrinsically maintained under disruptive selection (i.e., at a branching point) rather than having disruptive selection imposed on it. Presumably this should have no effect on the extent to which assortative mating (and the requisite linkage disequilibrium) can evolve, so it remains unclear why sympatric speciation appears to occur more readily in this recent theory. Further research examining the relationship between these results and those of earlier theory would help us understand where this difference comes from.

13.5.3 | Conclusions: Déjà Vu or Something New?

As will be clear by now, it is my opinion that the most recent techniques and approaches in theoretical evolutionary ecology do not represent a fundamental change in the way scientists are thinking about and modelling ecological-evolutionary feedbacks. Rather than being a paradigm shift, I believe it represents "normal" science. What it truly fascinating, however, is that the work on single-locus models, quantitative-genetic models, game-theoretic models, and adaptive-dynamic models has proceeded largely independent of one another, but these have led to what is fundamentally the same mathematical description of evolutionary change (compare the progression of results from the various modelling approaches: 13.13B, 13.18B, 13.19, 13.20, 13.31, and 13.41). This suggests that there is something fundamental and robust being described by the different approaches because they have all converged on similar answers from different starting points. Indeed, the main differences in these approaches stem from their difference in the as-sumption of a separation of timescales. Game-theoretic or adaptive-dynamic models make such an assumption, whereas single-locus and quantitative-genetic models do not (and therefore have additional dynamic equations governing the ecological dynamics in conjunction with the evolutionary dynamics). The ongoing union of ecology and evolutionary biology is proving to be a fertile enterprise, and the most important advances in the near future will likely continue to be refinements, developments, and extensions of the important groundwork laid over the past half-century.

REFERENCES

Abrams, P.A. (2001). Modelling the adaptive dynamics of traits involved in inter- and intraspecific interactions: An assessment of three methods. *Ecol. Lett.* **4**, 166–175.

Abrams, P.A., Harada, Y., & Matsuda, H. (1993). On the relationship between quantitative-genetic and ESS models. *Evolution* **47**, 982–985.

Abrams, P.A., Matsuda, H., & Harada, Y. (1993). Evolutionarily unstable fitness maxima and stable fitness minima of continuous traits. *Evol. Ecol.* **7**, 465–487.

Begon, M., Harper, J.L., & Townsend, C.R. (1986). "Ecology." Sinauer Associates, Sunderland, MA.

Bolnick, D.I. (2001). Intraspecific competition favours niche width expansion in *Drosophila* melanogaster. *Nature* **410**, 463–466.

Bolnick, D.I. (2004). Can intraspecific competition drive disruptive selection? An experimental test in natural populations of sticklebacks. *Evolution* **58**, 608–618.

Bolnick, D.I., & Doebeli, M. (2003). Sexual dimorphism and adaptive speciation: Two sides of the same ecological coin. *Evolution* **57**, 2433–2449.

Bolnick, D.I., Svanback, R., Fordyce, J.A. *et al.* (2003). Comparative approaches to intra-population niche variation. *Integrative and Comparative Biology*, **43**, 1078.

Bolnick, D.I., Svanback, R., Fordyce, J.A., Yang, L.H., Davis, J.M., Hulsey, C.D., & Forister, M.L. (2003). The ecology of individuals: Incidence and implications of individual specialization. *Am. Nat.* **161**, 1–28.

Bolnick, D.I., Yang, L.H., Fordyce, J.S. *et al.* (2002). Measuring individual-level resource specialization. *Ecology*, **83**, 2936–2941.

Brown, J.S., & Vincent, T.L. (1987a). Coevolution as an evolutionary game. *Evolution* **41**, 66–79.

Brown, J.S., & Vincent, T.L. (1987b). A theory for the evolutionary game. *Theor. Popul. Biol.* **31**, 140–166.

Brown, W.L.J., & Wilson, E.O. (1956). Character displacement. *Syst. Zool.* **5**, 49–64.

Bruno, J.F., Stachowicz, J.J., & Bertness, M.D. (2003). Inclusion of facilitation into ecological theory. *TREE* **18**, 119–125.

Bulmer, M. (1994). "Theoretical Evolutionary Ecology." Sinauer Associates, Sunderland.

Case, T.J. (2000). "An Illustrated Guide to Theoretical Ecology." Oxford University Press, Oxford.

Charlesworth, B. (1971). Selection in density-regulated populations. *Ecology* **52**, 469–474.

Charlesworth, B. (1990). Optimization models quantitative genetics and mutation. *Evolution* **44**, 520–538.

Christiansen, F.B. (1991). On conditions for evolutionary stability for a continuously varying character. *Am. Nat.* **138**, 37–50.

Cockburn, A. (1991). "An Introduction to Evolutionary Ecology." Blackwell Science, London.

Crow, J.F., & Kimura, M. (1970). "An Introduction to Population Genetics Theory." Harper & Row, New York.

Day, T. (2000). Competition and the effect of spatial resource heterogeneity on evolutionary diversification. *Am. Nat.* **155**, 790–803.

Day, T., & Young, K.A. (2004). Competitive and facilitative evolutionary diversification. *BioScience* **54**, 101–109.

Dieckmann, U., & Doebeli, M. (1999). On the origin of species by sympatric speciation. *Nature* **400**, 354–357.

Dieckmann, U., & Law, R. (1996). The dynamical theory of coevolution: A derivation from stochastic ecological processes. *J. Math. Biol.* **34**, 579–612.

Doebeli, M. (2002). A model for the evolutionary dynamics of cross-feeding polymorphisms in microorganisms. *Pop. Ecol.* **44**, 59–70.

Doebeli, M., & Dieckmann, U. (2000). Evolutionary branching and sympatric speciation caused by different types of ecological interactions. *Am. Nat.* **156**, S77–S101.

Eshel, I. (1983). Evolutionary and continuous stability. *J. Theor. Biol.* **103**, 99–111.

Eshel, I., Motro, U., & Sansone, E. (1997). Continuous stability and evolutionary convergence. *J. Theor. Biol.* **185**, 333–343.

Felsenstein, J. (1981). Skepticism towards Santa Rosalia, or why are there so few kinds of animals? *Evolution* **35**, 124–138.

Ferriere, R., & Fox, G.A. (1995). Chaos and evolution. *TREE* **10**, 480–485.

Fox, C.W., Roff, D.A., & Fairbairn, D.J., Eds. (2001). "Evolutionary Ecology: Concepts and Case Studies." Oxford University Press, Oxford.

Geritz, S.A.H., Kisdi, É., Meszéna, G., & Metz, J.A.J. (1998). Evolutionarily singular strategies and the adaptive growth and branching of the evolutionary tree. *Evol. Ecol.* **12**, 35–57.

Grant, P.R. (1972). Convergent and divergent character displacement. *Biol. J. Linn. Soc.* **4**, 39–68.

Hartl, D.L., & Clark, A.G. (1989). "Principles of Population Genetics." Sinauer Associates, Sunderland.

Hendry, A.P., & Kinnison, M.T. (1999). Perspective: The pace of modern life—Measuring rates of contemporary microevolution. *Evolution* **53**, 1637–1653.

Hofbauer, J., & Sigmund, K. (1988). "The Theory of Evolution and Dynamical Systems." Cambridge University Press, New York.

Iwasa, Y., Pomiankowski, A., & Nee, S. (1991). The evolution of costly mate preferences: II—The "handicap" principle. *Evolution* **45**, 1431–1442.

Kassen, R. (2002). The experimental evolution of specialists, generalists, and the maintenance of diversity. *J. Evol. Biol.* **15**, 173–190.

Kimura, M. (1965). A stochastic model concerning the maintenance of genetic variability in quantitative characters. *Proc. Natl. Acad. Sci. USA* **54**, 731–736.

Lande, R. (1976a). The maintenance of genetic variability by mutation in a polygenic character with linked loci. *Genet. Res.* **26**, 221–235.

Lande, R. (1976b). Natural selection and random genetic drift in phenotypic evolution. *Evolution* **30**, 314–334.

Lande, R. (1979). Quantitative-genetic analysis of multivariate evolution applied to brain: Body size allometry. *Evolution* **33**, 402–416.

Lande, R., & Arnold, S.J. (1983). The measurement of selection on correlated characters. *Evolution* **37**, 1210–1226.

Lawlor, L.R., & Maynard Smith, J. (1976). Coevolution and stability of competing species. *Am. Nat.* **110**, 79–99.

Levins, R., & Lewontin, R. (1985). "The Dialectical Biologist." Harvard University Press, Cambridge, MA.

MacArthur, R.H. (1970). Species packing and competitive equilibrium for many species. *Theor. Popul. Biol.* **1**, 1–11.

MacArthur, R.H. (1972). "Geographical Ecology." Princeton University Press, Princeton, NJ.

Maynard Smith, J. (1978). Optimization theory in evolution. *Annu. Rev. Ecol. Syst.* **9**, 31–56.

Maynard Smith, J. (1982). "Evolution and the Theory of Games." Cambridge University Press, Cambridge.

Maynard Smith, J., & Price, G.R. (1973). The logic of animal conflict. *Nature* **246**, 15–18.

Metz, J.A.J., Nisbet, R.M., & Geritz, S.A.H. (1992). How should we define "fitness" for general ecological scenarios? *TREE* **7**, 198–202.

Orr, H.A. (1998). The population genetics of adaptation: The distribution of factors fixed during adaptive evolution. *Evolution* **52**, 935–949.

Orr, H.A. (2003). The distribution of fitness effects among beneficial mutations. *Genetics* **163**, 1519–1526.

Parker, G.A., & Maynard Smith, J. (1990). Optimality theory in evolutionary biology. *Nature* **348**, 27–33.

Pianka, E.R. (1974). "Evolutionary Ecology." Harper & Row, New York.

Proulx, S.R., & Day, T. (2001). What can invasion analyses tell us about evolution under stochasticity in finite populations? *Selection* **2**, 1–15.

Provine, W.B. (2001). "Origins of Theoretical Population Genetics." University of Chicago Press, Chicago.

Rainey, P.B., Buckling, A., Kassen, R., & Travisano, M. (2000). The emergence and maintenance of diversity: insights from experimental bacterial populations. *TREE* **15**, 243–247.

Rainey, P.B., & Travisano, M. (1998). Adaptive radiation in a heterogeneous environment. *Nature* **394**, 69–72.

Real, L.A. (1994). "Ecological Genetics." Princeton University Press, Princeton, NJ.

Reed, J., & Stenseth, N.C. (1984). On evolutionarily stable strategies. *J. Theor. Biol.* **108**, 491–508.

Roughgarden, J. (1971). Density-dependent natural selection. *Ecology* **52**, 453–468.

Roughgarden, J. (1972). Evolution of niche width. *Am. Nat.* **106**, 683–718.

Roughgarden, J. (1983). The theory of coevolution. *In* "Coevolution" (D. Futuyma & M. Slatkin, Eds.). Sinauer Associates, Sunderland, pp. 33–64.

Roughgarden, J. (1996). "Theory of Population Genetics and Evolutionary Ecology: An Introduction." Prentice Hall, Upper Saddle River, NJ.

Slatkin, M. (1980). Ecological character displacement. *Ecology* **61**, 163–177.

Taper, M.L., & Case, T.J. (1992). Models of character displacement and the theoretical robustness of taxon cycles. *Evolution* **46**, 317–333.

Taylor, P.D. (1989). Evolutionary stability in one-parameter models under weak selection. *Theor. Popul. Biol.* **36**, 125–143.

Taylor, P.D. (1996). The selection differential in quantitative genetics and ESS models. *Evolution* **50**, 2106–2110.

Taylor, P.D., & Day, T. (1997). Evolutionary stability under the replicator and the gradient dynamics. *Evol. Ecol.* **11**, 579–590.

Travisano, M., & Rainey, P.B. (2000). Studies of adaptive radiation using model microbial systems. *Am. Nat.* **156**, S35–S44.

Turner, P.E., Souza, V., & Lenski, R.E. (1996). Tests of ecological mechanisms promoting the stable coexistence of two bacterial genotypes. *Ecology* **77**, 2119–2129.

Vincent, T.L., Cohen, Y., & Brown, J.S. (1993). Evolution via strategy dynamics. *Theor. Popul. Biol.* **44**, 149–176.

Waxman, D., & Gavrilets, S. (2005). Target review: 20 questions on adaptive dynamics. *J. Evol. Biol.*, in press.

Werner, T.K., & Sherry, T.W. (1987). Behavioral feeding specialization in *Pinaroloixas inornata*, the "Darwin's Finch" of Cocos Island, Costa Rica. *Proc. Natl. Acad. Sci. USA* **84**, 5506–5510.

Whittaker, R.H. (1977). Evolution of species diversity in land communities. *Evol. Biol.* **10**, 1–67.

Wright, S. (1935). Evolution in populations in approximate equilibrium. *J. Genetics* **30**, 257–266.

Wright, S. (1969). "Evolution and the Genetics of Populations: The Theory of Gene Frequency," Vol. 2. University of Chicago Press, Chicago.

14 | THE ELUSIVE SYNTHESIS

Kim Sterelny

14.1 | SOURCE AND CONSEQUENCE LAWS

In *The Nature of Selection*, Elliott Sober distinguishes between evolutionary theory's *source laws* and *consequence laws* (Sober 1984). Much of evolutionary theory, and in particular population genetics, has typically been concerned with the *consequences* of fitness differences. Population-genetic models specify the changes in populations of genes and organisms that are the consequences of fitness differences at the initial state of the system. These models probe the effects of population structure, genetic variation, migration, mating systems, and fitness differentials on evolutionary trajectories. Day (Chapter 13) reviews this tradition, showing how these models have been able to successively drop idealizations about populations and their structure. There has been the development of models with more realistic assumptions about genotype-phenotype relations; models of multispecies interactions, models of the effects of frequency dependence, and models of fitness effects that prevent the establishment of strategies that would be stable were they established. Nonetheless, the ecological parameters of these models—demographic growth rates, carrying capacity, and fitness values—are assumed by the models rather than explained within them. So these are consequence models. Work in this tradition abstracts from the causes of fitness differences, taking them as given. But without an account of the source of these fitness differences, evolutionary biology is radically incomplete. Arguably, its status as a genuinely empirical science is at risk. From one view, that status depends on the distinction between expected and actual fitness. If we have no way of specifying fitness differences except by measuring actual survival and reproduction, we cannot explain differential success by appealing to differential fitness. Fitness would not be an explanatory or even a predictive property of organisms or genes. Evolutionary theory needs source laws explaining the bases of fitness differences and consequence laws explaining their upshot.

The obvious home of source laws is ecology. Some fitness differences among organisms are not specifically tied to their environmental situations: selection will penalize genes that compromise basic metabolic or reproductive functions in almost any circumstances. But in many cases it is the relations between an individual organism and other organisms in the same population, its relations with members of different populations, and its relations to the abiotic environment that explain why some phenotypic traits confer a fitness advantage and others are handicaps. For example, when populations of large mammals are isolated by sea level changes, there is a persistent evolutionary trend for them to evolve to much smaller sizes. That is explained by the changed ecology they find themselves in and in particular by the sharp limits on food supply such populations face. As Holt notes, although Dobzhansky was surely right to insist that nothing in biology makes sense except in the light of evolution, likewise nothing in evolutionary biology makes sense except in the light of ecology (Chapter 12). This is not news. Darwin ended *On the Origin of Species* with a wonderfully vivid and justly famous description of the ecological interactions that drive evolutionary change (Darwin 1964, pp. 489–490):

> It is interesting to contemplate an entangled bank, clothed with many plants of many kinds, with birds singing on the bushes, with various insects flitting about, and with worms crawling through the damp earth, and to reflect that these elaborately constructed forms, so different from each other, and dependent on each other in so complex a manner, have all been produced by laws acting around us. These laws, taken in the largest sense, being Growth with Reproduction; Inheritance which is almost implied by reproduction; Variability from the indirect and direct action of the external conditions of life, and from use and disuse; a Ratio of Increase so high as to lead to a Struggle for Life, and as a consequence to Natural Selection, entailing Divergence of Character and the Extinction of less-improved forms. Thus, from the war of nature, from famine and death, the most exalted object which we are capable of conceiving, namely, the production of the higher animals, directly follows.

Evolutionary biology needs ecology to provide a systematic account of the fitness differences on which selection acts; likewise, community ecology, in seeking to explain the composition of local communities and the abundance of its members, needs evolutionary biology. It does so partly for the obvious reason that the characteristics that explain the fate of population in local communities have evolutionary explanations. Moreover, as Holt points out, evolutionary biology is important to community ecology because the timescale of evolutionary change overlaps the timescale of ecological interaction. Within the one community, there are organisms whose generation roll over in years to tens of years, and others—the myriads of microbes and small invertebrates—that turn over their generations much faster. These are evolving when the larger and longer-lived organisms are merely reproducing. Host-pathogen interactions are ecological, affecting both the presence and the abundance of pathogen and host, and evolutionary, changing the phenotypes of pathogens and host, as the rabbit history of Australia indicates. So if researchers are interested in (for example) the population dynamics of the mountain ash *Eucalyptus regnans* (the world's tallest hardwood), they cannot treat the phenotypes of those populations with which it interacts as fixed. Likewise, in considering disturbance and other nonequilibrium processes, the evolutionary characteristics of a population—population-level variation and individual plasticity—are often crucial. Migration from outside can have purely demographic rescue effects. But it can have evolutionary effects, too, allowing a local population access to genetic variation in the metapopulation of which it is a part. Just as we cannot think of populations as phenotypically fixed, in such circumstances we cannot think of them as phenotypically uniform, either. Holt's chapter is packed with examples showing the interpenetration of ecological and evolutionary timescales and the ecological importance of within-population variation (Chapter 12).

So, evolutionary biology and community ecology need one another. Yet the integration of evolutionary biology with community ecology has been at best patchy and partial. Much of evolutionary biology has simply idealized away from the problem of identifying fitness differences, although there are deservedly famous exceptions (Grant 1986, Schluter & Price 1993, Grant & Grant 1995, Schluter 1996). Equally, Holt's brief review of episodes from the history of community ecology shows how patchily evolutionary theory has penetrated ecology (the same message comes through from more systematic historical treatments; see Kingsland 1985 and Golley 1993). The normal factors of tractability and simplification partly explain the tendency of ecologists to bracket off evolutionary considerations. If you can, it is much simpler to treat populations as fixed and uniform, and this is sometimes a legitimate idealization. But there are more fundamental factors that have made a true synthesis of community ecology and evolutionary biology hard to build.

In the next section, I outline a standard community ecology picture of the integration of ecology and evolution. They are unified through an equilibrium model of local communities, a model in which competition plays the key role. On ecological timeframes, competition explains the presence and abundance of species in local communities and determines the selective environment experienced by the members of populations in such communities, and hence their relative fitness. If a population of rabbits is controlled by a shortage of safe burrows in some local community, the fittest rabbits will be those who compete best for the burrows. This picture faces three major challenges, and meeting those challenges changes the picture both of ecology and of its relation to evolutionary biology.

14.2 | THE LIMITS OF EQUILIBRIUM

Equilibrium models have never been an unchallenged paradigm in community ecology. Greg Cooper has shown that there have always been sceptics: thus, the classical equilibrium views of Nicholson and Lack found a strong sceptical response in those of Andrewartha and Birch. But he also shows that there have been times in the history of community ecology when equilibrium models of community organization were dominant (Cooper 2001, 2003). According to these conceptions, membership and abundance were held close to equilibrium by density-dependent interactions and especially by competition. According to this idea, unless communities are regulated, their obvious year-by-year qualitative stability would be miraculous. They are stable: for the most part, a species found one year will be found the next; species among the most abundant one year will be so the next. Stable patterns need explanation, because disturbances act independently of abundance. The force of a storm blowing through a nesting colony of herons is independent of the number of nests present; hence, the mortality it causes is density independent. Thus, local communities are regulated: the forces that act on local populations are sensitive to the size of those populations, affecting them more strongly when they are larger and less strongly when they are smaller. In local communities in the Canberra region, three species of parrots are often found together: eastern rosellas, red-rumped parrots, and crimson rosellas. These all live in open woodland environments and they all nest in tree hollows—a relatively rare resource (especially because other elements of the fauna use them) and a relatively fixed resource (because eucalypts develop suitable hollows slowly). Competition for this resource imposes a stable selective pressure on these local parrot species (thus, it may be no coincidence that interspecific competition is damped down by size differences; see Chapter 13). So, the thought goes, regulation defines a relatively stable selective environment that acts on each member of the community. Competition is crucial at ecological timescales in regulating communities and on evolutionary timescales in selecting phenotypes.

In this picture, competition has this dual role of structuring local communities and defining selective environments. Thus, these equilibrium-based models of the organization of local communities fit naturally with a particular version of evolutionary biology—namely, the version of evolutionary biology that takes adaptation to be a population's *accommodation* to its environment. On this view, the change in a population over time is an evolutionary response to the demands the environment makes on that population. The environment has causal effect on the lineage but not vice versa. Features of the environment explain features of the organism through selection. Organisms are shaped to their environments. The stabilized structure of communities is the basis of the source laws evolutionary biology needs. Stable patterns of interactions within a particular community explain why, say, rather dull male sticklebacks are overall fitter than brighter (hence, more vulnerable) males, despite female preference for those brighter fish (Schluter & Price 1993).

This may never have been a tenable view of the ecology-evolution interface because, as Kingsland and Cooper make clear, there has always been a strand of ecological thought that insisted on the importance of external disturbance and on the exceptions to stability (Kingsland 1985, Cooper 2003). But even if it was once a sound picture of the interface between ecology and evolution, it is so no longer. I shall describe three pressure points for competitively regulated community conception of this interface.

14.2.1 | The Grain Problem

The central project of evolutionary biology is to explain speciation and adaptation. Ecological interaction in local communities does not explain speciation and adaptation; ecological interaction and evolutionary response take place at different spatial scales. There is a disjunction between the spatial scale of local ecological interaction and the spatial scale of entrenched evolutionary change. Local populations do adapt to their specific circumstances, but local populations are often short lived, being fused with others through migration and local ecological change and thus losing their distinctive gene complexes. Their distinctive evolutionary responses are entrenched only when these populations acquire isolating mechanisms. Thus, the unit of evolution is the species. Yet species are typically distributed through many different local communities; hence, the local populations into which a species is divided are often subject to different selective environments. Some sticklebacks live in clear and predator-infested waters, whereas others are not subject to predation or are subject to predation under circumstances less favourable for the predators. For widely distributed species, the variation in environmental circumstances can be extreme. The common brushtail possum is found in cool temperate New Zealand rain forests, inner Sydney suburban gardens, and eucalypt woodlands. There is no single set of selection pressures acting on the possum population as a whole, and although these possums are unusually widespread in space and circumstance, many species are geographical and ecological mosaics. As Eldredge in particular has argued, species are typically ecologically fractured (see Eldredge 1989, 1995, 2003 and Sterelny 1999, 2001). So although competition may structure local communities, equilibrating them, competition-driven events in a local community do not explain the evolutionary trajectories of the species represented in those communities.

This is one significant difference between the perspective of this chapter and that of Holt (Chapter 12). He thinks the key problem is the contingency of local systems. Their dynamics are sensitive to so many factors that community change is difficult to predict. No doubt it is true there are no relatively simple general models of the trajectory of local communities. But even if there were predictive models of local communities, they would not be at the right scale to understand evolutionary change. Conserved evolutionary change takes place in metapopulations, not in local communities. It is the scale mismatch between community ecology and evolutionary biology that makes the local equilibrium model fundamentally limited for integrating evolutionary biology with ecology, rather than the empirical intractability of community ecology.

14.2.2 | Organisms Do Not Merely Experience Environments, They Change Them

Much evolutionary biology takes as its working assumption the idea that environments are independent of the lineages evolving within them. Thus, the Australian climate became hotter and drier for reasons independent of the Australian biota. That biota had to adapt to the new conditions or go extinct. Although some examples fit this model of lineages responding to a fixed or independently changing environment, many do not. At neither ecological nor evolutionary timescales do organisms merely respond to their environment. Organisms are not intentional or deliberative agents, for the most part, but they are agents. They interact with their environments, sometimes changing them to their advantage. Organisms may affect their environment in profound and surprising ways: thus Hamilton has suggested that clouds might be mechanisms made by microbes to assist in their own dispersal (Hamilton & Lenton 1998). Neither the abiotic features of the local environment nor the characteristics of other organisms should be regarded as ecological filters that sieve community membership independently of and before the organism's own activities. Although in principle almost everyone concedes this point, in practice biologists continue to idealize from the active role of organisms as agents. For example, community formation is often conceptualized as the outcome of a series of filters. First, the regional species pool specifies a set of potential community members. Second, that set is filtered by the abiotic features of a particular patch of habitat, which specifies the set of potential members of a community at that patch. Then biological interactions among those potential members exclude others, and this further filter determines the actual community pool (see Ricklefs & Schulter 1993 and Belyea & Lancaster 1999). This conceptual model idealizes both from the active role of organisms in making their world and from the evolutionary responses of populations to their local circumstances: it treats both environment and phenotype as fixed. Competition-driven equilibrium models are limited in ecology by understating the range of potentially structuring interactions among organisms. They are limited in evolutionary biology by assuming that the environment is a fixed background against which lineages evolve.

14.2.3 | Ecological Agents

If ecology is about organisms and populations, then the agents that drive ecological processes are also agents visible to selection. Those agents are individual organisms. But within ecology, individualism is a controversial doctrine. On some views, there are genuine ecological systems, and ecological theory is partly about the characteristics and behaviour of these ecological systems. For example, according to the diversity-stability hypothesis there are ensemble properties of communities that are causally important: species-rich communities are less sensitive to disturbance than species-poor ones. On this view, communities are real systems; they are not mere aggregates or mixtures of the species that happen to live in association in a particular habitat patch. But they are not *evolutionary units*. Communities do not form lineages (except perhaps in odd cases; see Wilson 1997). They do not found descendant lineages in their own image. Their components are evolutionary units, but they are not. So to the extent that ecological systems drive ecological processes, there is an extra layer of complexity in the ecology–evolution feedback. Interactions among organisms in a community, and between those organisms and their physical environment, will affect evolutionary units indirectly through their effect on the ecological systems of which evolutionary units are parts. Likewise, evolution will affect ecological processes indirectly: the evolved characteristics of the components from which ecological systems are built will constrain the behaviour of the system. Holt calls this the holism or top-down problem. He especially has in mind ecosystem ecology, which is strongly committed to the existence of distinctive ecological systems responsible for cycling matter and energy through the habitat. It is not surprising that, as Holt notes, Odum's text is light on evolutionary themes. Odum was an ecosystem ecologist; hence, the systems of interest to him were not evolutionary units. It will be difficult to integrate evolutionary biology and ecology if the two branches of biology are about different biological systems.

14.3 | THE GRAIN PROBLEM AND ITS MACROECOLOGICAL SOLUTION

The picture of local interaction determining the selective landscape of a local population (hence, explaining local fitness values) presupposes *local determinism*, as Robert Ricklefs calls it: community membership and abundance are determined by features of the local environment. Community ecology has typically taken community makeup to be a response to local properties (Ricklefs 2004). In general, even if communities are regulated, they are not autonomous in this way: there are important outside-inside influences on local communities (Gaston & Blackburn 1999, Lawton 1999). To the extent that regional influences affect local ecological interactions, the scale of ecological interaction becomes commensurate with that of speciation and adaptation. So the program of macroecology is important not just within ecology but also for the interface between ecology and evolution. One strand of macroecology emphasizes the influence of regional features on local communities. The program of macroecology offers one way of forging the links between ecology and evolutionary biology. If macroecologists are successful in finding robust causal factors operating over landscapes or regions, then the grain problem is less pressing. All or most of the communities through which a particular species is spread will be affected—and affected in a similar way. If, for example, species found at higher latitudes really have larger ranges, then regional factors must play a salient role in determining presence and abundance in many local communities. Gaston and Blackburn try to carry through a program of this kind in their macroecological analysis of the avifauna of Britain. Their local site is a wood, the Eastern Wood of Bookham Common, and they argue that (for example) the species richness of this local site is mostly to be understood in regional terms. Local factors matter—the size and the basic habitat structure of the wood are important. But the main variable is regional species richness and the access that regional pool has to Eastern Wood. They even interpret site size in relation to these regional factors: larger woods are easier targets for animals dispersing across a landscape, one reason size correlates positively with richness (Gaston & Blackburn 2000). Along similar lines, Ricklefs and others argue that community richness is mostly determined by external coarse-grained factors. Richness is not determined by the intricate pattern of biotic relations within a community. Rather, it is driven by regional species richness—the set of species available to make up the community—and the physical features of the habitat (Ricklefs 1989, Zobel 1997).

Lawton develops a similar picture. He distinguishes between *type I* and *type II* communities, arguing that if local determinism were true, communities even in rich regions would become saturated. The pattern of local activity—competition, predation, herbivory, and other crucial density-dependent interactions—would determine an upper bound for the community. Once a community had reached its richness limit, it would be full. Because this upper bound is set locally, it is independent of regional species richness. Communities with such internally determined limits are type II communities. To the extent that communities are not highly regulated by local interactions, local richness will increasingly be proportional to regional richness (Lawton 1999). Gaston and Blackburn argue that at least for avifauna, there is little evidence of saturation. In their view, although local events play a role, communities are closer to the type I end of the spectrum. Other theorists are more concerned with developing a balanced view of the role of local and regional factors. For example, Brown (with various co-workers) has developed models suggesting that species richness is itself an equilibrated characteristic of local communities. In his model, local factors are crucial: equilibration depends on the system as a whole being resource limited, so extinctions (say, as climate changes) open resource gaps that are likely to be filled. But it also depends on the local community being embedded in a rich regional community—one rich enough to contain species tolerant of new conditions as local environments change and to be able to use the full range of resources local communities offer (Brown *et al.* 2001, Ernest & Brown 2001).

The extent to which regional influences determine crucial local features is clearly an open empirical issue and likely varies by region, time, and kind of factor. *A priori*, a factor like widespread climate change seems likely to affect a large number of different communities in qualitatively similar ways, imposing similar pressures on most of the populations out of which species are built. This expectation gets some support from Whittaker's attempt to predict the basic shape of plain communities from simple climatic factors. He argues that the formation of plant community types and their geographic distribution is largely controlled by mean annual rainfall and temperature. Furthermore, the members of these communities are often adapted to these coarse climatic variables rather than to specific features of particular communities. Desert plants wherever they are found have a similar suite of physiological adaptations. Thus, they often have deep and wide-ranging roots for effective water uptake, water storage tissues, and leaves adapted to reduce water loss (Whittaker 1975, pp. 167–169). Likewise, such large-scale and pervasive biotic changes as those involved in the Great American Interchange, when the biota of North and South America mixed, would have induced similar invasive changes in many communities. The success of this form of macroecology is not yet known, but to the extent to which these regional models of ecological effects are vindicated, they improve the fit between the spatial scale of ecological processes and that of evolutionary ones. To the extent that fitness landscapes are shaped the same way in a series of local communities by regional factors, then although species are metapopulation ensembles, each component of the metapopulation will be under similar selective pressures and is likely to be evolving in parallel. Thus, patterns at regional scales are likely to have evolutionary upshots.

14.4 | NICHE CONSTRUCTION AND ITS CONSEQUENCES

On the simplest picture of the ecology-evolution interface, ecological interactions determine the fitness values of phenotypes. The fitness values of phenotypes affect the rate of replication of genes for phenotypic characteristics; hence, they drive changes in population gene pools over time. On this simplest model, evolutionary change is the downstream consequence of new fitness values set by ecological interactions. The environment changes, the community finds its way to a new equilibrium, and the population phenotypes finds their way to new local optima—without those ecological and evolutionary responses having reciprocal effects on the environment. A new species migrates in, or the environmental changes, and the fitness values associated with existing phenotypes (and genotypes) change. The consequence is an environmental trajectory to a new equilibrium, a trajectory driven by the new selective environment. That selective environment is not itself changed as lineages respond to it. The idealizations involved in this conception have always been recognized, and in the special case of social behaviour, there is no assumption that the environment is independent of the lineage. It is obvious that the fitness of many traits is frequency dependent. Day (Chapter 13) shows that evolutionary game theory—originally developed to model such paradigmatically social behaviours as cooperation—can been extended to other ecological interactions. For example, foraging has a frequency-dependant aspect. All else equal, the more my foraging diverges from other agents, the more abundant my resource target will be.

The niche construction perspective, however, rejects the assumption that the environment is independent of the evolving lineage in a much more fundamental way. For one thing, it makes facilitation and other non-competitive interactions more central to the ecology-evolution interface, and these change the environment as they themselves change. Classic community ecology has often treated competition as the crucial interaction-structuring communities. But populations within a community can be linked through niche construction networks. Litter recycling is the cleanest example. Plants produce litter as a by-product of their life: fallen leaves, twigs, bark. A host of organisms live by consuming the litter, and as a consequence of these actions, they return crucial materials to the soil. This is absorbed by the vegetation, which in turn produces more litter (Odling-Smee, Laland, & Feldman 2003, pp. 318–322). One population can influence another by changing important features of the physical environment. Trees buffer the wind, modulating the effect of storms and providing shelter to many organisms (Jones, Lawton, & Shachak 1997). In many ways the resource envelope is biotically regulated: think of nitrogen fixation, soil stabilization, and litter recycling. Soils are made by organisms. Yet it is unmade by them, too. Competition for resources might degrade those resources: cropping grasses beyond the point of recovery or eroding soils. Populations act on one another through the physical changes they induce. As Odling-Smee and his colleagues argue, an important benefit to ecological theory of this change in perspective is a (partial) unification of community ecology with ecosystem ecology. The distinction between community ecology and ecosystem ecology begins to break down once community ecologists recognize the niche-constructing role of organisms and populations (Odling-Smee, Laland, & Feldman 2003). Organisms do not just eat, breed, and die; they also reorganize their environment. Hence, an explanation of the presence, abundance, and activities of local populations will also explain the biotically caused flow of materials and energies through that local system.

Perhaps even more fundamentally, a consequence of the niche construction framework is a realization that change does not result in one-step alteration of fitness values whose effects then ramify through the populations of a community. Organisms respond to change by altering environments, phenotypes, and fitness values. Thus, to take the apparently simplest case, abiotic factors do not influence fitness independently of biological factors: rather, their importance is often a consequence of biological response on ecological and evolutionary time frames. A particular plant population may be exposed to frost or storm-based mortality because competition has restricted it to exposed habitat. Many birds nest on cliffs and other physically exposed places to avoid the even more unwelcome attention of predators. Organisms change the physical conditions in which they live and impose a biased sample of the physical environment on others. In short, ecological and evolutionary responses to local environments change these environments and thereby change the fitness consequences of agent's phenotypes.

Lewontin deserves credit for being the first to explicitly point to the importance of organism-environment interactions for both ecology and evolutionary biology (see Levins & Lewontin 1980 and Lewontin 1982, 1983). But these ideas have been taken much further by those working on niche construction (as its called in evolutionary biology) and ecological engineering (as its called in ecology). In particular, they have pointed out that this interaction between organism and environment sometimes results in a second inheritance mechanism. Many organisms engineer the developmental environments of their offspring. For this reason, Odling-Smee and his colleagues regard niche construction as a system of ecological inheritance. To the extent that organisms control the environment of their young, the environment can become a stable, repeatable background context in which varying developmental inputs can consistently have varying effects. This allows the selection of one in favour of another. Parental ecological engineering systematizes developmental contexts and allows rival alleles to be exposed to selection. Genes have more consistent phenotypic effects because developmental environments will be more standardized. For the same reason, their fitness consequences will be more consistent. If nests, trees hollows, beach scrapes, or cliff edges are reliably part of the causal context in which genes influence development, then the phenotypic differences those distinct environments induce will be visible to selection. The alleles that generate phenotypic differences in those environments will do so typically rather than occasionally (see Jones, Lawton, & Shachak 1997 and Odling-Smee, Laland, & Feldman 2003). For a review, see Sterelny (in press).

In summary, it seems to me that, in treating organisms as actively involved in the creation of their own world rather than as merely accommodating to the world in which they are born, an enormously important change in perspective has taken place simultaneously in both ecology and evolutionary biology. In ecology, the consequence of this perspective change includes the recognition of a much richer set of interactions that organize communities and improved prospects for integrating community ecology with ecosystem ecology. In evolutionary biology, niche construction changes both our conception of selective landscapes and our conception of a key evolutionary mechanism: heritability. The term *paradigm shift* is both vague and subjective. But as I understand it, a paradigm is a set of fundamental, interconnected assumptions about the basic causal processes that characterize a domain. So understood, the two shifts, from local determinism to a regional perspective and from seeing organisms as accommodating their world to seeing them as partly makers of that world, are jointly a good candidate for a paradigm change in ecology, one linked to a fundamental change in evolutionary biology.

14.5 | THE EMERGENT PROPERTY HYPOTHESIS

So far, the news has been good. The local equilibrium concept is not a good model of the ecology-evolution interface. But transformation of the spatial scale of ecological focus, and recognition of the active role of organisms as agents, seems well motivated in both ecology and evolutionary biology. That double shift is a promising basis both of an ecological account of evolutionary biology's source laws and of the importance of evolutionary change to ecological process. For example, the niche construction perspective emphasizes the importance of coevolutionary interactions in ecology—such as among the various members of the litter recycling functional group.

In the final section of this chapter, I want to describe a serious complication for this picture: Holt's top-down concern. I have so far assumed that evolutionary biology and ecological theory both characterize the same agents. Detrivore populations both recycle and evolve, and each process modifies the other. But that may be mistaken. Organisms and populations certainly play important roles in ecological activity. But they may not be the only agents that drive ecological processes. There may be distinctive ecological systems: composite ecological agents whose properties drive ecological processes. In particular, local communities may be more than composites or aggregates; they may be ecological agents. That is, these systems may have properties, properties that are causally important in driving ecological and hence evolutionary processes. I shall describe this idea by exploring a family of famous hypotheses that link the diversity of a community to its stability. The crucial claim of the emergent property hypothesis, as I see it, is that these emergent properties are causally important: they drive ecological processes. It is not enough, in other words, to identify or define ensemble properties of systems, such properties as biomass, trophic complexity, or species richness. It must also be shown that these ensemble properties are causally important. The diversity-stability hypothesis is one attempt to show this.

The idea that diversity adds stability to a community or ecosystem has enormous intuitive plausibility. Diversity adds redundancy and hence allows that community to survive fluctuations in the fortunes of its members. If only one population in Canberra's Black Mountain community pollinates the gum tree, *Eucalyptus rossii,* and if those pollinators were to suffer serious local declines, *Eucalyptus rossii* would be in trouble, unable to recruit new plants into its population. However, if there is a suite of eucalyptus pollinators, such a fluctuation would not ramify through the community. Redundancy buffers disturbance, and diversity adds redundancy. Despite the appeal of this picture, it was apparently undermined by theoretical work by Robert May, who showed that *particular populations* are less stable in more diverse communities (May 1973). However, Tilman and others have argued that *community-level properties* are more stable in more diverse communities—in particular, total productivity. The biomass of more diverse communities is more stable than that of less diverse ones. Community-level properties stabilized by diversity include total biomass, increased resistance to invasion, and, in some views, biodiversity itself. Tilman's crucial theoretical idea is that of compensation. If one population declines, another population, using somewhat similar resources, expands and hence stabilizes the overall productivity of the community. Importantly, the idea that populations compensate for one another's fluctuations does not depend on strong, controversial ecological assumptions. Community-level stability occurs despite population-level volatility because individual populations have somewhat overlapping resource requirements but different environmental tolerances. These tolerance differences explain why populations fluctuate out of synchrony, although the resource overlaps explain how one population can expand as a result of another contracting (and by contracting leave resources unused). The overall effect is to partially stabilize the overall productivity of the community. Compensation so understood requires only weak biological assumptions: it does not demand that the expanding population causes the contracting population to contract; hence, it does not presuppose strong competitive interactions in the communities. This covariance effect depends only on resource overlap. Compensation leads to negative covariation between two populations, which in turn leads to stabilization (Lehman & Tilman 2000, p. 536). There is empirical evidence that supports this cluster of ideas, too. Species-rich plots resisted drought better, overall biomass varied less in species-rich plots, and species-rich plots returned to the predrought biomass more rapidly than species-poor plots (Tilman 1996). Tilman's own empirical work concentrates on Minnesota grassland plots, but he thinks African data leads to similar conclusions (Tilman 1996, 1999).

The case for ecological systems looks promising. The diversity-stability hypothesis has empirical support and is based on undemanding theoretical assumptions. There are, however, problems. The empirical case for the connection between diversity and stability is not so strong at second glance. It mostly depends on measuring the plants within a community. It is their biomass that is measured. There are serious doubts about whether this ensemble relationship holds when considering the interactions between plants and animals and those between animals. When attention shifts to herbivores and those that eat them, resource exploitation efficiency is by no means an obvious stabilizing mechanism. On the contrary, enhanced resource use can cause overexploitation; hence, productivity collapses (Loreau *et al.* 2001, p. 807). In short, it is not a settled fact of descriptive ecology that more diverse communities are more stable (see also McCann, Hastings, & Huxel 1998 and Naeem 2002).

There is a second reason for scepticism about the diversity-stability hypothesis. Even if more diverse communities are more stable, it is not clear that they are more stable *because* they are more diverse. Diversity may be a symptom of causally relevant properties of individual populations rather than a causally important property of ensembles. Let me explain. Suppose stability depends on redundancy effects. The overall productivity of the community depends on a set of key processes. These include the acquisition of energy by primary producers, the flow of minerals to and from the abiotic substrate, the decomposition by the detrivores, and the flow of organic material from organisms to organisms through predation, herbivory, and similar activities. Ecosystem function depends on these key processes, and ecosystems that are more diverse, and hence have a variety of species with different tolerances that can compensate for one another while driving these processes, are thereby more stable. If this is the right story, diversity itself is genuinely causally important (Naeem 1998).

However, there is an alternative possibility: the *sampling effect*. Species-rich communities have more tickets in the relevant biological lotteries. Thus, a diverse community is more likely to have species that resist disturbance. They are more likely to have drought-resistant members; likewise, rich Serengeti plots are more likely to contain grazing-resistant members. More generally, to the extent that resistance to disturbance, productivity, resistance to invasion, and other community properties depend on the presence of *specific taxa* from the regional species pool, richer communities are more likely to manifest these properties just because they have more tickets in the relevant lottery. Diversity-stability relationships may be explained individualistically by the responses of individual organisms to their environment (Wardle 1999). Tilman notes, "Diversity itself is both a measure of the chance of having certain species present in a system and a measure of the variation in species traits in an ecosystem" (1999, p. 1470). The second of these properties—variation in traits in the community—is a genuine property of the community. Thus, if a rich community has *a set of species* that can use the available resource envelope with maximal efficiency and is more able to resist invasion by lowering the level of resources below the point at which potential invaders can sustain a viable population, then variation in the community is causally crucial to its resistance to invasion. This is not true if diversity just increases the chance that an already-dominant species is present and it is the presence of *that key species* that explains resistance to invasion.

In short, if the sampling effect can explain diversity-stability relationships, the door is still open to the possibility that the agents of ecological interaction are also agents of evolutionary interaction. There would then be a direct interface between ecology and evolutionary biology.

Patterns of ecological success and failure would translate back into evolutionary change, and evolutionary change would feed back into ecological interaction. Understanding this coupled interaction would still present formidable problems; nonetheless, the interaction would be direct. This would not be the case if top-down approaches to ecology are right: if important ecological processes are driven by composite, nonevolving ecological systems. The functional group of litter consumers in a particular Australian woodland is not an evolutionary unit: it is not a lineage. So if nutrient recycling is genuinely an effect of this functional group rather than just an aggregate effect of all of the members of the group, then the ecology-evolution interaction is indirect. To the extent that ecological systems drive ecological processes, we will face extra complexity in the ecology-evolution feedback. The effect of ecological events on evolutionary units will then be indirect through their effect on the ecological systems of which evolutionary units are parts. So, too, evolutionary changes will affect ecological processes indirectly. Ecological systems are built from evolved components—organisms—and the traits of those components will constrain the behaviour of the system.

So here is the state of play: the idea that local communities are stabilized and organized by competition among their components was never quite a paradigm. There were always serious dissenters. But it was certainly an influential view, and one that gave a clear view of the ecology-evolution interface. However, this view of local communities is almost certainly false: macroecology and niche construction revise it in substantial ways. The scale and kind of ecological interaction is expanded greatly. With this change in our model of ecology there is a correlated and independently plausible change in our view of evolutionary mechanisms and of their relation to ecology. But holist views, if vindicated, would make it hard to smoothly integrate ecology with evolutionary biology. Classical community ecology and evolutionary biology at least talked about the same biological systems: organisms and populations. Holist community ecology would not do so. However, holism is hard to establish. To do so, researchers have to show that an emergent property—such as diversity—covaries robustly with some alleged effect (such as stability) and that the connection is genuinely causal. So this may be an extra complication we do not have to face. Ecological interactions may well be driven by individual agents, not by collectives with ensemble properties. Let us hope so: understanding the interface between ecology and evolution will be hard enough without having to take into account indirect effects on and from ecological ensembles.

REFERENCES

Belyea, L., & Lancaster, J. (1999). Assembly rules within a contingent ecology. *Oikos* **86**, 402–416.

Brown, J.H., Morgan Ernest, S.K., Parody, J.M., & Haskell, J.P. (2001). Regulation of diversity: Maintenance of species richness in changing environments. *Oecologia* **126**, 321–332.

Cooper, G. (2001). Must there be a balance of nature? *Biol. Phil.* **16**, 481–506.

Cooper, G. (2003). "The Science of the Struggle for Existence." Cambridge University Press, Cambridge.

Darwin, C. (1964). "On the Origin of Species: A Facsimile of the First Edition" (1859). Harvard University Press, Cambridge, MA.

Eldredge, N. (1989). "Macroevolutionary Dynamics: Species, Niches, and Adaptive Peaks." McGraw-Hill, New York.

Eldredge, N. (1995). "Reinventing Darwin." John Wiley & Sons, New York.

Eldredge, N. (2003). The sloshing bucket: How the physical realm controls evolution. *In* "Evolutionary Dynamics: Exploring the Interplay of Selection, Accident, Neutrality, and Function" (J.P. Crutchfield & P. Schuster, Eds.). Oxford University Press, Oxford.

Ernest, M., & Brown, J.H. (2001). Homeostasis and compensation: The role of species and resources in ecosystem stability. *Ecology* **82**, 2118–2132.

Gaston, K.J., & Blackburn, T.M. (1999). A critique for macroecology. *Oikos* **84**, 353–368.

Gaston, K.J., & Blackburn, T.M. (2000). "Pattern and Process in Macroecology." Blackwell Science, Oxford.

Golley, F.B. (1993). "A History of the Ecosystem Concept in Ecology: More Than the Sum of the Parts." Yale University Press, New Haven, CT.

Grant, P.R. (1986). "Ecology and Evolution of Darwin's Finches." Princeton University Press, Princeton, NJ.

Grant, P.R., & Grant, B.R. (1995). Predicting microevolutionary responses to directional selection on heritable variation. *Evolution* **49**, 241–251.

Hamilton, W.D., & Lenton, T.M. (1998). Spora and Gaia: How microbes fly with their clouds. *Ethol. Ecol. Evol.* **10**, 1–16.

Jones, C., Lawton, J., & Shachak, M. (1997). Positive and negative effects of organisms as physical ecosystems engineers. *Ecology* **78**, 1946–1957.

Kingsland, S. (1985). "Modeling Nature: Episodes in the History of Population Ecology." University of Chicago Press, Chicago.

Lawton, J.H. (1999). Are there general laws in ecology? *Oikos* **84**, 177–192.

Lehman, C., & Tilman, D. (2000). Biodiversity, stability, and productivity in competitive communities. *Am. Nat.* **156**, 534–552.

Levins, R., & Lewontin, R. (1980). Dialectics and reductionism in ecology. *Synthese* **43**, 47–78.

Lewontin, R.C. (1982). Organism and environment. *In* "Learning, Development, and Culture" (H.C. Plotkin, Ed.). Wiley, New York, pp. 151–170.

Lewontin, R.C. (1983). The organism as the subject and object of evolution. *Scientia* **118**, 65–82.

Loreau, M., Naeem, S., Inchausti, P., Bengtsson, J., Grime, J.P., Hector, A., Hooper, D.U., Huston, M.A., Raffaelli, D., Schmid, B., Tilman, D., & Wardle, D.A. (2001). Biodiversity and ecosystem functioning: Current knowledge and future challenges. *Science* **294**, 804–808.

May, R.M. (1973). "Stability and Complexity in Model Ecosystems." Princeton University Press, Princeton, NJ.

McCann, K., Hastings, A., & Huxel, G. (1998). Weak trophic interactions and the balance of nature. *Nature* **395**, 794–798.

Naeem, S. (1998). Species redundancy and ecosystem reliability. *Conserv. Biol.* **12**, 39–45.

Naeem, S. (2002). Biodiversity equals instability? *Nature* **416**, 23–24.

Odling-Smee, F.J., Laland, K., & Feldman, M.W. (2003). "Niche Construction: The Neglected Process in Evolution." Princeton University Press, Princeton, NJ.

Ricklefs, R.E. (1989). Speciation and diversity: The integration of local and regional processes. *In* "Speciation and Its Consequences" (D. Otte & J.A. Endler, Eds.). Sinauer Associates, Sunderland, MA, pp. 599–624.

Ricklefs, R.E. (2004). A comprehensive framework for global patterns in biodiversity. *Ecol. Lett.* **7**, 1–15.

Ricklefs, R.E., & Schulter, D. (1993). Species diversity: Regional and historical influences. *In* "Species Diversity in Ecological Communities" (R.E. Ricklefs & D. Schulter, Eds.). University of Chicago Press, Chicago, pp. 350–363.

Schluter, D. (1996). Ecological causes of adaptive radiation. *Am. Nat.* **148**, S40–S64.

Schluter, D., & Price, T. (1993). Honesty, perception, and population divergence in sexually selected traits. *Proc. Roy. Soc. Lond. B* **253**, 117–122.

Sober, E. (1984). "The Nature of Selection: Evolutionary Theory in Philosophical Focus." MIT Press, Cambridge, MA.

Sterelny, K. (1999). Species as ecological mosaics. *In* "Species: New Interdisciplinary Essays" (R.A. Wilson, Ed.). MIT Press, Cambridge, MA, pp. 119–139.

Sterelny, K. (2001). "The Evolution of Agency and Other Essays." Cambridge University Press, Cambridge, pp. 152–178.

Sterelny, K. (in press). Made by each other: Organisms and their environment. *Biol. Phil.*

Tilman, D. (1996). Biodiversity: Population versus ecosystem stability. *Ecology* **77**, 350–363.

Tilman, D. (1999). The ecological consequences of changes in biodiversity: A search for general principles. *Ecology* **80**, 1455–1474.

Wardle, D.A. (1999). Is "sampling effect" a problem for experiments investigating biodiversity–ecosystem function relationships? *Oikos* **87**, 403–407.

Whittaker, R.H. (1975). "Communities and Ecosystems." Macmillan, New York.

Wilson, D.S. (1997). Biological communities as functionally organized units. *Ecology* **78**, 2018–2024.

Zobel, M. (1997). The relative role of species pools in determining plant species richness: An alternative explanation of species coexistence? *TREE* **12**, 266–269.

VI | ECOSYSTEM ECOLOGY

15 | THE LOSS OF NARRATIVE

T.F.H. Allen, A.J. Zellmer, and C.J. Wuennenberg

15.1 | INTRODUCTION

Paradigms lost? These are times of an unrelenting mechanist binge in ecology, and our discipline is unhealthy for these excesses. The collective vision is blurred and the body is worn down by an unbalanced intellectual life style. Unable to kick the habit, ecology cannot see beyond the next mechanism. Somehow the sober, larger vision of whole ecological systems has been lost. We will attempt to talk the discipline down and offer rehabilitation. We will see if we cannot get down to having just an occasional mechanism. The issue of too local a treatment of scientific investigation, by insisting on mechanism, is particularly pressing when ecology tries to do something useful, such as consider ecological systems with a large human presence. Our lost paradigm is analogy and narrative as crucial tools for giving meaningful context to lower-level devices in ecology, such as model or mechanism. Models and sometimes mechanisms are more local in the discourse and should serve as slaves to the narrative. The modernist agenda is to find the truth about the real world. Our agenda here is to challenge the modernist project in ecology. We substitute a postmodern, post-normal approach where understanding is constructed in the observer by interaction with aspects of observation. This forces observers to take responsibility for their decisions in narratives, and that lets us deal with complexity at face value.

Narrative is a complex scaling device that transforms rate-dependent process into rate-independent, but time-dependent, events. Our focus on narrative derives from the historical ecologist Bill Cronon (1992). An earthquake is sudden and widespread in its effect. A drought is widespread and persists for a long time. The arrival of a new species of insect is small in scale but might bring a feedback that changes a landscape indefinitely. All these differently scaled events involve dynamics and processes that are so different in amplitude, frequency, duration, and spatial extent as to be incommensurable. Yet the insect, drought, and earthquake can be stably juxtaposed as events in a narrative. Although narratives can handle disparate components, models cannot, quickly becoming chaotic, then failing. Narratives, not models, are the device scientists use to make complex systems simple.

As we expand the ideas of model, analogy, and narrative in the body of this chapter, we will quickly define what we mean because our usage is quite specific. Mechanism gets a longer treatment because it is likely to be taken for granted by the audience we wish to influence. Analogy identifies equivalence between two systems on the basis of some characteristics, actively ignoring the large number of non-equivalences that also pertain. For instance, a common analogy makes the flow of electricity down a wire equivalent to the flow of water down a pipe. Voltage becomes water pressure, amperage becomes the width of the pipe, and watts become the quantity of water flowing. Of course, water is not electricity and is only like electricity in the specific analogy. Electricity is not wet, and water is not necessarily negatively charged. Analogies have a linguistic aspect that is rate independent. Note that although water flows through a pipe, a pipe does not have a width at a rate, only a width. The equivalences just are; they do not have dynamics even if the equivalence is a dynamic, such as flux. Analogy shares with narrative the freedom to choose relationships. There is nothing necessary about either analogy or narrative. They rather amount to what the users or creators have chosen because they find it useful.

Narratives also are rate independent in that they collapse dynamics, such as seismic waves, to events or structures, such as an earthquake. Disparate events, identified to a class of event by analogy, are then put together in a sequence. The power of narratives is that they can relate in a coherent way contrasting types of things from different scales. Narratives are the device people use to grasp large ideas. Models can be used to calibrate things, and even improve the quality of narratives, but models cannot work with the scope natural to a narrative.

According to Rosen (2000), scientific models say that if the world worked in such and such a way, then there would be this particular outcome from these initial conditions. The world would behave that way, too, but only if it worked in the specified manner. All models are abstractions, and cannot capture every aspect of the world, without actually being the world (whereupon they would not be models). So all models are in that sense false. At that point veracity is beside the point because there are not degrees of being right on a single issue. Models necessarily invoke symbols, so they are representations, not approximations. Approximation is always in the same terms as that which is approximated; that is indeed a matter of degree, but that is a different issue. Because models are not in the same terms as that which they model, it makes no sense to say that they approximate the truth. Measurements do approximate the value that they attempt to capture, but that, too, is a different matter.

Formal models allow situations to be made equivalent with all asserted as being the same between the model and the thing modelled except for the scale (Rosen 1991). Formal models are more specific than what we mean by analogy. For instance, the allometric equations are a formal model for growth of organisms. They take the strength of bone and sinew, and the weight of flesh, as givens. The allometry then defines relationships between size and proportion based on arguments that the weight to be supported goes up as a cubed function and the cross section of the legs that support that weight increases only as a squared function. Accordingly, the legs of ants are proportionally thinner in an ant compared with the legs of an elephant. Trees, like elephants, are big, hence the analogy of "tree-trunk legs". The allometric equations say how much thicker elephants' legs must be than ants' legs by making up the difference between the cubed mass and the squared cross section with terms that cancel out the respective exponents. Models are thus statements of relative scale and so apply independently of any particular scale or size. The formal allometric equation says that mice and elephants are the same except for size. The same is true for the equations of aerodynamics, explaining why paper darts stall at slower speeds than airliners.

If, as in the allometric equation or the laws of aerodynamics, it is possible to encode two structures into a particular equation and then decode them back out again, something wonderful happens. Locked together by the formal model embodied by the equation, the two material systems become models of each other (Fig. 15.1). Clearly, a hobbyist's toy model of a DC-10 is a model, but it works the other way around, too. The actual airliner is also the model for the toy. It is upon these equivalences that experimentation rests, such as when tests on a flask with algae growing in it are taken as an indication of what researchers might expect to happen in a whole lake, as in Jim Drake's (1992) assembly rule experiments. All experimental systems assert this equivalence, although often it is taken as implicit and then forgotten. Models are carefully fixed at a particular level of analysis, and it is this that allows competing models to be set on a level playing field. They can be compared without semantic arguments muddying the issue so that one or other

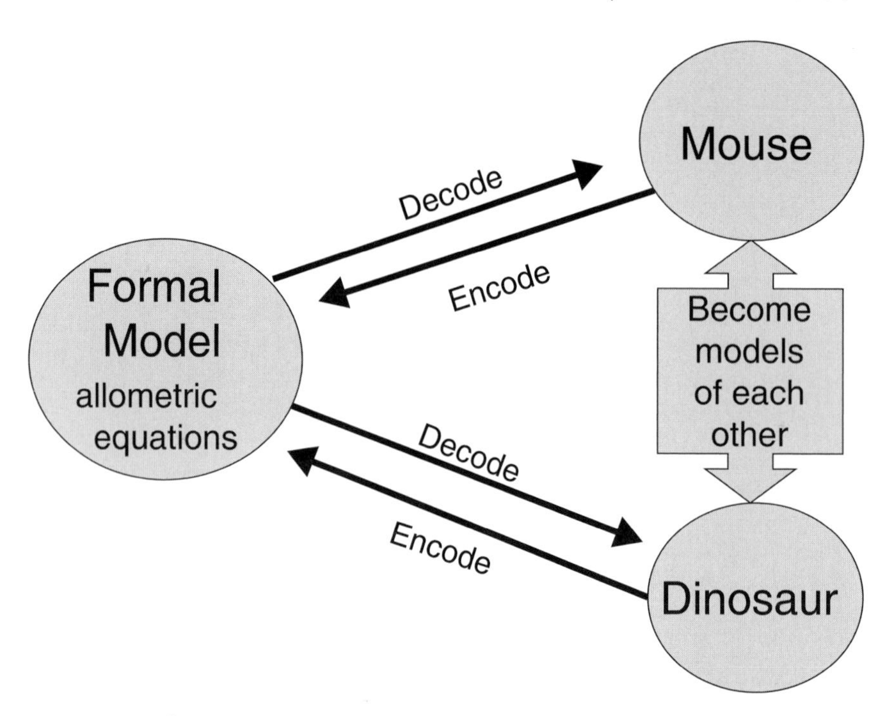

FIGURE 15.1 | Rosen (1991) identifies that a formal model is a set of scale relative statements that are therefore scale independent in themselves. Observable systems may be encoded into and decoded out of the equations of the formal models, such as when a paper dart and a DC-10 may be related to the laws of aerodynamics. When two systems can be encoded and decoded in this manner with regard to the same formal model, the two observable systems become models of each other. The formal model states that all is the same between the two systems except scale. That is not true, but it is the way that science uses models. Experimental approaches depend on this principle.

model can win as their respective outcomes match observations better or worse.

We see mechanism in large-system ecology as being a model for reduction of some upper-level entity or pattern to a lower-level set of linked explanatory devices. The links in a mechanism follow in a simple chain of causality. Not all models do this. For instance, the allometric equations in themselves do not embody a chain of causality. Nevertheless, scientists can invoke the allometric equations in a mechanistic model that grows animals in a sequence of hormones acting to modify the effects of genes and so on. Mechanism does not generally deal well with multiple causality; indeed, its beauty is reduction to an unambiguous causal explanation that passes monotonically through time.

Mechanisms are achieved by narrowing the focus so that ignoring enough of the context puts the causes in the model in a series or some other simple construction. The system is described as a linear approximation, which only works in a particularly narrow discourse. Approximate linearity can be achieved, but only if most of the curved line on the graph is ignored. An example of the dramatic cost of linearization occurs in the standard treatment of muscles and skeletons in biomechanics of joints. The favoured mechanism is the lever, such as when the bicep pulls on the forearm to raise the wrist. Levin (2002) points out that such accounts of all joints are achieved by narrowing the focus. The forearm is pictured as a free body, the shoulder is excluded, and the wrist joint is frozen. Why does this matter? The bicep is connected not to the upper arm but to the shoulder at one end and the forearm at the other. The upper arm is free and, in fact, floating. So shortening the bicep should lift the whole arm upwards more than bend it. Therefore, a person needs to put tension on the triceps, also connected to the forearm and shoulder, to hold the arm down. Now the system is in feedback and should oscillate, but it does not, only because the whole system is set in a web of tension (Levin 1997) in which the bicep lever is a small component. The bicep in the standard representation is arbitrary to the point of being capricious; the lever is denied the critical anchor at one end. Archimedes said he could move the world with a lever, but everyone forgets that he also said he needed a place to stand. If you want to say, "Well, close enough", then Levin (2002) explicitly calculates that the bones, muscles, and cartilage needed to work the levers are not even close to strong enough. Furthermore, there is insufficient energy to run the muscular-skeletal system as a set of levers. All models are wrong, so that is not the objection, but they are so in-your-face wrong and capricious that the bicep and forearm as a lever is beyond the pale. Ecological mechanisms are often no less inadequate; it is just harder to see them as such.

Mechanisms offer a set of rules that link lower-level components. Mechanism is a model, so we see the notion of a real mechanism in the real world as an oxymoron. As John Searle, philosopher and psychologist, is fond of saying, a psychologist may have rules that simulate animal behaviour, but his dog does not have rules for its joy and enthusiasm, even though the psychologist's rules for dog behaviour may be predictive and offer a good mimic. For some issues, some mechanism may be the best model, but it does not follow that the world in reality is mechanistic. The bicep lever is not the real mechanism; it is just the easiest, intuitive way to linearize the system. We cannot fathom what *real mechanism* could mean, let alone what it does mean.

The understanding that mechanisms sometimes offer encourages overindulgence, in which the whole that is being explained is asserted as *nothing but* the mechanism. Even if mechanism is the best model, the thing is never *nothing but* a mechanism. Chicken-and-egg issues are beyond the ken of mechanism because there are several levels of explanation there. Although a scientist can identify the mechanism in feedbacks, loops of causality do not sit easily in mechanistic explanations (see Bateson 1980 on steam engine governors). Mechanism works well if summation of the parts from one lower level is feasible, if mythical. Like all models, mechanism is explicit in the level of explanation to which it gives privilege, but that privilege does not ever make the whole *nothing but* a mechanism. Even a steam engine has meaning, uses, and roles beyond its mechanism.

15.1.1 | The History of the Problem

The first moves towards mechanism in the ecology of large systems were promising. After all, by the 1980s, the descriptive multivariate methods had yielded their insights, and their use was degenerating into a ritual that community ecologists were expected to perform. It was forgotten that multivariate analysis is a hypothesis-generating approach, not a means of hypothesis testing. How different from the middle decades of the century. From those days, I (Allen) remember Peter Greig-Smith insisting that a thesis in Bangor, North Wales, should have some experimental tests that address the insights from the gradient analysis. Community ecology in Bangor was all a matter of focus on a particular system. How sensible were Eddy van der Maarel's gentle appeals for more mechanistic understanding to be part of community ecology. Colin Prentice and Eddy van der Maarel had arranged a splendid meeting in July 1985 in Uppsala, Sweden, with new players welcomed into vegetation science. The newcomers were distinctly mechanistic, generally doing experiments on organismal interactions or the physiological basis for specific and focused community phenomena. In the preface to the proceedings, published in *Vegetatio*, Prentice and van der Maarel (1987) listed "key aspects of modern vegetation science". They were aware of ranges of timescale and of vegetation-environment relations in time and across space, both emerging issues. But Prentice and van der Maarel were also specific in their inclusion of "the functional basis of vegetation in terms of the individual plants and plant populations that it comprises". Mechanistic reduction of vegetation to lower-level components was recognized as a key aspect of the larger science, and it was a good idea.

But by then the mechanistic excesses had already begun elsewhere. Instead of tests directed at local patterns, some experimentalists were already searching for general mechanistic explanations for large abstractions, such as species diversity or stability in principle. The descriptive multivariate methods went into disrepute, and instead population equations were introduced as proof of mechanism. Like ordinations, population equations are only hypothesis generators, the proof in them being mathematical, not of material observables. But that was forgotten by, or otherwise lost on, those enjoying the new rigor. Soon naïve realists were searching for the one true mechanism—if they could only find it.

Alfred Korzybski made famous the idea that the map is not the territory (Bateson 1980). The system itself is not a mechanism. Mechanistic accounts are fixed within a narrow range of explanation. Models are most effective when they are put in a proper context. Overreach the context, and the model no longer applies. Overreaching the applicability of a mechanistic model comes sooner rather than later. It has definitely come soon enough in mechanistic accounts of diversity, stability, coexistence or sustainability, and the host of other large issues that have mechanistic accounts as the agenda of modern ecology. The bottom line is that we need more narratives and fewer mechanisms for those large concepts.

15.1.2 | A Postmodern View of Ecology

Rejecting a realist take on mechanism, the view expressed in this chapter is instead more measured, using a constructivist philosophy and method. There is an old guard that is loud and energetic in its rejection of postmodernism and its constructivist philosophy. As in all paradigm shifts, the merely modernist old guard attacks without understanding what the new paradigm says. No, constructivism does not imagine that it constructs the truth of what happens in the material system. It is observers who are constructed in what they understand to be happening in their respective continuous *sense experience*. As postmodernists we use analogy and narrative with a focus that is unusual. Narratives are not blunt instruments that offer the easy, inexact solutions. We choose to use them as tools for precision.

Our champion in this paradigm fight, postmodern before his time, from the halcyon days and volume of the Clements-Gleason fracas, is no less than the great knight himself, Sir Arthur Tansley (1926, pp. 686–687; his emphasis and quotation marks):

> I mean we must always remain alive to the fact that our scientific concepts are obtained by "abstracting from the continuum of sense experience," to use philosophical jargon, that is, by *selecting* certain sets of phenomena

> from the continuum and putting them together to form a concept which we use as an apparatus to formulate and systematize thought. This we must continually do, for it is the only way in which we can think, in which science can proceed. What we should not do is to treat the concepts so formed as if they represented entities which we could deal with as we should deal, for example, with persons, instead of being, as they are, mere thought-apparatuses of strictly limited, though of essential, value.

Of climax community, Tansley said (1926, p. 687; his emphasis and quotation marks):

> [It] is a particular aggregation which lasts, in its main features, and is not replaced by another, for a certain length of time; it is indispensable as a conception, but viewed from another standpoint it is a mere aggregation of plants on some of whose qualities as an aggregation we find it useful to insist . . . But we must never deceive ourselves into believing that they are anything but abstractions which we make for our own use, partial synthesis of partial validity, never covering *all* the phenomena, but always capable of improvement and modification, preeminently useful because they direct our attention to the means of discovering connections we should otherwise have missed, and thus enable us to penetrate more deeply in the web of natural causation.

If one imagines that analogy is soft when reduction and mechanism are hard, one would do well to remember the keystone physicist, William Hamilton. Physicists have been driven to use analogy and are quite secure and self-conscious about it. Ecologists are less self-confident than physicists, and Joel Cohen (1971) has pointed out that ecology has physics envy. Unfortunately, it is Newtonian tidiness that they envy, not the penetrating logic of quantum mechanics. Fearing that they may be viewed as somehow soft, ecologists are loathe to admit that they use devices such as analogy. They still analogize, but they will not readily own up to it. Rosen (2000) points out that no lesser man than Maxwell tried to reduce optical geometry to Newtonian particle mechanics, and he failed. Some decades earlier, Hamilton achieved a successful link between the two facets of physics through analogy. He wrote a sort of dictionary to translate between them. Having used analogy successfully, Hamilton went on to reduce optical geometry to wave mechanics. If he had only thought to do the same on the Newtonian particle side of the analogy, using essentially the same equation, he would have discovered, in the manner of Erwin Schrödinger a century later, quantum mechanics (Rosen 2000). Who says analogy is soft? Therefore, ecologists need not feel that analogy is soft, because some tough guys have used it to great affect in the very field that ecologist feel is as hard as they would wish to be themselves.

15.1.3 | Analogy in Ecology

By insisting on reduction of biology to physics, scientists end up with impotent intellectual devices in which models are assumed closed and close enough to equilibrium to use it as the reference state. Almost all physics, even that which uses analogy, still needs closure and equilibrium for the bookkeeping to work. It is not just that the assumptions of modern physics are wrong; all assumptions are that. In biology you cannot get away with those assumptions and still be able to predict significant biological and ecological systems. Biologists can use diffusion models or mass balance from physics but only in very local biological systems.

Schrödinger's *What Is Life?* (1967) is not the pillar of orthodoxy that everyone thinks it is today. First published in 1944, that heterodox work stated that we needed a new physics for biological systems, and we still do. There must be a regular ordered physics of systems that is widely open and far from equilibrium, even if physicists do not often study it. Physics must be reliable because organisms continually depend on it being predictable for their survival moment to moment. For instance, enzyme pathways are predictable but far from equilibrium. That is how they function to do what they do. An enzyme pathway at or close to equilibrium would not process anything. That is what happens when an organism dies—its enzymes pathways become blocked and product piles up, causing the enzymatic reactions ahead and behind to reach equilibrium. Far from equilibrium does not have to mean unpredictable; it just cannot be predicted with reference to the equilibrium. The small amount of physics and chemistry studied far from equilibrium, such as Bénard cells and Prigogine's fluctuations (Nicolis & Prigogine 1977), are perfectly predictable, and that work has earned Nobel Prizes. But even that is too limited to serve biology with a physics that is properly useful to answer large questions.

Biology and ecology should not be always touching the forelock to physicists. It should be the other way around. Analogy is more than respectable. We need it as the primary tool of ecology. By taking the organism, through analogy, into physics, it should be possible to create the new physics of which Schrödinger spoke and for which we still have a pressing need (Rosen 2000).

Narrative works like analogy, but it operates at a higher level of symbolism. Narrative was in Tansley's day particularly useful for dealing with things physically larger than the ephemeral human form. The community concept often applies to physically large systems, as does the derivative notion of ecosystem. In ecosystems scientists find the entry of systems analytic endeavours into ecology. With systems approaches, ecology gropes up the scale to deal with the big issues, the ones that take ecology into the realm of significant application at a societal level. More than ever we need narrative and analogy as devices for precision treatment of human ecological affairs. But the discipline at large has forgotten how narratives and analogy work and is not fully aware of when we need them.

This chapter begins the counterattack of the narrator. An ecological description is at best incomplete, and at worst impotent, unless a narrator has taken responsibility for what is part of the story and what is not. We fear that if young ecologists were to give narrative a thought, their comments too often would be, "It is just a matter of stories, whereas I have to make ecology properly scientific." We will attempt to rehabilitate anyone with that view. Narrative gives us courage to begin and end, and it makes scientific rigor meaningful.

15.2 | THE PARADIGM OF NARRATIVE

Paradigms are intellectual contexts shared by scientists who work with an unspoken set of rules. Written down explicitly or not, paradigms are the context in which scientists work to find and reinterpret useful ideas. Many paradigms are so much part of the furniture that they rest overstuffed and unseen by those who sit in them in gentleman's clubs.

15.2.1 | A History of the Ecosystem Paradigm

As to paradigms in ecology, our discipline has several large ones. One such paradigm is the ecosystem. That way of looking at the world had to be explicitly invented, although it is not clear that Tansley (1935) fully understood the implications of the term he coined. All he was trying to do was clean up the mess left by the burgeoning terminology of community ecology. When the physical environment is put inside the system, along with the biota, organisms melt away and become connectors of, say, primary productivity and the detritus compartment. Tansley had named a conception of ecological systems that is fairly independent of community ecology, working largely without evolution as an organizer. With sufficient computational power to achieve simulations of nutrient flows, some of the International Biological Program models in the 1970s broke the ecosystem concept from community ecology, and the ecosystem paradigm stood on its own feet. Paradigms do not have to be titanic, as were the paradigms created by Newton and Einstein. Ecology, as in most science, has a host of smaller notions and practices that are examples of more modest paradigmatic frameworks.

There is a tradition of narrative in ecology. Some of that comes naturally from ecology being often a historical science. There is some reverence for traditions of ecological storytelling by the great narrators, such as Aldo Leopold and Wallace Stegner (both from our University of Wisconsin). These are not just fanciful accounts, because Leopold (1949) tells a tale of atoms in biological material as they move around what was, at the time, the new conception of ecosystem, after Tansley (1935). Narrative, much in the sense of literally telling stories, has a venerable tradition in ecology. But it appears tradition is not enough for narrative to hold its own as a device, which is one reason we decided to write this chapter. Narrative appears to need defending to the extent that Cronon (1992) argues for "a place for stories", as if the narrative were under siege.

In the narratives traditional in ecology is a deep sense of the wholeness of function in ecological systems. This sense sometimes leads to flaccid statements that everything is linked to everything else. It may be true with regard to gravity, but as a generalization it ignores that the science of ecology is about how some connections are enormously more important than others. Beyond that, the importance of each connection changes with each conception of the issue at hand. Take a set of organisms in a place and cast them in community terms, and the narrative turns on evolutionary issues, not material fluxes and mass balance. Take that same set of organisms in that same place and cast them in ecosystem terms, and the organisms disappear in the ecosystem melting pot. In an ecosystem conception, evolution almost ceases to have explanatory power, and the narrative is about mass balance and cycles (Allen & Hoekstra 1992). That is why narrative matters so much, because it can deal with changing connections and changing times. Communities and ecosystems are merely different narratives about the same ecological materiality.

15.2.2 | Scientific Paradigms Versus Humanitarian Commonplaces

Narratives are devices commonly used in the humanities, and a comparison between humanities and science shows why that is the case. In the humanities, the approach to scholarship rests not on data but on the documents and the narratives they tell. The social structure of the community in the humanities is different from that in the sciences. That cultural divergence between these two parts of academe is reflected in the patterns of publication and period of residence in graduate school. A humanities student works with remarkable independence from the major professor. The humanities student arrives in graduate school and spends a long time reading the literature of the field and little time interacting with the advisor, by the standards of a scientific laboratory. Whereas scientists employ paradigms, humanitarians reference their discourses with devices they call *commonplaces*. A paradigm defines what is in the area of discourse, although a commonplace only says where to start. Normal scientists fight to achieve the best model within the paradigm. It is a knock down, drag out fight. Each laboratory is a platoon of collaborators all working on the task of winning out. The patterns of publication reflect this in that most scientific papers have multiple authors. In the humanities, by contrast, graduate students rarely publish with their advisors. This reflects their use of commonplaces instead of paradigms. Both paradigms and commonplaces are agreements about common concerns. Paradigms say what is in, whereas commonplaces say where to start.

A commonplace is the agreement as to what is common about, say, Hamlet. It recognizes a play in which the Prince of Denmark is the picture of indecision, and a tragedy follows from that. From the point of the commonplace, each scholar then departs to develop new interpretations and meanings about Hamlet's latent homosexuality or some other such theme. The journey takes the student from the activities of the advisor, and the patterns of publication reflect that. Commonplaces may be used to tie treatments of the play into the body of literary criticism. By contrast, nobody leaves the arena of a paradigm—that is, not until they have tenure. They must stay and face combat in the arena in science, whereas students of the humanities have not shown sufficient creativity unless they move away.

It is entirely misplaced to imagine that the humanities are a lesser activity. The humanities appear insecure in the face of the funding and kudos that science commands, but that insecurity is misplaced. Many practitioners do not notice, but science as an endeavour is at its limits. It is not quite up to the job society needs and expects it to do. Science would benefit greatly if it would liaise with the humanities and begin to do more synthesis using the device of narrative and the expansion of meaning. The humanities can do more than collaborate as junior partners; they may be able to save the day.

At the interface of science and history, Cronon (1992) shows the stark difference between the devices of the humanitarian and those of the scientist. Whereas modernist science attempts to corner some truth, Cronon points out that narratives, the device of humanitarians, are never true, so there is no singular truth as the prize. Narratives are neither right nor wrong. Narratives can be false with regard to known happening or accepted understanding, but stories that are not false are still not true stories to the exclusion of other narratives. A given situation may be a given account as a tragedy or a heroic saga without one or other being the correct version of events. Narratives are not about reality; they are about the narrator's interpretation of the narrator's experience. Who are we to assign verity or otherwise to that? Truth does not reside in telling literally everything that happened. Not only is it impossible to achieve a complete chronology of everything that happened in the past, but if the narrator could do it, the account would not be a narrative. A narrative requires a narrator to take a point of view. The interpretation is the reality.

Cronon speaks of two accounts of the dust bowl and the human-directed ecology of the Great Plains. Bonnifield (1979) gives a heroic account of Euro-Americans learning from disaster and returning to overcome patterns of drought in the Midwest. Published in the same year, Worster (1979) covers generally the same time period but with a tragic narrative. There was a balanced ecological situation on the Great Plains, but Euro-American technology stressed the system that then collapsed. How can the two stories agree, as they do, on the facts and yet have one be heroic and the other tragic? If a piano being hoisted into an upper-story window breaks loose and falls on your head, then that is most unfortunate, but it is not a tragedy. A tragedy would be, "He did not know she was his mother." Once Oedipus has married his mother, then it all unfolds inevitably. It is the inevitability of it all that makes the tragedy. Cronon refers to yet another tragedy, as he quotes the story of Plenty Coups, chief of the Crow, published in the 1930s. Poignantly, Plenty Coups ends his tale saying, "When the buffalo went away, the hearts of my people fell to the ground, and they could not lift them up again. After that nothing happened" (Linderman 1962, p. 311). Cronon (1992, p. 1366) points out that the Crow still survive on their reservation, but "for Plenty Coups their subsequent life is all part of a different story. The story he loved best ended with the buffalo. Everything that has happened since is part of some other plot, and there is neither sense nor joy in telling it." What followed, as told by Bonnifield and Worster—the great drought, the depression, and the triumph of technology—did not happen for Plenty Coups, and all those events are lost in the "nothingness at the end of Plenty Coups's story" (Cronon 1992, p. 1367). Narratives require a narrator to take responsibility for "once upon a time," when the characters "lived happily or not ever after," and everything between.

15.2.3 | Paradigms, Complexity, and Narratives

The distinction between complex and simple systems is normative. Human devices and decisions regularly convert complexity into simplicity. It is worth making a distinction between systems that are complex and those that are merely complicated (Allen, Tainter & Hoekstra 1999). Complexity arises from uncertainty about system specification. A good story always has the tension and the uncertainty between the focal and the tacit levels of analysis (Needham 1988), so it is a good device for facing down complexity. Simple systems may be complicated because they have many parts of many types. They may have many internal relationships. As long as those relationships are unambiguous, complication does not necessarily lead to complexity. Ambiguity is removed by decisions about system specification. Models remove uncertainty about system specification, but this means that models cannot address complexity in the way narrative can. Needham (1988) makes much of the competition between focal and tacit attention, emphasized by Michael Polanyi. But you have to plump for the focal attention, and that leaves the remainder in the realm of tacit attention ambiguous and tangentially treated. Narratives can handle that uncertainty, but models precisely remove it by specifying the context so that it is no longer tacit and becomes the focal-specific initial conditions.

Chaotic systems are more about simplicity than complexity by virtue of involving equations that are short to capture infinite complicatedness in the states of the system (undisturbed chaotic systems never repeat behavioural states). Chaos is something of a special case in that at the bottom it addresses systems that cannot be resolved by definition, because the strange attractor is never complete. In that sense, chaos remains complex. Before chaos theory, complicated behaviour was assigned to complexity and was presumed to have complex explanations. Some complicated behaviours are not explained by low-dimensional chaos, and they remain complex on any criterion or for any question you might ask. At one level, paradigms are the story you choose to tell or hear to the end to make a class of systems simple. Accordingly, when May started to work on chaos, it was a matter of complexity in several ways. Now there is a normal science of chaos, and it addresses simple, if complicated, systems because the story has been told in the specialist literature. The lay version of the story particularly took an obviously narrative form in James Gleick's (1987) treatment.

If you have a paradigm for a system, aspects of it may be complicated, but it is at its bottom simple, not complex. Normal science only deals with simple systems. That is why mainstream scientists do not readily take to complexity; it belongs in the system specification phase that has passed even before the normal scientist gets to work. If scientists are dealing with a system that may have only a few parts, but for which they have no paradigm, then the system is complex. Kuhn (1962) does not cast paradigms in our terms, but what he says (shared methods, vocabulary, and concepts for system organization) translates to our criteria for distinctions that arise in making a paradigm. Kuhn does not disagree with us, but our notion of paradigm is only implicit in his definition. For us a paradigm defines the system on a small number of general principles.

For a system of study, then, a paradigm is a narrative that specifies what is:

1. *Continuous and therefore readily quantifiable as opposed to discrete and therefore inviting a qualitative description.* Whereas the number of eggs in a nest is an integer, and is in that sense discrete, it is a variable number for which only a special case of variation is counted. Three eggs is almost as many as four, which cannot be said about being pregnant. A woman is either pregnant or not, with "almost pregnant" having no meaning. Pregnancy is discrete any way you look at it, whereas three being almost four has meaning. Although three is not four, it is closer to four than, say, two. But closer or further from pregnant is like the sound of one hand clapping, nonsense fashioned from conflicting definitions. One could change the level of analysis from pregnancy to pregnancy rates, but that is a different discourse.
2. *Structural and emergent as opposed to behavioural and dynamic.* Structures emerge, whereas a structure-changing state is dynamic behaviour. Change of state is not the structural all-or-nothing that pertains to a new structure coming into existence, or at least achieving observability or recognition.
3. *A significant difference as opposed to an incidental difference.* Acid rain causing acidification is associated with a change in pH beyond a certain threshold at which significance arises. Daily change in pH because of photosynthesis and respiration rates is only an incidental difference if the description is of acid rain damage.

4. *Rate dependent and process-oriented as opposed to rate independent and linguistic.* A ball may fall at a rate, but it is not a ball at a rate; it just is a ball.
5. *The whole as opposed to a part.* This distinction fixes the level on analysis.
6. *The type of the whole.* To what equivalence class has the whole been chosen to belong?

Once all those decisions have been made, and there is agreement by paradigm adherents, then the system is simple because its organization is defined, or may be easily characterized. The preceding decisions about how to bound and characterize the system plump for a particular level of analysis. Specification of the level of analysis occurs even if the paradigm adherents often are not explicitly aware that they have a level of analysis, let alone what theirs is. Naïve realism fosters such mistakes by encouraging a belief that our paradigm is true, or darned close to it, and who needs to care about level of analysis when you have truth in the bag. Until all the preceding distinctions have been made you do not know what you are talking about, so you face a complex system. Complexity scientists are in the business of finding good and productive ways to define systems, at which point they have a prescription of some sort of workable simplicity.

In a paradigm the narrative is fixed, so nobody needs to tell it. Thus, the paradigm is often invisible to its adherents. The paradigm fixes arbitrary decisions, such as a decision that a difference is big enough to say there is a new structure, not just a change of state of some old structure. The invisibility of the paradigm to its adherents causes what is arbitrary to be mistaken as necessary. This does no harm for work within the paradigm, but it does freeze the mindset and so causes resistance to paradigm change or significant paradigm modification. Resistance to paradigm change is usually a good thing, because it maintains memory in the day-to-day practice of science. Free acceptance of new paradigms would spawn new dialects in scientific vocabulary, and the whole endeavour would degenerate into a shambles of bright ideas.

Paradigms mostly operate by stubbornly not changing. Progress is not naked change but is rather lack of change in every direction except one. Editors need to be conservative, and having only a slavish understanding helps them adopt that posture. The review process is rarely fully competent, but it always seems to improve the paper because it forces a translation of something new into terms more likely to be understood. If the review process were highly competent with regard to big new ideas, science would not work like it does now and probably would not work as well as it does. The truculence of paradigms amounts to system memory in science, and every information-processing system, including science, needs its memory to make it largely unchanging.

We would go further than to say narrative has a place; we think it should be the ultimate product of science. Ecology started to take models seriously with the equilibrium school of MacArthurian equations, which was quite late in the game. Unfortunately, models are now mistaken to be the whole purpose of the enterprise. No, we need models to specify assumptions, and to make them operational in, say, an experiment, but that is at best the opening gambit. We use the models to challenge and thus improve the quality of ecological narrative. For a nuanced description of quality see Pirsig (1992). An improved, higher-quality narrative is the point.

We say that quality of narrative is the copestone of science. We hasten to note that quality has been assessed in different ways in the classical world, the modern world, and now the postmodern world (Funtowicz & Ravetz 1992). In the classical world, quality was identified in a relatively democratic fashion. Although the duke might have commissioned the painting, the assessment of quality in the time of Titian was done by the externality of public opinion among courtiers. They knew what they liked, as do people with untrained eyes today. In the modern world, democracy goes to the wall, and it is an elite group that makes the decisions. Modern painting, as with all modernism, uses reality as the external criterion for assessing quality. For a modernist, higher-quality work gives a better account that is closer to reality. The reason a Picasso cubist painting has faces with noses off to the side is that the full face misses what you would see from a profile. You see the full face, but the profile is always there to be seen as well. By mixing the full face and profile, and using an interpretation only understood by an elite world of art criticism, the cubists can claim to be getting closer to the reality of the whole face (Allen *et al.* 2001). Reality is also the benchmark for quality in modernist science. Success in modernist terms can say, "My model is better because it is more real." In a postmodern world, there is an explicit acknowledgment that we do not have access to reality and we are not sure what it might be, let alone what it is. If a model indeed hit right on the nose of reality, we would not be able to tell it from a model that had not yet been shown wrong, so reality has serious limits as a reference.

In a postmodern world, there are no external criteria for quality. Popular opinion and external reality are viewed as insufficient. Quality resides in the very activity under way, be it science, technology, or art. Because science relies on observation, there is no irrefutable logical reason to suppose that science gets closer to reality independent of observation. Even so, science is still the best game in town despite proximity to reality being denied as an adequate criterion. In a postmodern world, high quality resides and is assessed in the very process of doing science. Quality is internal to scientific activity.

There are two sides to quality: structural and dynamic. By becoming a craftsman and getting all the bits to fit just right, the craftsman achieves structural quality. In science, structural quality is a matter of precision and reliability in the execution of the prevailing paradigm. Dynamic quality is at odds with structural quality, but a craftsman needs both. Dynamic quality attacks the premises on which high structural quality is set. Without dynamic quality, all that happens is more precision within an established framework. Too much bias in favour of structural quality gives more precision than is necessary. Dynamic quality is the antidote in that the changes it proposes question the premises on which more precision would be based. Dynamic quality is progress beyond the prevailing view; it is acceptance of new premises that move the enterprise forward. Both structural and dynamic quality are required for good science and good art (Funtowicz & Ravetz 1992).

High quality in science comes from a proper tension between dynamic and structural quality. Models are a device to achieve both dynamic and structural quality. Models can improve the quality of narratives in two ways. First, models are often quantitative and deliver structural quality. Second, models challenge the premises of the primitive narrative and so raise dynamic quality. But the bottom line is an improved narrative. We start with some narrative that may have good credentials, such as, "It came from Darwin himself." We then challenge the narrative with models that will often depend on technology not available in the mid-nineteenth century. But it is not over until a new, higher-quality narrative is told.

15.3 | HIGHER DIMENSIONALITY IN NARRATIVES

15.3.1 | Essences, Models, and Observables

Being purposely confrontational, Rosen (2000) speaks of essence and realization of essence to account for complex organized structures, such as organisms. Essence is the contextual intangible that lies behind complex organized systems. The presidency of the United States lies behind the actual, realized presidents. A US president is usually realized through the process of an election. Role and essence are related concepts. Deoxyribonucleic acid (DNA) is not the essence of, say, a dog; it is just a device for realization of an essence. Behind every dog is the essence dogginess. That is what allows the word for dog to arise in different languages and still refer to the same types of animals. People rarely make mistakes calling something a *dog, chien, canis, perro,* or whatever. In Chinese there are several words for dog, one being a dog that belongs to the emperor. You probably do not want to kick a dog with that name. We use essence as the explanation for the consonance of experience all people have with regard to dogs. Here we wish to use the notion of essence in an intuitive fashion, realizing that it is a rich concept that has received a larger treatment elsewhere (Rosen 2000, Allen, Giampietro, & Little 2003). The nub of an essence is that it is the target of our understanding of biological systems. We use models and eventually narratives to construct understanding of essences in an observer. Essences are the causes of what makes things equivalent in an equivalence class. Much as what is seen in an observation cannot be completely independent of arbitrary observation protocols; essences are the context of models but do not exist independent of the model for equivalence in the equivalence class in question.

Biological systems are different from physical systems in that organisms appear to have models of the world in which they live. Complicating the situation further is that the world in which an organism lives has other organisms in it and they, too, have models, including models for each other. Realized from an essence, the organism enters the world with a model, and it tells narratives that reflect the model. For a dog that model of the world has a lot of smells in it. The organism presents that model as a narrative to the world. The success or failure of the organism's model often turns on how its narrative plays out in exchanges with other organisms and their narratives. Success is transmitted back to the essence, in this case through natural selection. In our social science example, Richard Nixon changed the presidency, obviously not with DNA and natural selection but by changing the political context of subsequent elections and the view of the populus as to what it will not tolerate in a president.

15.3.2 | Dimensions of Narratives

If a bug has a model that enhances survival and reproduction, the quality of the essence is improved as to how its realizations fit their context. The changed essence then realizes itself in another version of the bug, which has a modified model of the world. The improvements in fit may be considered an added dimension to the narrative that the second and subsequent realizations of bugs tell the world. Perhaps the new narrative is one of improved mimicry. There is a first story that is a one-dimensional account of the organism to the world, and it represents the homology that the organism shares with its brethren. In the improvement in a mimic there is a second narrative, an analogy that hides the first. "No, I'm a ladybug", says the cockroach as it lies to the bird. Mimicry is a two-dimensional narrative.

The organism with the improved narrative, like all organisms, does not have self-knowledge. There are other organisms with their models of the world. Self-knowledge comes from how the models of the other organisms respond to the first organism's narrative that comes from its model. The mimic lies to the other organisms. The response comes from the narratives they tell using their respective models. The lie constitutes a two-dimensional narrative. Self-knowledge comes from how those other organisms respond to that two-dimensional narrative with their own models and narratives. Self-knowledge is therefore a reading and interpretation of the two-dimensional mimic narrative and is accordingly a three-dimensional narrative.

So if self-knowledge is a three-dimensional narrative, modification of self-knowledge therefore adds another dimension to make a four-dimensional narrative. All the time through the reinforcement phase, the realized organisms are telling narratives to the essence. The essence is altered by these reports of what is happening to the realizations.

15.3.3 | The Observer-Observation Complex

McCormick, Zellmer, and Allen (2004) lay out a protocol for working with observations, types, and essences. Their particular topic was of salmon sustainability in the Columbia River. In Figure 15.2 there are four levels in the study and analysis of an observed object. The concrete thing of primary experience, perhaps a dog, anchors the diagram at the bottom in level N. There is a metaobserver who sets the context by defining the arena of discourse, in this diagram at level $N+3$. Pattee (1978) identifies a set of restrictions that are universal to a particular discourse. He calls these restrictions laws. Laws are independent of structures and pertain to relative rates, as do the laws of physics. Biology, in its local discourse, has its universals that are not those in physics. In McCormick, Zellmer, and Allen's conception, the laws reside at level $N+2$. Inside the purview of the laws are rules, which represent the local, arbitrary, structures used by the scientist in defining models. Structures are by nature rate independent. The rules define some equivalence class, and they reside at level $N+1$.

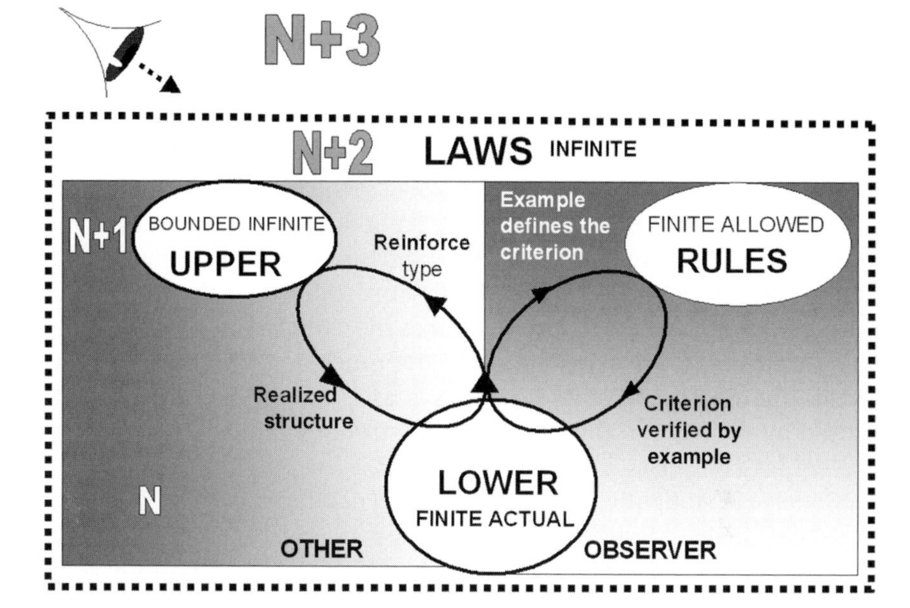

FIGURE 15.2 | The object of direct investigation lies at level *N*. Laws are local universals, and they reside at level $N + 2$. They offer possibilities for the system at hand. The upper is either an essence, role, or relational function residing at level $N + 1$, the level above the realized structure at level *N*. The upper represents what generates the patterns seen across the realized structures that arise for inspection at the bottom of the diagram. The upper resides in the other, the part of observation that arises above observer decisions. The rules pertain to the equivalence class to which the lower-level structure is assigned. Also at level $N + 1$, but on the side of observer decisions, is a cycle of definition and verification that amounts to a modelling operation. That cycle touches another cycle in the other at the point of the realized structure. Realized structures modify essences or roles, and essences and roles generate realizations or incumbents.

Although observers recognize things, and make decisions as to protocol, there are aspects of observation that happen beyond observer decisions. The *other* is a device used by constructivists, who make much of how the access is to data and observation, not external reality. The other is that which is other from observer decisions but still part of observation. The preceding roles or essences reside at level $N + 1$ but on the side of the other in the observer–observation complex. Although tied to models and protocols, essences and roles apply beyond observer decisions about observation. The point of modelling is to find suggestions from observation about what essences and roles might be. A scientist investigates dogs to gain a better handle on dogginess or salmon to address their role in the ecosystem.

The side of the other is usually about behaviour and dynamics of the thing at level N that is the focus of observation. Perhaps the observed complex system at level N is an organism. It is realized by the essence and it plays a role, but it is also assigned by observer decisions to the equivalence class defined by the rules at level $N + 1$. The observed complex system, at level N, links the model to the roles and essences (Fig. 15.2). On both the side of the observer's model and the essence is a cycle that exchanges between the level $N + 1$ and the complex organized structure at level N. On the side of the essence there is realization of the complex system, which in turn reinforces the essence through its actions and according to its experience. On the modelling side of the pair of cycles, the observed complex system is assigned to class. That loop is closed by verification that the complex structure is the special case of the equivalence class. For instance, "Yes, it is a salmon in distress".

Models are so tightly prescribed that you can think of them as degenerate narratives. Models and experiments are contrived to remove as much semantic uncertainty as possible, whereas proper narratives are semantically rich. Proper narratives leave room for alternative interpretations, even though they describe the same situation, such as when one story is the heroic creation of a dam on the Columbia River and another is the tragic loss of wild salmon populations. McCormick, Zellmer, and Allen (2004) address multiple approaches to salmon restoration. In the conclusion of their report, they insert a series of different modelling loops on the right-hand side of Figure 15.2 to give Figure 15.3. They string together the conflicting alternative models of fisheries belonging, respectively, to ecologists, fisheries experts, Native Americans, and toxicologists. The conflicts and uncertainties as to how the models of those different groups may contribute to salmon restoration are aggregated into a narrative. Even if the model is flexible enough to have variation in a stochastic component, there is a fixed distribution from which the stochasticity is drawn. So in that sense, the stochastic model is fixed. At some level all models have no slack and cannot be made to apply outside their own particularly specified universe.

15.3.4 | Dimensionality in Science

We are now in a position to press together the preceding arguments about the dimensionality on organismal narratives and the role of narrative in science. Scientists cannot require a tragic account to be superior to a heroic one, but they can insist that the narrative passes through the zero-dimensional points defined by the models and their implications for the narrative. As scientists raise the quality of the narrative by

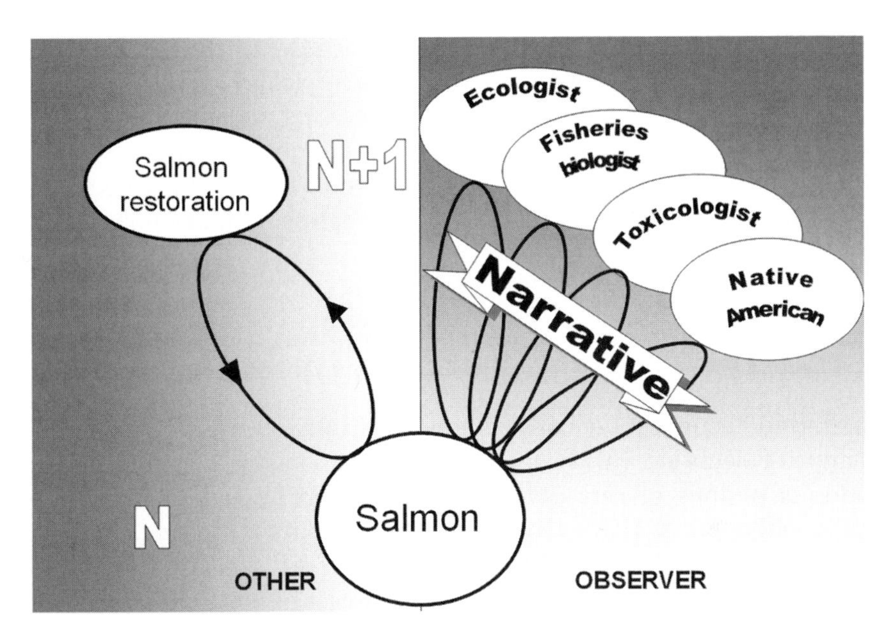

FIGURE 15.3 | As a modification to Fig. 15.2, multiple models are offered by different players in addressing the problem of sustaining salmon in the Columbia River, the situation described by McCormick, Zellmer, and Allen (2004). The ecologist might model ecosystem properties including persistent toxic chemicals. The Native American might see the issue as a lack of respect for the land with a prediction for seven generations. The fisheries biologist might model populations, and the toxicologist might model everything that might have an effect that is not caused by his industry clients but might cause them to be blamed in error. Each player monitors an aspect of fish sustainability. All the various models of player–stakeholder may be brought to the table and juxtaposed in a narrative. Only narrative can rise to the level of integration, the respective models being locked into their local discourses and paradigms.

challenging it with models, the models inform a fully one-dimensional narrative. The models say that if such and such applies then this particularity must follow. As the narrative becomes more sophisticated, like the mimic, it begins to invoke lies. Sophisticated models underpinning an experiment lie to the world, saying, "No, really, it is as simple as the experiment prescribes". Scientists in some way always manipulate the world in which the science is being conducted. The dimensionality of the scientists' narrative increases to become a two-dimensional story, a lie. The world then responds to the narrative, as it does to the mimic, by buying the story or not. The behaviour seen under the experimental conditions informs the scientists whether or not the assumptions fit the observations. The experimental results take the two-dimensional lie of the experimental contrivance and tell the scientists about their assumptions. This adds another dimension to the narrative, as did the response of the bird to the ladybug mimic. The scientists achieve self-knowledge. Science informs the scientists about themselves and the meaning of their experiences. Science is thus a three-dimensional narrative told back to us by the materials that we manipulate. Progress in science adds another dimension, making scientific progress a four-dimensional narrative.

15.4 | THE COMPLEMENTARITY OF NARRATIVES

In the nineteenth century, one of the issues in physics was about the very nature of electrons. This issue was premised upon science being able to address notions of *very nature*. Such a project implies a modernist view, where better science achieves higher quality by getting closer to the truth. However, the more insistent the physicists were in trying to determine the nature of electrons, the clearer it became that there was something wrong with the root question. It appears that the question of the very nature of electrons being either particles or waves cannot be settled. Unfortunately for the modernist view, both particles and waves are indicated by the slot experiment. In that experiment, individual electron hits on a screen (implying particle) accumulate to make a wave-interference pattern (implying wave) but only if a second slot is open. So, did the electron that made the hit split and go through both slots, or did it go though only one slot, somehow "knowing" that the other slot was open? Neither option is appealing. We seem to need both wave dynamics and particle structure to explain what we see, even if they play contradictory roles in explaining what we see. The trick is to abandon a unified view and accept a complementarity one in which electrons are both waves and particles, but that keeps the notions separate to avoid contradiction.

It is possible to opt for a unified treatment, but then contradiction inserts itself, and scientists can get any result they want by choosing this as opposed to that side of the contradiction. That is why the Tao is muddled. The master said, "I achieved complete enlightenment while sweeping the porch." Excitedly, the student inquired, "So what did you do then?" The reply: "I kept on sweeping the porch." Surely such an event as complete enlightenment is earth-shattering for the individual, but then complete enlightenment levels everything, including complete enlightenment. Completely enlightened the master keeps sweeping the porch; otherwise, he would not have achieved the ultimate insight. Here Russell's set of all sets gets underfoot. In the Buddhist school of archery you must try to hit the target, but without noticing. Notice that you have done it, and you need to do it again, to make sure, and then you are back where you started. Buddhist cycles of reincarnation are not viewed favourably as keeping hold on life, as we might feel in the West. Rather, they are something that complete enlightenment will ease so precisely that the enlightened can avoid having to do it all again. For all this, we still prefer to keep the duals apart so that we can continue in the tradition of logic where, as scientists, we feel comfortable.

Pattee (1978) raises the issue of complementarity in social and biological science. The harder science presses its models and data, the more often the issue of duality comes to the fore. He recommends dealing with the contradictions in duality by recognizing that two separate descriptions appear to be necessary. One description is law based, by which he means an account of the situation in dynamic terms as to what is possible. The other description is rule based and reflects what is allowed by the arbitrarily asserted names and structures, given that a self-assertive observer must be part of any observation.

Using Rosen's (1991) description of the use of formal models, we have compared the two cycles presented in the previous section of this chapter with Figure 15.2. We found that comparison between the two cycles in Figure 15.2 requires a complementarity. In a general statement, Rosen (1991) develops a cycle of scientific activity in which there is an encoding into the formal model and a reciprocal decoding back to the observed system (Fig. 15.4). Consider just the levels N and $N + 1$ in Fig. 15.2 and imagine a third cycle that links the two sides of level $N + 1$ in a cycle. Start by using one arm of each of the three cycles to travel anticlockwise around the triangle of upper, lower, and rules. Then do the same thing using the respective other arms of the three cycles to link the same triangle clockwise.

The essence links to the realized structure through its integrative phase. Using Koestler's (1967) word for realization, connect the upper and the lower in a process of *integration*. The individual structure relates to its equivalence set by acquiring a definition. It could be a human, a primate, or a graduate student, depending on the equivalence class to which it belongs. McCormick, Zellmer, and Allen (2004) might see a salmon as a member of a class, a small year class of returning adults in a failing fishery. The failing fishery is recognized in a set of rules for the equivalence class. The equivalence class relates to the essence through a process of decoding in Rosen's (1991) terms (as in Fig. 15.4), and the triangle is complete.

In the clockwise cycle, you might start with the essence at the upper. The essence is encoded into the model of the equivalence class (as in

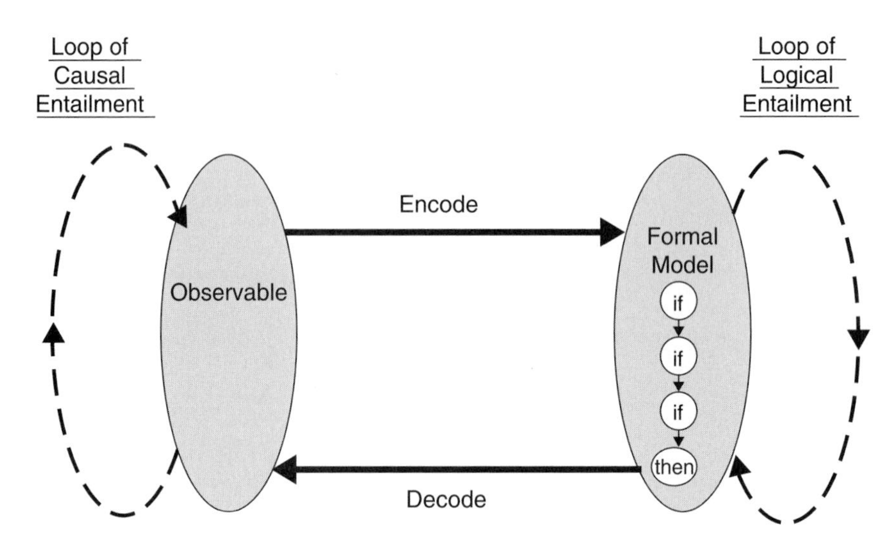

FIGURE 15.4 | Rosen (1991) notes that formal models are logically entailed and map variously well to areas of interest in observable systems. We presume that observable systems are entailed with some sort of material causality. The use of models is an encoding and decoding process among entailed systems as we look for anomalies, which are used to improve the understanding of the modelled system.

Rosen's diagram in Fig. 15.4). The coherence of the equivalence class and of the mapping of the structure to the class is captured in the act of verification that the structure meets the criteria of the equivalence class. Yes, there are only a few salmon in the year class you have identified as failing. The verified structure then asserts itself (Koestler's 1967 terminology) in an observable way. Poor year class does lead to failure of salmon sustainability (as in Fig. 15.3) The lower observable becomes self-assertive with regard to the essence, as was Nixon on the US presidency.

We are now in a position to reconstruct the cycles as two-way patterns, and we will obtain some significant improvement of clarity (Fig. 15.5). The cycle of model building is linguistic. The structure is assigned, recognized, and then labelled as fitting in the pattern of equivalence of the equivalence class. It fits the rules. On the other side is the cycle of the modified observable and how it contributes beyond the observer's models as to what is observed. That cycle is dynamic. Perhaps the salmon adapt to the new conditions, or perhaps their genetic diversity goes down as wild salmon are swamped by hatchery-raised fish. Without defined structure, on this side of the diagram is a loop of pure dynamics. Continuing the journey, this leaves the last side of the triangle across

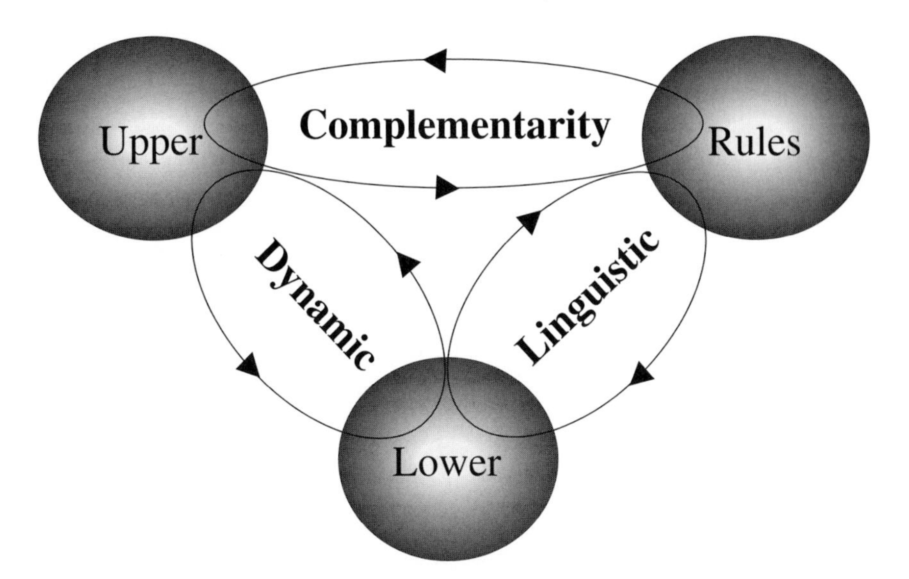

FIGURE 15.5 | When the two cycles of Figs. 15.3 and 15.4 are mapped onto each other, three cycles emerge. On the left are found dynamics. On the right are found the linguistics of definitions and models. Across the top is the cycle that links the model to the upper-level observed patterns, presumably arising from causal entailment. Because the class is defined and discrete, whereas the causes of the patterns in observables relate to dynamics, the link across the top creates a complementarity, such as in the wave-particle duality of electrons. You cannot know exactly where a particle is (rules) and know its velocity and direction (upper-level dynamics). The model and the material observables can never map without ambiguity and contradiction. The harder you try to make the mapping exact, the more in your face are the contradictions of the complementarity.

the top. This is the loop of assessing the model in the light of the observed changing realizations under influence of a shifting role or essence. We cannot see the whole change in the essence of salmon sustainability, but the scientist can go round the linguistic loop again and check for expected changes under a new model. However, a problem is likely to arise. If the model is linguistic and the essence is dynamic, then putting the essence and the rules in juxtaposition creates a duality. The intrinsic incompatibility of pure dynamic accounts and linguistic accounts forces a duality that is a complementarity. Complementarities are not helpful alternatives; they embody an intrinsic potential for contradiction. There is no resting place in the triangle because the whole process embodies a necessary struggle to which there can be no resolution. Ecologists are often muddled because they attempt to model a unity across the duality. The map is not the territory; the model maps only with difficulty to and from the essence, and the process is never over. A model is not an approximation of the essence. It does not approach the truth of the situation. Truth here is beside the point because it is the struggle that counts. Progress is not towards truth but is the way that earlier parts of the struggle precede and affect later rounds. And the fight does not stop with a winner at 15 rounds.

The only way to handle this situation to achieve periods of temporary respite for pedagogy and contemplation is narrative. Narrative rises above the models and allows the narrator to get out of the ecological muddle by deciding upon a story. Scientists can get predictions, but they are always bound by the rest of the seething mass of materiality mixed with conception. Narrative rises above models to deal with true complexity. Rosen (2000) says that a system is complex if you cannot write a model for it. He has said in several places that if you cannot write a single model that accounts for the system as experienced, then it is complex. With complementarity tied inexorably into the essence, you indeed cannot write a single model. Furthermore, this is a statement coming from principles, not local exigencies. Simple systems rely on laws and structures to inform the model. But complexity runs and hides in the essence. One cannot model your way there to get predictions, but one can tell stories about it. Analogy and narratives are vehicles for you to take responsibility in the process that allows science to work. Analogy and narrative are indeed the point of it all.

15.5 | WHY IT MATTERS IN APPLIED SYSTEMS

The disdain that normal scientists have for narrative and analogy is captured in the contemporary usage of the word *cartoon* for a graph that gives a general impression with ambiguous axes. For those who use the term, there are real graphs and then there are cartoons. Although formal graphs need to be explicitly scaled to convey relationships among explicit variables in models, the so-called cartoons tell the story of the whole picture.

The big ideas about breakpoints in adaptive management are captured nicely in Holling's (1986) figure-eight diagrams, which tell narratives more than offer a model to be tested. In the Holling diagrams, the system builds and then loses capital and organization (Fig. 15.6). I (Allen) have had vociferous discussions with other theorists in which I had to defend the worth of that diagram, the attacks complaining that the axes are at best ambiguous. The diagram starts in the lower left where capital is small and organization is weak. The system

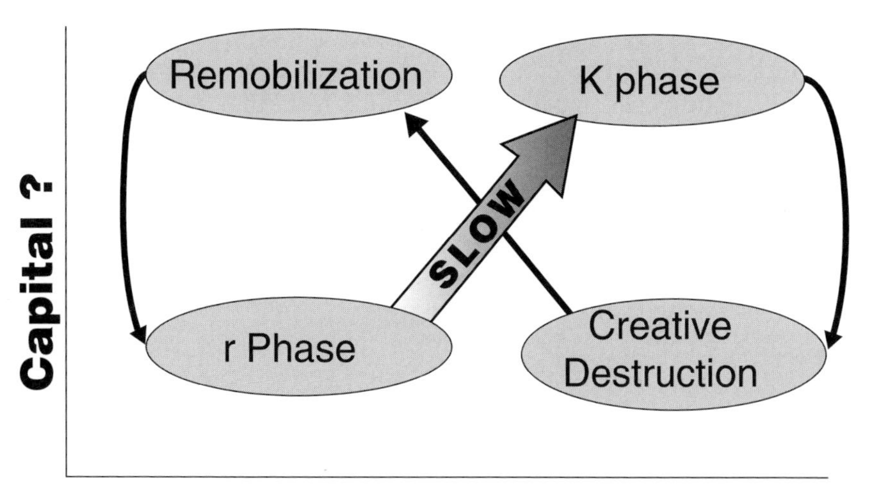

FIGURE 15.6 | Re-creation of Holling's (1986) figure-eight diagram, emphasizing the tentative nature of the labelling of the axes. This is not really a graph; rather, it tells a narrative. In narratives different types of things are linked together, so the axes cannot apply fully to all parts of the figure. Disparagingly, such narratives are often called "cartoons", as if they were inferior. Technical graphs, with their insistence on internal consistency, are often mistaken as superior devices. No, narratives are the point of the scientific exercise, and they are not in any way lesser than models and graphs.

then climbs to the right, taking a long time to build the capital that it defends with increased organization. In the end the ensuing rigidity puts the system in a configuration of an accident waiting to happen. When the system rapidly loses capital, it is the same tight connections of the capital-rich organized phase that drag the system down. It is a case of normal functioning in an unworkable posture that causes collapse. Without living, growing capital, the system loses organization. It discovers the capital from before, but that capital is not now in a vibrant state. Perhaps that capital is dead trees that used to be living forest in capital-rich organized phase. Without organized conservation, the system burns through the capital, such as when fungi consume dead trees, in contrast to the living forest that made more living trees. The system returns to the original phase of little capital and poor organization. The message is that there will be periods of growth to maturity followed by decline and a return to a new beginning. The four phases are a cautionary tale that warns against fighting collapse, because resistance will only make the inevitable fire or pestilence worse when it does come.

In the figure-eight diagram for ecological cycles, capital is destroyed by wind, fire, pestilence, senescence, or some other catastrophic event in what Holling calls "creative destruction". Note that collapse is not a pejorative here; it is the precursor for a remobilization phase that takes the system to a new springtime. Tainter (1988) comes to the same conclusion in the loss of complexity in societal collapse. After the collapse of the Western Roman Empire, production actually increased in southern Gaul when lands were only taxed, postempire, at a fraction of imperial levels. People could again afford to farm. The decisions of the elite that drive societies to the bottom of diminishing return curves are rational. So are the decisions of the rank and file to abandon societal complexity, because at that time the burden of complexity is not worth it. After the collapse phase, the system capital reemerges but under a regime with little organization. Both Attila the Hun and the fungi after a budworm outbreak think that the collapse is just the right thing.

But notice here that there is ambiguity in the axes of the figure eight. I (Allen) have asked Holling to clarify what the axes should be. Holling preferred to ask my opinion rather than clarify it in his own terms. That was just fine because there are valid alternative interpretations. One can question the degree of organization appropriate to characterize the creative destruction. The figure eight defines that phase as highly organized but with lost capital. True, there are still a lot of connections in the collapse phase, as fire or pestilence enters the positive feedback phase, but you could question whether that really is a highly organized phase. Then there is the matter of the capital, say, the dead trees after a budworm outbreak, reappearing immediately as carbon for fungus to decay in the redistribution phase. Did the capital ever disappear in the creative destruction phase?

It takes a narrative to deal with the large issues, such as the coming crisis in resources facing the industrial world. That narrative is captured in a chart that appears in various versions in Allen, Tainter, and Hoekstra (2001) and Allen *et al.* (2001). That diagram is contradictory in the manner of Holling's figure-eight diagram. The authors in those works employed a graph of marginal return curves over time, where at the beginning of a period of exploitation more effort gives increasing returns but later greater effort and elaboration return less and less. Locally, marginal return increases at first, but later diminishing returns cause declines over time. The chart ordered these convex curves in a sequence that ended up with highly organized societies. If it was marginal return on the ordinate, then the cycles of marginal returns should as a whole sequence fall, not rise. This decline in marginal return would be because there is greater effort put into increasingly complex societies and so poorer marginal returns. However, the grand pattern in their graph of successive cycles on marginal return ascends, because I (Allen) and my respective colleagues wished to capture not only the local cycles of marginal return but also the increasing capital and organization over the long term. The local periods are marginal return curves over the short run, but the string of shifts in marginal returns are a matter of increasing capital and complexity.

Both Holling's figure-eight diagram and the string of marginal returns are narratives, not graphs. Being narratives, they offer the contradictions of a unified complementarity. But that is all right; both charts have given the story that needed to be told. The Allen *et al.* (2001) graph showed cycles of short bursts of self-organization emerging on a high-quality resource: hunting-gathering, imperial invasion, and fossil fuel. These phases alternate with longer periods of highly organized activity to exploit a lower-quality resource: agriculture, imperial taxation, and now a shift to renewable energy sources. Extinction of species under these human regimes appears in low-quality phases such as agriculture, not in high-quality phases such as use of fossil fuel, because to acquire low-quality resources, individuals must go everywhere to get them. Renewables will therefore not be the ecological nirvana that green politics suggests. Its proponents could never have come to that conclusion using only consistent models. The big issues need narratives if we are to address them effectively.

Applied ecology usually has to employ several devices in accounting for the problem and its various phases of solution. That is to say, one internally consistent model will not usually be sufficient. Applied ecology, taken to its conclusion, requires narratives to link the various phases of modelling. The strategy of adaptive management resembles the telling of a narrative. As one model fails, adaptive managers attempt to improve the narrative they tell of system dysfunction with a new model that is then tested in real time in ensuing management action. It is important to cast ecological systems in terms of consistent models, but the job is not properly useful until the narrative has been told.

15.6 | THE POSTMODERN PARADIGM IN ECOLOGY

The authors of this chapter are all comfortable with a postmodern ecology and see it as the coming revitalization of a somewhat tired discipline. In this volume, filled with reflections of seasoned ecologists at counterpoint to a set of young bloods, we, in this chapter, come in with a pincer movement. The senior author here is the old guy, whereas the junior authors are younger even than our young-blood counterpart (Peterson). We have outflanked resistance on both sides. Let me (Allen) establish my credentials as one of the geezers by reminiscing about the old days. How appropriate to begin the conclusion of this chapter with a narrative. The point of this particular story is to show that the hopes and expectations we had for the sixties did not come to pass but the sixties have had a lasting effect we did not expect at the time. It was the decade of the first stirrings of postmodernism, the notion of paradigm being part of that.

When I turned up to take my new position at the University of Wisconsin in Madison, the place was in some disarray. Only days before, the Army Math Research Center, housed in the building next door, had been blown up. It was the last hurrah of Madison radicals, because a graduate student in the building, working on something that was nothing to do with the army, had been killed in the blast. That sobered all but those who remain aging hippies to this day. Some 3 years later, on the first spring day of fair rioting weather, students running naked through the streets, instead of rioting, was the preferred display. It was called streaking. At that point we all knew the sixties were over. The radicals became doctors and lawyers. One outstanding radical became the fiscally responsible, business-friendly mayor of Madison. A terrified business community was bemused as his first months in office were spent going over the books to achieve at last the best credit terms with a triple A bond rating for the city. He was an excellent mayor, his greatest pride being no more radical than the bike path that goes around the lake. Even so, the sixties were heady times, and their bequest remains large, albeit not in the radical politics that were supposed to change everything. Rather, the legacy is in ideas about investigations of all sorts, including a postmodern ecology.

The change brought by the sixties was in intellectual posture, and the notion of paradigm has been part of that legacy. Undergraduates nowadays do not know who Marshall McLuhan was (I know because I check each year in my big class). That McLuhan turned up, in person, in a Woody Allen movie is lost on most of my charges because well over half of them have never seen *Sleeper* or *Manhattan*. The darlings of the sixties are hardly ever read today: Thomas Kuhn, Marshall McLuhan, Lynn White, Garrett Hardin, and Michael Polanyi. Perhaps Hardin deserves to be forgotten, but the rest do not. Kuhn (1962) along with McLuhan, was an early postmodernist from whom sprang post-normal science. These days undergraduates turn up already postmodern, and my mechanistic colleagues have to beat reductionist realism into them. Not so in the 1960s, because high school pupils then had not only been taught but also had accepted the modernist view that sanctifies science as the seeker of truth. It was part of the secularism of those days. But then Kuhn showed us, for the first time, how science has a socially constructed context that is undeniable. He used the word *paradigm* for these socially constructed basins of attraction. The word was borrowed from grammarians, who use it to signify the full context of a given word, every case of the root together. Although the radicals are now doctors and lawyers reaching retirement age, the sixties were not lost, because the notion of paradigm, and other ideas beyond the stifling modernist orthodoxy of the 1950s, survives in how we think about ideas today.

According to Kuhn (1962), paradigms are identifiable through common methods, vocabulary, and shared concepts. Paradigms function as a framework that defines what are valid and interesting questions. Individuals belonging to a paradigm understand, but reject as old fashioned and unimportant, the questions of paradigms that theirs replaced. Paradigm adherents are more vociferous, and energetically reject the new questions raised, when they get their comeuppance from a yet newer paradigm. Under siege, the old timers do not understand what is being queried and dismiss the vocabulary of new paradigms as pointless or even pretentious jargon. Old paradigms do not go away; they remain frozen in time, useful for what they did well before. We do not build bridges with quantum mechanics. Old paradigms are rejected as not asking the pressing questions. The old guard suffers much consternation because the young adherents to the new jargon simply refuse to ask what the old guard still regard as the telling questions. Anyone irritated at the pretentiousness of this chapter may wish to reflect to make sure that they are not an example of someone still using an old paradigm for what used to be important. We may be wrong, but we suspect that the reader who feels aggravated may exist only at the cutting edge of conventional wisdom. Within the social science discipline of history of science, Kuhn's notion of paradigm has been nitpicked to death. By contrast, we practicing scientists know all about those tacit agreements not to ask certain questions. There is indeed structure to scientific revolution, particularly in the little revolutions in which editors and cliques resist and champions of radical new findings fight on to acceptance of some new method. Dark humour making the rounds says that science moves forward one funeral at a time.

In resurrecting narrative, I may be a stuffy old guy who refuses to get with the times, but it should be no surprise that I do not see it that way and neither do my young, postmodern coauthors; I am not dead yet. Instead we see a certain irony in the modernist view of young postdoc mechanists being behind the times. The prevailing method in ecology is modernist, not postmodern. But some of us have been to pay homage to the Wizard of Oz and saw instead the modernist man behind the curtain. Losing faith in the self-deception of the project of the Enlightenment, we have preferred to move from modernist realism.

Ecologists, like many scientists, have not properly grasped Kuhn's meaning. They acknowledge that there was some politics in defence of old paradigms. Even so, many think their new paradigms replaced old ones because an unseen hand has guided science closer to the truth. Later Kuhn writings (e.g., the postscript in his second edition of 1970) appear to soften on the issue of realism and truth, but remember that Darwin's later editions of *On the Origin of Species* were downright Larmarkian. Kuhn must be allowed not to follow his own early hard line, but I think we should. For modernist ecologists, somehow, the old *they* had paradigms, but *we* now have a proper, or at least more dispassionate, view. No, we all have paradigms, noticed or not, and that certainly includes the authors here and other adherents of systems analytic approaches. Because paradigms come crashing down all the time, there is no reason to suppose that our paradigms have more of a corner on the truth. There is a progression in the truism that later paradigms are set in the context of previous paradigms, but that in itself does not mean later paradigms are truer paradigms. Ultimate truth has nothing to do with paradigms.

Social construction of scientific knowledge is still misunderstood by modernists. Strident modernism accuses the postmodernists of meaning that any of us can construct anything we want and that there is no discovery of anything any more. I and my colleagues (Allen *et al.* 2001) pointed to the irony of the inflammatory hyperbole used by defenders of normal, modernist science as they accuse postmodernism of excess. No, a postmodern view merely focuses on the quality that resides inside the scientific process. Quality should not be assessed by assertions of proximity to an undefined external truth. The quality internal to the scientific process itself arises in meticulous, repeatable work that raises structural quality of the enterprise. Another sort of quality derives from the constant challenge, which infuses science with creativity; that is called *dynamic quality*.

So for many ecologists who want to get on with saving species and restoring prairies, what we say here might sound like a bunch of esoteric philosophical niceties that have nothing to do with getting ecology done. We disagree. There are too many species, and too many sites, with too many nutrients for us to find the simple truths of the world by studying one more place. The methods of modernists are sound and there is the potential for much improved understanding, but we argue that normal science can be done for different reasons than a desire to catalogue everything and find ecological truth. We argue for a clearer vision as to context of ecology so that the normal science that hopes to save prairies is properly directed. We refer you to the following chapter by Peterson, which deals more explicitly with how a postmodern perspective is advancing conservation and management efforts with systems approaches.

The important ecological issues of today relate to the management of a gigantic human presence on this planet. That can come down to the effect on some local site of prairie diversity, and we applaud good work of that local sort. But even those small, concrete, real-time projects are meaningless unless they are properly embedded in the larger context. Even fancy metapopulation studies are moot if the context is not specified intelligently and creatively. The most elaborate nutrient cycling measurements mean nothing until the broader implications are set in place. Sooner or later we have to deal with complexity. At some point, exact specification is not possible. The way to tie the bits together and make them meaningful is narrative.

As the reader should be aware by now, narratives are not just simple stories. We have laid out some of their nuances. Particularly ecosystem studies, the ones that deal with the concrete consequences of intangible nutrient cycles, need to achieve meaning. Science is at the limits of what it can achieve in its present modernist form. The discipline needs help. That help may well come from those for whom narrative is a serious mode of expression. Postmodernism came as much from literary criticism as from anywhere. With precision use of ideas, the humanities can indeed help science keep control even as complexity and the need for upper-level meaning comes down on the project. Wuennenberg is a student of linguistics, that strange science that is as often as not housed in a humanities faculty. All of us are keenly interested in the interface among all the disciplines across the gamut of hard and soft science, art, and the humanities. In this chapter we have attempted to use the vehicle of ecosystem and community studies to put a larger vision back into ecology.

In conclusion, we wish to dedicate this effort to the life and work of James Kay, a friend and colleague who was taken from us recently and far too soon. He, more than anyone we have known, sat at the interface of science and meaning. His training was in hard science, he was a physicist engineer. But in his life and teaching at the University of Waterloo, he reached across to the social and soft sciences. He had grand visions of world sustainability, and he cast that creatively in the thermodynamic terms central to ecosystem science. Being a hard scientist by any standards, he had the confidence to look at the soft meaning of ecosystems science. Physics and engineering notwithstanding, he also worked with degraded physical and social environments in Bangladesh. Once the street cleaners were defined with social justice by the community, then the rest of the stakeholders could begin to take proper responsibility. Kay's work on the integration of the Huron Natural Area was masterful. It started as an undeveloped tract left behind by the development of the city of Waterloo in Ontario, Canada. He and his team turned it into a working place, with a grandfathered, small industrial component put in the context of a place in which biodiversity appears safe for a while. If you do not believe the analysis that we have presented here, living proof is to be found at Kay's Web site (http://www.jameskay.ca), an astonishing place on the Net, where complexity is taken seriously and is applied in practical terms.

ACKNOWLEDGEMENTS

This work was supported by the National Science Foundation awards NSF DEB-0083545 and NSF DEB-0217533, both administered through Limnology at the University of Wisconsin.

REFERENCES

Allen, T.F.H., & Hoekstra, T.W. (1992). "Toward a Unified Ecology." Columbia University Press, New York.

Allen, T.F.H., Giampietro, M., & Little, A. (2003). Distinguishing ecological engineering from environmental engineering. *Ecol. Engin.* **20**, 389–407.

Allen, T.F.H., Tainter, J.A., & Hoekstra, T.W. (1999). Supply-side sustainability. *Syst. Res. Behav. Sci.* **16**, 403–427.

Allen, T.F.H., Tainter, J.A., & Hoekstra, T.W. (2001). Complexity, energy transformations, and postnormal science. *In* "Proceedings of the Second Biennial International Workshop on Advances in Energy Studies" (S. Ulgiati, Ed.). SGE Ditoriali, Padova, Porto Venere, Italy, pp. 293–304.

Allen, T.F.H., Tainter, J.A., Pires, J.C., & Hoekstra, T.W. (2001). Dragnet ecology, "Just the facts ma'am": The privilege of science in a postmodern world. *Bioscience* **51**, 475–485.

Bateson, G. (1980). "Mind and Nature." Bantam, New York.

Bonnifield, P. (1979). "The Dustbowl: Men, Dirt, and Depression." University of New Mexico Press, Albuquerque.

Cohen, J. (1971). Mathematics as metaphor: Book review, "Dynamical System Theory in Biology" by R. Rosen. *Science* **172**, 674–675.

Cronon, W. (1992). A place for stories: Nature, history, and narrative. *J. Am. Hist.* **78**, 1347–1376.

Drake, J.A. (1992). Community assembly mechanics and the structure of an experimental species ensemble. *Am. Nat.* **137**, 1–26.

Funtowicz, S.O., & Ravetz, J.R. (1992). The good, the true, and the postmodern. *Futures* **24**, 963–976.

Gleick, J. (1987). "Chaos." Viking Penguin, New York.

Holling, C.S. (1986). The resilience of terrestrial ecosystems: Local surprise and global change. *In* "Sustainable Development of the Biosphere" (W.C. Clark & R.E. Munn, Eds.). Cambridge University Press, Cambridge.

Koestler, A. (1967). "The Ghost in the Machine." Gateway, Chicago.

Kuhn, T.S. (1962). "The Structure of Scientific Revolutions." University of Chicago Press, Chicago.

Kuhn, T.S. (1970). "The Structure of Scientific Revolutions," 2nd edition, enlarged. University of Chicago Press, Chicago.

Leopold, A. (1949). "A Sand County Almanac." Oxford University Press, Oxford.

Levin, S.M. (1997). Putting the shoulder to the wheel: A new biomechanical model for the shoulder girdle. *Biomed. Sci. Instrum.* **33**: 412–417.

Levin, S.M. (2002). The tensegrity-truss as a model for spine mechanics: Biotensegrity. *J. Mech. Med. Biol.* **2**, 375–388.

Linderman, F.B. (1962). "Plenty-Coups: Chief of the Crows." University of Nebraska Press, Lincoln, NE.

McCormick, R.J., Zellmer, A.J., & Allen, T.F.H. (2004). Type, scale, and adaptive narrative: Keeping models of salmon, toxicology, and risk alive to the world. *In* "Landscape Ecology and Wildlife Habitat Evaluation: Critical Information for Ecological Risk Assessment, Land-Use Management Activities, and Biodiversity Enhancement Practices" (L.A. Kapustka, H. Gilbraith, M. Luxon, & G.R. Biddinger, Eds.). ASTM International, West Conshohocken, PA.

Needham, J. (1988). The limits of analysis. *Poetry Nat. Rev.* **14**, 35–38.

Nicolis, G., & Prigogine, I. (1977). "Self-Organization in Nonequilibrium Systems: From Dissipative Structures to Order through Fluctuations." Wiley, New York.

Pattee, H.H. (1978). The complementarity principle in biological and social structures. *J. Soc. Biol. Struct.* **1**, 191–200.

Pirsig, R.M. (1992). "Lila: An Inquiry into Morals." Bantam, New York.

Prentice, I.C., & van der Maarel, E. (1987). Preface. *In* "Theory and Models in Vegetation Science" (I.C. Prentice & E. van der Maarel, Eds.). Junk Publishers, Boston, p. 3.

Rosen, R. (1991). "Life Itself." Columbia University Press, New York.

Rosen, R. (2000). "Essays on Life Itself." Columbia University Press, New York.

Schrödinger, E. (1967). "What Is Life? And Mind and Matter." Cambridge University Press, Cambridge.

Tainter, J.A. (1988). "The Collapse of Complex Societies." Cambridge University Press, Cambridge.

Tansley, A.G. (1926). Succession: Its concept and value. *In* "Proceedings of the International Congress of Plant Sciences" (B.M. Duggar, Ed.). Banta, Menasha, WI, pp. 677–686.

Tansley, A.G. (1935). The use and abuse of vegetational concepts and terms. *Ecology* **16**, 284–307.

Worster, D. (1979). "Dust Bowl: The Southern Plains in the 1930s." Oxford University Press, New York.

16 | ECOLOGICAL MANAGEMENT: CONTROL, UNCERTAINTY, AND UNDERSTANDING

Garry D. Peterson

16.1 | INTRODUCTION

Ecological management is an important and exciting topic for ecological research for practical and theoretical reasons. Practically, ecological management is important because human action is transforming the ecological services that underpin human well-being. Humans are transforming the face of the earth, eliminating and introducing species, accelerating biogeochemical cycles, and modifying the atmosphere at a rate and scale never before seen in human history (McNeill 2000). Inadvertent and frequently surprising ecological transformations have stimulated people to manage their actions to attempt to produce ecosystems that they desire rather than what they inadvertently have. As Stewart Brand wrote in the 1968 edition of the Whole Earth Catalog, "We are as Gods and might as well get good at it." Unfortunately, almost 40 years later, we still need to improve the competence, finesse, and elegance of ecological management. Ecological management theory has focused on developing management approaches for the controllable, well-understood situations rather than the difficult-to-control, uncertain situations people increasingly confront. Ecologists need to develop new, integrated, cross-scale theoretical frameworks for ecological management.

Studies of ecological management also offer an opportunity to improve ecological theory. Ecological management encourages researchers to work on large scales (Walters & Holling 1990), stimulates new ecological questions, and prompts exploration of interdisciplinary frontiers. Ecologists working on applied and management issues such as eutrophication, acid rain, and climatic change have fundamentally advanced the scientific understanding of nature—partly from being forced to look at nature in new ways (Groffman & Pace 1998). Ecological management exists at the interface of people and nature, and this interface is something fundamentally different from either of them alone. Nature constrains human ambitions, and people provide a dynamic reflexivity missing from ecological dynamics. The quest to understand this ever-evolving relationship continuously generates questions that demand new research methodologies and offer exciting challenges to scientists.

In Chapter 15, Tim Allen and others describe the theoretical aspects of the role of the human actor in ecosystems. In this chapter I address the role of the human actor in ecosystems in practice by examining changes in approaches to ecological management. I will sketch a history of ecological management that focuses on the role of control and uncertainty. I will describe traditional approaches that have been developed for controllable, low-uncertainty situations and newer approaches that attempt to cope with uncertainty and difficult-to-control situations.

16.2 | A HISTORY OF ECOLOGICAL MANAGEMENT

Ecological management, historically, has not been very ecological. Ecological managers have tended to view nature as a place where isolated, individual resources exist, and they have focused on the "optimal" use of these isolated resources. In particular this was done by attempting to achieve maximum sustained yield (MSY). This approach was developed in the twentieth century and used quantification and technical understanding, with bureaucratic control, to increase the production of particular resources. MSY has been a guiding philosophy of human-dominated ecosystems such as agriculture, forestry, hunting, and fishing. Ecological management based on MSY has been primarily developed in forestry and fisheries.

16.2.1 | Forestry

In nineteenth-century Germany, government forest managers sought to improve the productivity of state-owned forest land (Scott 1998). They did this by measuring and monitoring forest growth then modifying the forest to improve growth. Foresters used this approach to conceptualize forests as a stock of timber growing at a specific rate. To make forest management more effective, forests were carefully seeded, planted, and cut to create ecosystems that were easier to measure and manipulate. These techniques reduced diverse, complexly structured old growth forests into ordered rows of a single species of even-aged trees.

Initially these simplified ecosystems were successful. The even-sized and even-aged trees lowered the costs of management and logging, and they enhanced the value of timber produced by forested lands. However, this success had unanticipated ecological and social costs. Simplifying the ecosystem into an apparatus for converting sunlight into wood ignored the ecological functions of bushes, shrubs, and litter in sustaining the forest. Simplifying the forest to maximize wood production greatly reduced many non-provisioning ecosystem services, such as decomposition and pollination that maintained the forest, enhanced its growth, and protected it from the pest outbreaks. Ecological simplification reduced the cycling of nutrients and so reduced forest production. The regular and even-aged character of the forest increased its vulnerability to disturbance. German foresters attempted to solve these problems by using management to provide some of the services they had eliminated. They added nest boxes and reintroduced spider species. They encouraged schoolchildren to raise ant colonies for the forest. The net results were ecologically impoverished forests that required higher amounts of human inputs. Despite these failings, this model of scientific forest management was exported worldwide with a similar pattern of success followed by failure, a pattern described as the "pathology of natural resource management" (Holling 1986, Holling & Meffe 1996).

16.2.2 | Fisheries

A similar history of MSY can be found in fisheries. It was first applied to fisheries in the early twentieth century. Following World War II, it was formalized, extended, and extensively applied to stock assessment and management (Schaeffer 1954, Ricker 1975, Clark 1985). Fisheries management was largely based on fitting a curve to the relationship between fish reproduction and population size and thereafter managing to produce a harvest level that maximized the catch, the economic value, or some other aspect of the fishery. Difficulties in measuring fish populations, identifying stocks, and analyzing environmental variation all provide technical challenges to this technique (Hilborn & Walters 1992). Political interference in the process of stock analysis has also been a persistent problem (Hutchings, Walters, & Haedrich 1997).

A classic case of fisheries decline is the collapse of the Newfoundland cod. Cod were harvested for centuries until 1993, when the fishery collapsed and was then closed at a huge cost to Newfoundland coastal communities (Harris 1998). In this case, the fisheries management by Canada's Department of Fisheries and Oceans favoured large-scale offshore fisheries. The mobile and technologically sophisticated offshore fishers were able to maintain high catch rates despite declines in cod abundance. Furthermore, because it fished a larger area and was able to pursue many different groups of cod, the offshore fleet was able to maintain high catches of northern cod off Newfoundland when the less-mobile inshore fleet could not (Finlayson 1994). The persistence of high catches, and the focus of managers on MSY, led to overestimates of stock size despite requests of inshore fishers for greater regulation based on their perceptions of fisheries' decline (Walters & Maguire 1996). The focus on MSY-based management devalued information that did not fit stock assessment models, and the objective of maximizing the value of the fishery ignored conflicts between local, long-term fishers and economically larger, multinational fishers (Finlayson 1994).

Unfortunately, the idea of MSY has been more resilient than the fisheries it has been used to manage. Because of its many problems, fisheries scientists have been urging the abandonment of MSY approaches for more than 25 years. For example, prominent fisheries scientist Larkin (1977) argued that MSY should be abandoned: (1) to reduce risk of catastrophic decline of populations, (2) to recognize ecosystem interactions, and (3) because MSY is not necessarily economically desirable. MSY focuses management on the fisheries catch when often what people care about is how fishing interacts with an ecosystem. A fishery can kill non-fished species such as dolphins or turtles and, by altering the relative abundances of predators and prey, influence the growth rates of other species. Similarly, MSY provides little insight into issues such as the design of marine protected areas or the assessment of the effect of nutrient pollution on marine populations.

Despite these limitations, the use of MSY and other optimization-based approaches continues. Why? A simple answer is that it is a well-developed approach to environmental management that has no real alternative. Why there are no alternatives is a more complicated question—with several potential answers. First, in many ecological situations evidence of short-term success is present before evidence of long-term decline (Carpenter, Ludwig, & Brock 1999). Early success tends to lead managers to underestimate the chance of failure and the ease with which declines can be reversed—until it is too late. Second, the people who benefit from MSY management and those who lose from its failures are often different groups of people. For example, MSY fishery management often favours highly mobile modern fishing fleets. These fleets can relocate to another fishery following a collapse, but local fishers cannot. These mobile fishers are often more politically and economically powerful than the local fishers. In many cases the establishment of an MSY regime involves privatizing an ecological commons informally regulated by locals. In such a situation, larger, external users are provided with access to a local resource that is then quickly degraded, undercutting the ability of local users to persist (Berkes *et al.* 2001). Similarly, nineteenth-century German forestry excluded traditional users of nontimber forest products who used forests for grazing, as a source of bedding, and as medicine. Third, and most generally, simplistic standardized rules such as MSY are compatible with organizations that seek to avoid paying for the consequences of their actions. The negative effects of harvesting, (e.g., declines of other species and habitat destruction) are not explicitly considered by management that focuses on stocks of a single species, thus allowing organizations to benefit from extractive activities and someone else, or nature, to bear the costs of their actions (Scott 1998).

16.2.3 | Ecosystems and Ecological Management

Optimization approaches have had great success in engineering but have a poor track record in ecosystem management. This is probably because engineered systems are designed to behave predictably. Optimization approaches assume that the effects of an event can usually be treated as if they were local and immediate and can attenuate over time and space.

Ecologists know that ecosystems are neither linear nor atomized. Rather, they are full of indirect effects, thresholds, and cross-scale connections. These perspectives on the nature of ecological effects are illustrated in a simple figure that shows the distribution of effect in which causation is local versus a case in which effects are mediated through complex interactions, such as food webs (Fig. 16.1).

In some situations, Figure 16.1A is a good description of the world. For example, the effects of cutting a tree in a forest attenuates over time and space, thereby making its effects less predictable. However, in other situations, Figure 16.1B is a better model. For example, persistent organic pollutants, such as polychlorinated biphenyl and mercury, have high concentrations in Arctic animals, even though the Canadian Arctic is distant from the locations from which these pollutants are released.

These different models of how human action transforms ecosystems can be used to separate intended engineering of ecosystems from their

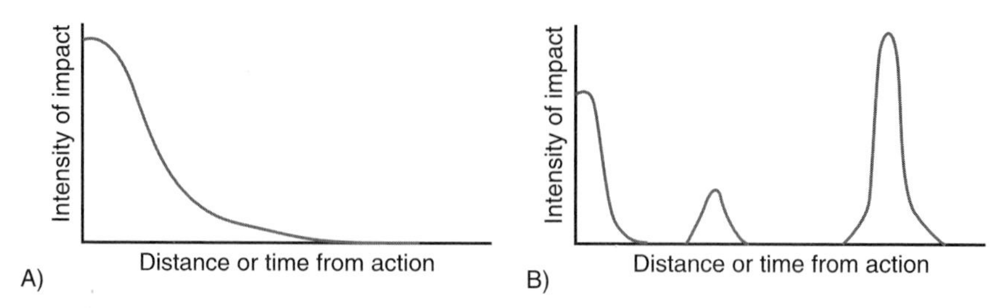

FIGURE 16.1 | The distribution of ecological effect because of human action. (A) Simple linear world. (B) Surprising world. Adapted from Holling (1978).

inadvertent transformation. People frequently alter ecosystems to purposefully alter the provisioning of ecosystems services. For example, farmers engineer agricultural ecosystems to emphasize the production of a desired set of ecosystem services, such as agricultural production, at the expense of others, such as the production of fresh water and wildlife. When such trade-offs are well understood, then the simple model of effects applies and optimization approaches such as MSY can be usefully applied.

Human action produces a lot of intentional ecological change. For example, although people understand that agricultural runoff will impair water quality, the costs of low water quality are usually borne by someone other than those benefiting from agriculture. These inadvertent changes can be local and direct, such as when agriculture pollutes a neighbouring lake, but frequently they are distant or indirect. For example, the large "dead zone" in the Gulf of Mexico is an area of low oxygen produced by fertilizer use practices on farms across the US Midwest that the Mississippi River aggregates and transports to the Gulf of Mexico (Foley *et al.* in press). In such situations, optimizing farm yield does nothing to address a distant but serious ecological problem. These inadvertent ecological changes are largely caused by the failure of institutions to prevent or punish people from harming the person or property of other people. As the scale and intensity of human action have increased so has the frequency of surprising, inadvertent ecological changes. A classical example is the destruction of the ozone layer by apparently safe chlorofluorocarbon. Surprises also occur as change percolates through a transformed ecosystem. For example, the reintroduction of wolves to Yellowstone National Park has changed elk foraging behaviour to the extent that riparian tree growth has increased with potential benefits for water quality (Ripple & Beschta 2003). The ubiquity of surprise and unwanted change in ecosystems suggests that although optimization approaches have a place in ecological management, they are dangerous to apply in many, if not most, situations.

16.3 | A THEORETICAL FRAMEWORK FOR ECOLOGICAL MANAGEMENT

The social-ecological situation in which ecological management is being attempted will determine what type of ecological management could be successful. Drawing on previous work in adaptive management (Holling 1978) and postnormal science (Funtowicz & Ravetz 1993), I argue that two key aspects of any social-ecological situation that influence the success of ecological management are the extent of uncertainty about a system's behaviour and the degree to which a system can be controlled.

Most ecological management theory has been developed for situations in which uncertainty is low and controllability is high. Although such situations exist, most current environmental problems, such as concern over the ecological effects of transgenic organisms or climate change, are situations in which control is difficult and uncertainty is high (Fig. 16.2).

Uncertainty and partial control are difficult to avoid in real ecological management situations because ecosystems are complex adaptive systems (Gunderson & Holling 2002; Gunderson, Holling, & Light 1995; Holling 1978; Lee 1993; Walters 1986). Ecosystems are complex adaptive systems whose behaviour emerges from the interaction of adapting components (Levin 1999). Their behaviour is frequently nonlinear, sometimes resisting large perturbations and other times transforming because of small perturbations. Social-ecological systems are even more complex (Westley *et al.* 2002), with surprises occurring frequently (Gunderson & Holling 2002). In the next sections, I explain what I mean by uncertainty and controllability.

16.3.1 | Uncertainty

Ecological management depends on making decisions that improve a situation. Consequently, making decisions requires some model of how present actions will alter the future. That is, ecological management must evaluate implicitly or explicitly the possible outcomes of potential actions.

Current approaches to ecological management often cope with uncertainty in prediction by fitting models to data and comparing competing models (Clark *et al.* 2001, Hilborn & Mangel 1997). The aim is to discover

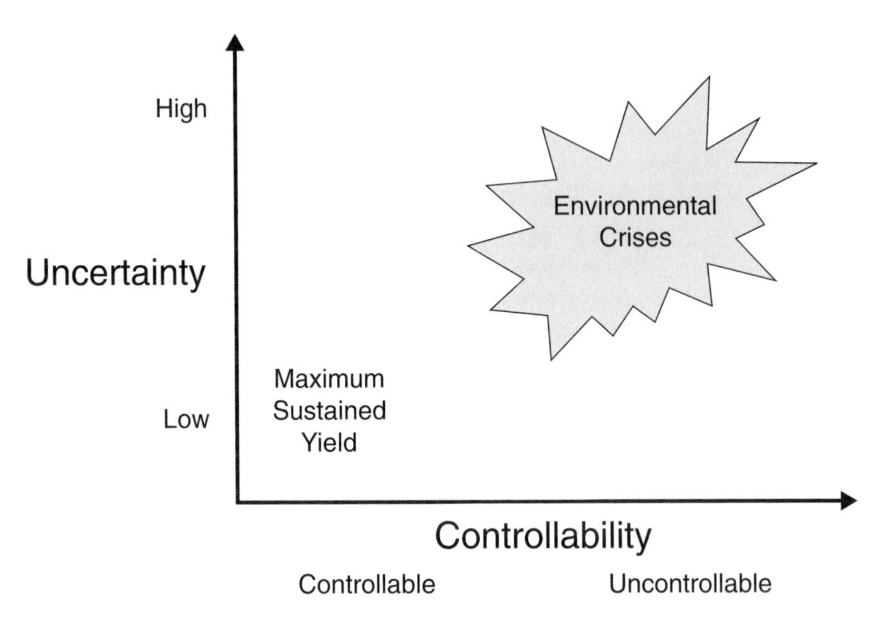

FIGURE 16.2 | An ecological management situation can be represented in a two-dimensional space defined by the uncertainty that surrounds an issue and the degree to which the system of management is controllable by management modified. Most ecological management theory, such as the concept of MSY, has been developed for situations that are relatively certain and controllable; unfortunately most environmental problems are poorly understood and only weakly controllable. These types of situations represent the frontier of research on ecological management.

which model or models appear to best forecast future behaviour of the ecosystem. Often the consideration of uncertainty focuses only on the prediction errors of a single model rather than on the credibility of the model structure. The credibility of a given model depends on the other models with which it is compared and on the data available. If the data represent only a subset of the potential behaviour of the ecosystem, then the model comparison may be biased, or the appropriate model may not even be discovered, because the behaviours of the ecosystem relevant to the appropriate model have not been observed. If an important model is omitted from the set under consideration, substantial errors can occur in assessing credibility, making predictions, and choosing management actions. Thus, model uncertainty has critical implications for ecological management, but the assessment of model uncertainty is limited by the range of ecosystem behaviours observed in the data and by the diversity of models created by the analyst. However, it is impossible to consider all possible or even all plausible models. Consequently, model uncertainty means that even well-considered management approaches should expect surprises (Peterson, Carpenter, & Brock 2003).

The future state of human drivers is even more difficult to predict than ecological dynamics because of rapid social change and the reflexive nature of people. Ecological policy is often based on existing technologies and values or on the continuation of existing trends. However, new inventions frequently change the way people affect ecosystems. For example, protected areas designed before World War II were not designed to incorporate jet skis or all-terrain vehicles. Now, many of these areas are struggling to develop new regulations to reduce the ecological effect of people using these machines. Furthermore, humans are reflexive and therefore consider the consequences of their, and other people's, future behaviour before making a decision. This reflexivity can make predictions self-fulfilling or self-negating, which makes predicting future behaviour more complex (Brock & Hommes 1997; Carpenter, Brock, & Hanson 1999).

16.3.2 | Controllability

Controllability of ecological processes by management action depends on both the nature of the ecological processes involved and the organization of the society being managed. Available knowledge and technology strongly influence management control of both these aspects of controllability.

Ecological processes vary in their ability to be controlled depending on three factors: novelty, visibility, and connections across spatial and temporal scales. First, the uniqueness of a given ecological situation makes control more difficult because control techniques are more easily transferred between analogous situations. For example, there are fewer analogues to the Florida Everglades than to a small North Temperate lake. Second, the visibility of ecosystem processes depends on the ease with which ecosystem functional relationships can be separated from noise, enabling an understanding of system dynamics. Visibility is low when ecological change occurs slowly, when there are long lags between management and environmental change, or when ecological processes are obscured beneath the ground or water. Examples of each of these situations are, respectively, the accumulation of phosphorus in soil (Bennett *et al.* 1999), the effect of deforestation on groundwater dynamics (Clarke *et al.* 2002), and the difficulties of monitoring oceanic sea turtle populations (Hays *et al.* 2003). Third, the controllability of a system being managed is decreased if processes external to it strongly influence the system's dynamics. For example, attempts to restore salmon to a section of river will be influenced by dam management, runoff from surrounding land, changes in fishing techniques, and the ocean temperature patterns in the North Pacific (Peterson 2000).

Along with these ecological features of controllability, society strongly influences what type of control is possible in a given location. The degree to which people are able to agree on how to manage a shared resource influences the degree of control people can exert over ecological dynamics. Attributes of ecological management institutions can help or hinder the ability to respond to change (Michael 1973, Röling & Wagemakers 1998, Westley 1995). Four key attributes of institutions that affect ecological effectiveness appear to be shared ecological understanding, match of scales of organization and ecological processes, effectiveness of collective action, and effectiveness of past conflict resolution. Shared understanding among actors of how the natural world works is important because it determines whether people can agree on an intervention in the system (Ostrom 1990). The match between institutional and ecological scales represents the degree to which management institutions function at the scales at which important ecological change occurs (Folke *et al.* 1998). Effectiveness of collective action is the degree to which organized groups of people can implement new policies (Bromley 1989). The legitimacy of resolutions of past conflict determines the ability of an institution to address new issues. Governments have frequently expropriated resources from local people then had attempts at management fail because of passive or active resistance to management policies from these people (Scott 1998). Consequently, the controllability of an ecosystem is determined largely by the group of people using and managing it.

Technology, with social organization, determines the controllability of an ecosystem. Depending on the technology available to managers, different ecological processes are more or less easy to control. For example, with little technology, it is easier to control access to an island than it is to an offshore fishery. Similarly, it is easier to monitor and regulate a stream than to do so for groundwater. Clearly, technology can shift these relationships. For example, satellites and transponders lower the cost of monitoring and enforcing fishing, increasing the controllability of an offshore fishery. Such systems have been used increasingly since the 1990s to monitor fishing in large, geographically isolated areas, such as the Pacific.

By working to increase controllability and reduce uncertainty, people make social-ecological situations more manageable. Increasing social agreement can increase the ability of people to control a system, as can ecological engineering interventions. Similarly, social learning processes, such as comanagement, and complex systems approaches that allow the useful simplification of complex situations can increase understanding.

16.4 | CURRENT APPROACHES TO ECOLOGICAL MANAGEMENT

Ecological management approaches were initially developed based on assumptions of certainty and control. Although techniques to manage controllable, well-understood situations are well established, techniques for coping with situations that are uncertain or difficult to control represent the frontier of ecological management (Fig. 16.3).

Ecological management approaches, such as MSY, were originally developed to deal with understood, controllable situations. For more than a century approaches to optimize the production of various provisioning ecosystems services have been developed in agriculture,

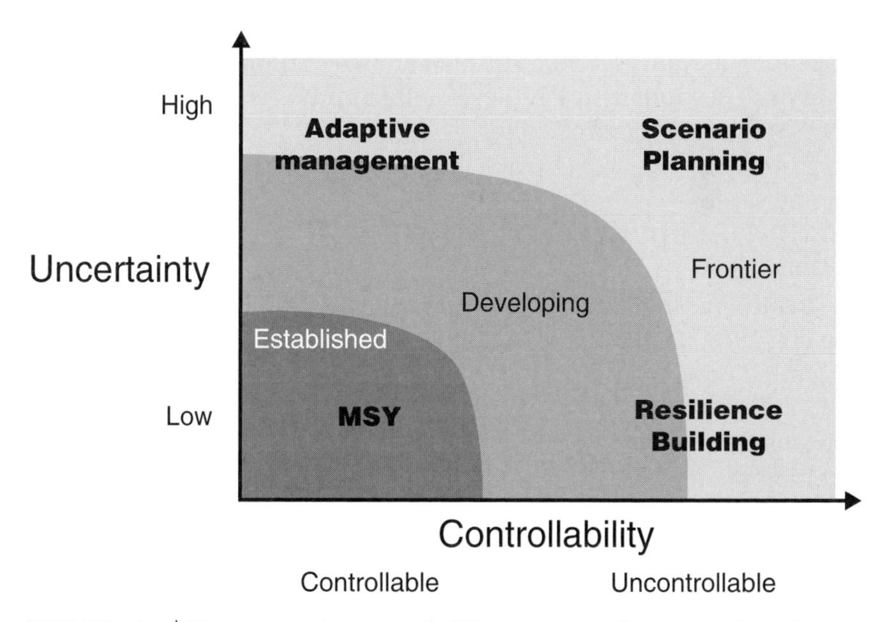

FIGURE 16.3 | The appropriateness of different approaches to ecological management varies according to the relative uncertainty and controllability of an ecological management situation. MSY: maximum sustained yield. Adapted from Peterson, Cumming & Carpenter (2003).

forestry, wildlife, fisheries, conservation biology, and economics. These techniques are applicable in many situations today and continue to be developed.

Ecological management techniques to cope with situations that lack either control or certainty represent an actively developing area of research. There has been more than 25 years of research developing experimental approaches to management to cope with situations of high uncertainty. In these situations management approaches such as adaptive management are appropriate because they use ecological manipulation to hedge against uncertainty and incorporate learning from experience.

There is also a body of research conducted over past decades that focused on increasing the ability of ecosystems to cope with uncontrollable shocks or stresses. In situations in which uncertainty is low but controllability is limited, approaches that increase resilience, or take insurance-type approaches, are sensible.

There has been relatively little development of approaches that allow managers to cope with situations that are both uncertain and uncontrollable. Although this represents the frontier of ecological management research, there are techniques that can be used, such as scenario planning. Scenario planning aims to make sense of complex situations by creating narratives. I have already outlined optimization approaches; in the next sections I present developing approaches to situations that lack certainty and controllability.

16.4.1 | Adaptive Management

Adaptive management is an approach to management in which policies are treated as hypotheses and management actions as experiments (Holling 1978, Walters 1986). Adaptive environmental management is a structured process of "learning by doing" that aims to reduce the social and ecological costs of management experiments and to increase the opportunities for learning. It aims to facilitate social learning by using a combination of assessment, computer modelling, and management experimentation to identify critical uncertainties facing managers. During this process people develop alternate hypotheses that address these uncertainties. Management plans are developed to evaluate these hypotheses by using the human manipulation of ecological processes to strategically probe the functioning of ecosystems.

An adaptive management process typically has three phases: assessment, modelling, and management experimentation. Assessment integrates existing experience, data, and theory. This integrated understanding is then embodied in dynamic computer models that attempt to make predictions about the effects of alternative policies. The modelling step serves three functions. First, it clarifies the problem and provides a common forum for people to discuss management issues. Second, the model can be used to eliminate policies unlikely to be effective. Third, the modelling process identifies which gaps in existing knowledge have the largest consequences for policy. Often these gaps involve poorly understood large-scale processes that cannot be investigated by small-scale experiments. If this is the case, this discovery leads to the design of large-scale management experiments to resolve these key uncertainties at the temporal and spatial scales relevant for management.

Despite being widely advocated in the last decade, and being initially proposed more than 25 years ago, adaptive management has not been widely practiced. Barriers to the implementation of adaptive management include the unwillingness of managers or decision makers to confront uncertainty, the desire of vested interests to avoid the change that experimentation may produce, and the cost of monitoring and experimentation (Walters 1997). They are produced by the social and political context in which adaptive management takes place. These barriers indicate that adaptive management is unlikely to be successful if it is imposed on a social-ecological system rather than developed with participants in that system. Adaptive co-management attempts to solve this problem by treating adaptive management as part of a collaborative social learning process (Berkes, Colding, & Folke 2003). Adaptive co-management integrates practical and theoretical insights from the study of institutions with technical insights of adaptive management. It aims to strengthen the adaptive aspects of local management by linking them to the maintenance of ecological knowledge and the folklore, rituals, and ceremonies that maintain local management institutions (Berkes, Colding, & Folke 2000, 2003).

16.4.2 | Resilience Building

Resilience building, which requires increasing the ability of a system to cope with stress or surprise, is an approach that has been advocated in situations in which control is difficult but in which there is some understanding about how the system works. The basic idea is to manipulate the system to increase the likelihood of it persisting despite disturbance. Building resilience is equivalent to hedging bets or purchasing insurance for unlikely but possible outcomes (Costanza *et al.* 2000). A simple example is the preservation of mangroves in tropical areas to reduce the vulnerability of coastal areas to flooding. For example, one factor that resilience depends on is response diversity—the variation of responses to environmental change among species that contribute to the same ecosystem function (Elmqvist *et al.* 2003). Increasing response diversity, for example, by allowing the recovery of populations of algae-grazers on a coral reef, can build system resilience (Bellwood *et al.* 2004).

An ecological example of building resilience is provided by the manipulation of lake food webs. Many agricultural areas have experienced large increases in soil phosphorus. Lakes in these regions are vulnerable to eutrophication from the runoff from this soil phosphorus. Controlling much of this runoff is difficult. Consequently, one approach to coping with this increased stress of lakes is to work to increase their resilience to phosphorus loading (Beisner, Dent, & Carpenter 2003). One way of doing this is to ensure that lakes have a robust food web that includes substantial populations of fish-eating fish (Carpenter & Kitchell 1993). These fish decrease the likelihood of increased phosphorus loadings tipping the lake into a state in which undesirable algal blooms occur. Through a trophic cascade, an increase in piscivorous fish increases populations of the large herbivorous zooplankton that prey on lake algae. Similarly, the vulnerability of coastal zones to the undesirable consequences of eutrophication appears to have been increased by declines in the population of large predatory fishes and marine mammals (Jackson *et al.* 2001). There are not general well-developed approaches to building resilience in ecosystems, but it is an area of active ecological research (Berkes, Colding, & Folke 2003).

16.4.3 | Scenario Planning

Scenario planning is a systemic method for thinking creatively about possible futures in which uncertainty is high and controllability is low (Peterson, Cumming, & Carpenter 2003; van der Heijden 1996; Wack 1985). Rather than attempting to predict a specific future state, scenario planning considers a diversity of possible futures that include many of the important uncertainties in the system. A scenario planning process begins with the identification of a central issue or problem. This issue is then used to define the key actors, linkages, and ecological attributes of the system used to identify the aspects of the systems whose future state is both uncertain and important. These key uncertainties provide a skeleton around which a set of scenarios is built. Quantitative and qualitative data and model output are used to construct a set of plausible scenarios.

Scenarios are not defined in terms of probabilities; rather, they are contrasted against one another to provide a tool for thinking about the relationships among today's decisions, social-ecological dynamics, and alternative futures. There are a variety of approaches to scenario planning. The appropriateness of any approach depends on the information available and the goals of the analysis (van Notten *et al.* 2003). The plausibility and relevance of a set of scenarios can be tested against models, and empirical data, before the scenarios are used to evaluate alternative policies. In addition, scenarios can provide an opportunity to create inspiring visions that encourage people to work together to create futures that they desire and avoid futures they do not. Scenario planning has been used to explore global sustainability (Raskin *et al.* 1998), biodiversity loss (Sala *et al.* 2000), and carbon dioxide emissions (Nakicenovic & Swart 2000). Increasingly, scenario planning has been used in an ecological management context; recent examples include tropical forest management (Wollenberg, Edmunds, & Buck 2000) and development of a temperate lake district (Carpenter *et al.* 2003, Peterson *et al.* 2003). The Millennium Ecosystem Assessment is using scenario planning to explore the future of ecosystem services (Bennett *et al.* 2003).

16.5 | FRONTIERS OF ECOLOGICAL MANAGEMENT

The development of methods to cope with the uncontrollable, uncertain realities of ecological management is vitally needed and represents the frontier of ecological management research (Board on Sustainable Development 1999). Researchers and practitioners are working to develop integrated frameworks and practical methods that build on the work of adaptive management, building resilience, and scenario planning. A variety of approaches build on different traditions in ecosystem ecology. Some approaches build on complexity theory (Allen, Tainter, & Hoekstra 2003; Allen *et al.* 2001; Kay *et al.* 1999; Waltner-Toews *et al.* 2003). Others use integrated approaches from management and community development, such as soft systems in organizations (Checkland 1981), participatory management (Borrini-Feyerabend, Kothari, & Pimbert 2000; Röling & Wagemakers 1998), learning organizations (Senge 1990), participatory modelling (Bousquet *et al.* 2001), and participatory geographic information systems (Craig, Harris, & Weiner 2002). Furthermore, others integrate insights from the institutional study of ecological management with adaptive management (Berkes, Colding, & Folke 2003).

16.5.1 | An Approach: Resilience Analysis

One approach being developed to manage low controllability/high uncertainty situations is resilience analysis (Walker *et al.* 2002). Resilience analysis integrates adaptive management with scenario planning and resilience building. This approach has been developed through dialogue among a group of researchers affiliated with the Resilience Alliance (http://www.resalliance.org), who are attempting to understand resilience in an international set of regional case studies. These case studies emerge from a diverse history of theoretical and empirical work on the dynamics of regional ecosystems and ecological management that have focused on management problems that are difficult to control or understand (Gunderson & Holling 2002; Gunderson, Holling, & Light 1995; Holling 1978; Walters 1986).

Resilience analysis is a process aimed at building a shared understanding of what generates and reduces resilience in a socio-ecological system (SES). An SES is defined as a set of actors (the people, groups, corporations, or other entities that play a substantial role in structuring a system), ecosystems, and the linkages among them. Resilience analysis is a participatory process that can be used to both evaluate existing policies and suggest new ones. The effective involvement of various stakeholder groups in this process has the potential to produce a shared understanding of how the system functions. This shared understanding can clarify trade-offs and conflicts among actors.

Resilience analysis iterates through several processes: scoping, exploring what an appropriate definition of the SES could be, imagining the system and its dynamics in detail, and finally exploring how these dynamics produce and destroy resilience (Fig. 16.4). The phases of this process are described in more detail in the next sections.

Phase 1: Scoping

Scoping is a process of identifying the system and issues of interest. Its goal is to examine a system from a variety of perspectives before deciding on a set of key issues and a working definition of an SES.

FIGURE 16.4 | An outline of a process of resilience management. Stakeholders define a problem, and this problem is used to analyze a social-ecological system in a fashion that examines interactions between key social and key ecological parts of a social ecological system with the aim of discovering vulnerabilities and opportunities for positive transformation. This understanding can be used to develop new policies and institutions.

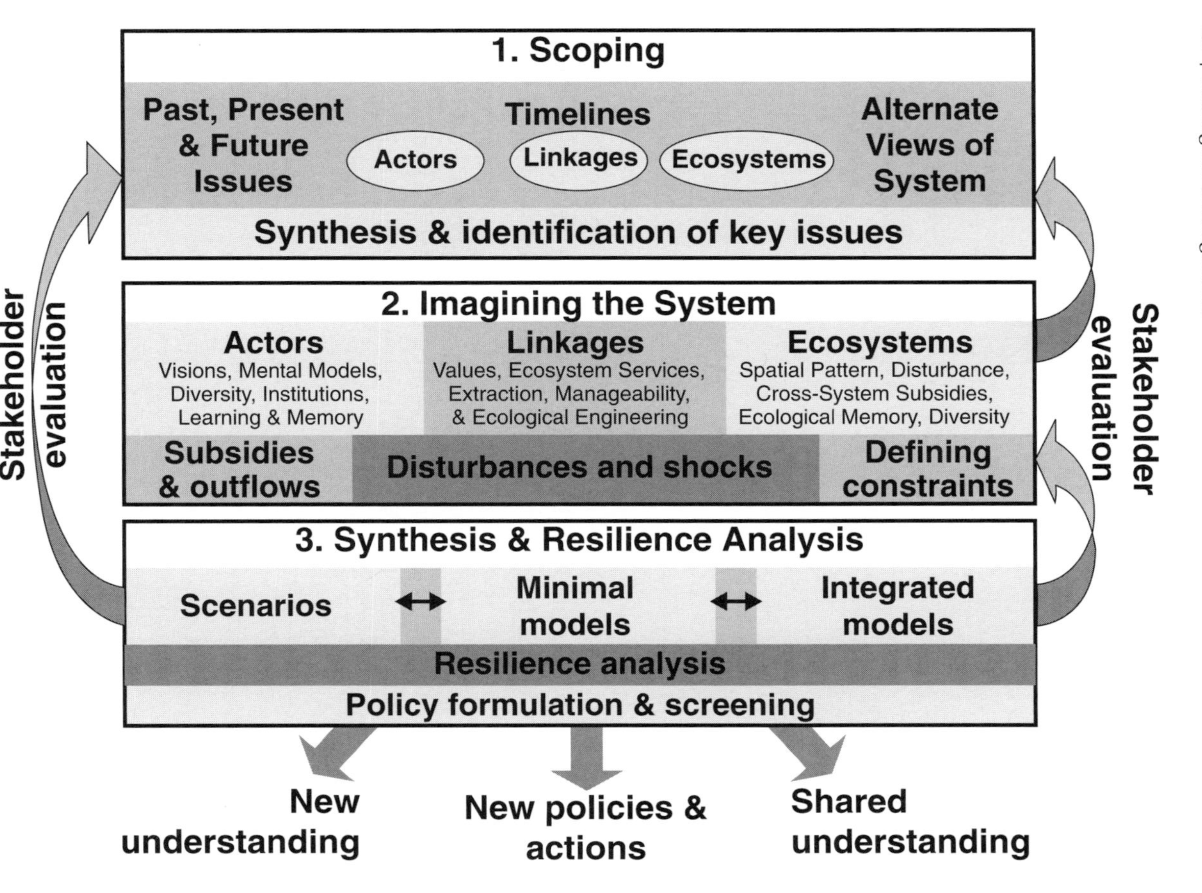

1. Scoping
Past, Present & Future Issues
Timelines
Actors
Linkages
Ecosystems
Alternate Views of System
Synthesis & identification of key issues
2. Imagining the System
Actors
Visions, Mental Models, Diversity, Institutions, Learning & Memory
Linkages
Values, Ecosystem Services, Extraction, Manageability, & Ecological Engineering
Ecosystems
Spatial Pattern, Disturbance, Cross-System Subsidies, Ecological Memory, Diversity
Subsidies & outflows
Disturbances and shocks
Defining constraints
3. Synthesis & Resilience Analysis
Scenarios
Minimal models
Integrated models
Resilience analysis
Policy formulation & screening
New understanding
New policies & actions
Shared understanding
Stakeholder evaluation
Stakeholder evaluation

Defining a system is a subjective process. To be useful a system should be conceptualized with a question or issue in mind. However, often apparent problems are merely symptoms of more fundamental problems. Stepping back from the contested issue of the moment is useful. One effective way of doing this is analysing the historical dynamics of the system. Describing the complex events of the past forces simplification. Attending to how the actors, ecosystems, and linkages among them have changed over time (especially attending to periods of crisis and reorganization) can help organize thinking about problems and issues confronting a system.

Frequently, in such a process it is useful to conceptualize or imagine a system in alternative ways. There are many alternative histories of a place. What they emphasize and ignore will depend on both the knowledge and the questions that motivate the historian. Analysing the same period using different questions can prove illuminating in determining what perspectives reveal or conceal about a situation. For example, Allen, Zellmer, & Wuennenberg's (Chapter 17) description of narratives mentions alternative histories of the Great Plains of the United States produced by environmental, agricultural, and Native American perspectives.

This process should aim to define a set of key issues and an SES. An SES can be defined by its spatial boundaries and a temporal horizon. Its key components should be identified: who are the actors, what are the ecosystems, and how are they linked? This rough definition of the system provides the essential basis for phase 2. It defines what aspects of ecosystems or society are of concern, because it is resilience of these features that is to be assessed (Carpenter *et al.* 2001).

Phase 2: Imagining the SES as a System

Scoping explored a diversity of concepts and issues to define a rough outline; phase 2 adds flesh to this outline with the minimal complexity that allows the defined set of issues facing the SES to be explored. The process of imagining the SES as an integrated system delves into structure, dynamics, and functions of the actors and ecosystems and into the linkages among them to the external world.

The behaviour of actors will be determined by the interaction of their goals, or visions; their mental models of how the world works; and their interactions with forces external to the system. Their behaviour will be constrained by the institutions that regulate interactions with ecosystems. Memory of the success or failure of past actions provides ideas for future action, and the ability of actors to learn from the consequences of their actions will shape how their behaviours will change.

Similarly, ecological dynamics are produced from interaction of diverse species and ecosystems across a landscape. These interactions will be transformed by disturbances, such as flooding or fire. The ecosystems that reorganize following these disturbances will depend on the legacies, spatial pattern, and mobile links preserved by the system.

Ecosystems and actors interact. People rely on the predictable supply of ecosystem services, such as clean air and water. Actors will choose what they extract and how they engineer ecosystems based on their cultural and aesthetic values. However, their ability to learn about and manipulate will depend on how well their technology is able to sense and influence ecosystems (e.g., it is easier to count trees than fish).

By examining the components of an SES and the connections of the SES to its environment, this stage of resilience analysis should identify what an integrated conceptualization of the system needs to include. Synthesis of these components into an integrated system is performed in the next step of the analysis.

Phase 3: Synthesis and Resilience Analysis

Compressing the complicated details of a real-world system into a useful form is at the heart of resilience analysis. Three separate, yet interrelated, analytical approaches are scenario planning, minimal modelling, and integrated modelling. Each approach focuses on a few aspects of a social-ecological system, providing each with its own strengths and weaknesses.

Scenario planning is a qualitative approach to resilience analysis that can be useful in situations high in uncertainty. As described earlier, scenario planning typically uses three to five structured accounts of plausible futures that include a range of possible outcomes to assess vulnerability and resilience of specific aspects of an SES. Scenarios can be used to test the robustness of policies or systems to different futures. Scenario planning is accessible and allows situations with many unknowns and uncertainties to be addressed, but its qualitative nature can be a barrier to analysis (Peterson, Cumming, & Carpenter 2003).

An alternate approach is *minimal* modelling of key components of the integrated system. Minimal models highlight the significance of processes operating at different timescales, especially the importance of relatively slow processes that shape ecological organization. Including the reflexive behaviour in social and ecological dynamics both enriches and constrains system behaviour (Janssen 2003). Minimal models allow aspects of a system to be examined in detail, but they have a limited ability to capture the dynamics of a system as a whole.

Integrated models capture the key aspects of an SES in a model that is rich but manageable. These models forgo the analytical precision of minimal models for a richer holistic representation of a system. Many such regional models have been developed and used in adaptive management exercises, for example, models of the Everglades (Walters, Gunderson, & Holling 1992), Eglin Air Force Base (Hardesty *et al.* 2000, Peterson 2002), and Grand Canyon (Walters *et al.* 2000). Integrated models surrender analytical rigor but are more suited to evaluating and exploring alternative policies.

Using one or, preferably, a set of these approaches allows researchers to assess what components of the system contribute to resilience and to identify vulnerabilities within and opportunities for increasing resilience or for transformation. These approaches can be used to examine the robustness of existing policies and to determine whether there are new policies that could reduce vulnerabilities or profit from opportunities.

16.6 | CONCLUSIONS

Ecological management approaches have gradually evolved as ecologists and managers have been confronted by the limits of existing approaches. Approaches have moved from a narrow technical focus on the management of single provisioning services to more integrated ecological management approaches, which now often explicitly consider human dynamics. Ecological management techniques initially assumed that managers had understood and could control the ecosystems in which they intervened. As ecologists have better understood the functioning of ecosystems, and people have demanded more say over decisions that affect their lives, this type of command and control management has become less feasible, leading to the development of management approaches, such as adaptive management, that explicitly deal with uncertainty and social learning.

The context in which ecological management is practiced strongly influences the success of different management approaches. Although most ecological management decisions are made in situations of high uncertainty and low controllability, the theoretical background most commonly used in ecological management is based on MSY, which is only appropriate for low-uncertainty and high-controllability situations. Other approaches, such as adaptive management and resilience building, have been developed, but researchers are only beginning to develop methods to cope with problems that are both highly uncertain and difficult to control.

Ecological management methods should aim to remove situations from high uncertainty, low controllability situations. Processes that either increase the controllability or reduce the uncertainty can allow better developed approaches to ecological management to be used. Changing the situation of a social-ecological system can occur through either technical or social processes. Increasing social agreement can increase the ability of people to control a system, as can ecological engineering interventions. Similarly, social learning processes, such as comanagement, and complex systems approaches that allow the useful simplification of complex situations can increase understanding. We, ecologists and humanity, need to develop better methods to understand and control the ecosystems we are haphazardly altering. We need to get good at it so that we can produce a world that we want rather than the world that we will otherwise get.

ACKNOWLEDGEMENTS

This chapter was improved by comments from Elena Bennett and Beatrix Beisner. Most of the ideas expressed in this chapter developed from my collaboration with members of the Resilience Alliance, in particular Buzz Holling, Fikret Berkes, Steve Carpenter, Graeme Cumming, Carl Folke, and Carl Walters.

REFERENCES

Allen, T.F.H., Tainter, J.A., & Hoekstra, T.W. (2003). "Supply-Side Sustainability." Columbia University Press, New York.

Allen, T.F.H., Tainter, J.A., Pires, J.C., & Hoekstra, T.W. (2001). Dragnet ecology, "Just the facts, ma'am": The privilege of science in a postmodern world. *Bioscience* **51**, 475–485.

Beisner, B.E., Dent, C.L., & Carpenter, S.R. (2003). Variability of lakes on the landscape: Roles of phosphorus, food webs, and dissolved organic carbon. *Ecology* **84**, 1563–1575.

Bellwood, D.R., Hughes, T.P., Folke, C., & Nystrom, M. (2004). Confronting the coral reef crisis. *Nature* **429**, 827–833.

Bennett, E.M., Carpenter, S.R., Peterson, G.D., Cumming, G.S., Zurek, M., & Pingali, P. (2003). Why global scenarios need ecology. *Front. Ecol. Environ.* **1**, 322–329.

Bennett, E.M., Reed-Andersen, T., Houser, J.N., Gabriel, J.R., & Carpenter, S.R. (1999). A phosphorus budget for the Lake Mendota watershed. *Ecosystems* **2**, 69–75.

Berkes, F., Colding, J., & Folke, C. (2000). Rediscovery of traditional ecological knowledge as adaptive management. *Ecol. Appl.* **10**, 1251–1262.

Berkes, F., Colding, J., & Folke, C. (2003). "Navigating Social–Ecological Systems: Building Resilience for Complexity and Change." Cambridge University Press, Cambridge.

Berkes, F., Mahon, R., McConney, P., Pollnac, R.C., & Pomeroy, R.S. (2001). "Managing Small-Scale Fisheries: Alternative Directions and Methods." International Development Research Centre, Ottawa, Canada.

Board on Sustainable Development (Policy Division), National Research Council. (1999). "Our Common Journey: A Transition Toward Sustainability." National Academy Press, Washington, DC.

Borrini-Feyerabend, G., Kothari, A., & Pimbert, M.P. (2000). "Comanagement of Natural Resources: 'Learning by Doing' Throughout the World." IUCN ROCA, Cameroon.

Bousquet, F., Le Page, C., Bakam, I., & Takforyan, A. (2001). Multiagent simulations of hunting wild meat in a village in eastern Cameroon. *Ecol. Mod.* **138**, 331–346.

Brock, W.A., & Hommes, C.H. (1997). A rational route to randomness. *Econometrica* **65**, 1059–1095.

Bromley, D. (1989). Economic interests and institutions: the conceptual foundations of public policy. Basil Blackwell, New York.

Carpenter, S., Brock, W., & Hanson, P. (1999). Ecological and social dynamics in simple models of ecosystem management. *Conser. Ecol.* **3,** 4. URL: http://www.consecol.org/vol3/iss2/art4.

Carpenter, S., Walker, B., Anderies, J.M., & Abel, N. (2001). From metaphor to measurement: Resilience of what to what? *Ecosystems* **4,** 765–781.

Carpenter, S.R., & Kitchell, J.F. (1993). "The Trophic Cascade in Lakes." Cambridge University Press, Cambridge.

Carpenter, S.R., Levitt, E.A., Peterson, G.D., Bennett, E.M., Beard, T.D., Cardille, J.A., & Cumming, G.S. (2003). "Scenarios for the Future of the Northern Highland Lake District." University of Wisconsin, Madison, WI.

Carpenter, S.R., Ludwig, D., & Brock, W.A. (1999). Management of eutrophication for lakes subject to potentially irreversible change. *Ecol. Appl.* **9,** 751–771.

Checkland, P.B. (1981). "Systems Thinking, Systems Practice." Wiley, New York.

Clark, C.W. (1985). Bioeconomic modelling and fisheries management. John Wiley & Sons, New York.

Clark, J.S., Carpenter, S.R., Barber, M., Collins, S., Dobson, A., Foley, J.A., Lodge, D.M., Pascual, M., Pielke, R., Pizer, W., Pringle, C., Reid, W.V., Rose, K.A., Sala, O., Schlesinger, W.H., Wall, D.H., & Wear, D. (2001). Ecological forecasts: An emerging imperative. *Science* **293**, 657–660.

Clarke, C.J., George, R.J., Bell, R.W., & Hatton, T.J. (2002). Dryland salinity in southwestern Australia: its origins, remedies, and future research directions. *Austral. J. Soil Res.* **40**, 93–113.

Costanza, R., Daly, H., Folke, C., Hawken, P., Holling, C.S., McMichael, A.J., Pimentel, D., & Rapport, D. (2000). Managing our environmental portfolio. *BioScience* **50,** 149–155.

Craig, W.J., Harris, T.M., & Weiner, D. (2002). "Community Participation and Geographic Information Systems." Taylor & Francis, London.

Elmqvist, T., Folke, C., Nystrom, M., Peterson, G., Bengtsson, J., Walker, B., & Norberg, J. (2003). Response diversity, ecosystem change, and resilience. *Front. Ecol. Environ.* **1**, 488–494.

Finlayson, A.C. (1994). "Fishing for Truth: A Sociological Analysis of Northern Cod Stock Assessments from 1977–1990." Institute of Social and Economic Research, Memorial University of Newfoundland, St. John's, Newfoundland.

Foley, J.A., Kucharik, C.J., Donner, S.D., Twine, T.E., & Coe, M.T. (in press). Land use, land cover, and climate change across the Mississippi Basin: Impacts on land and water resources. *In* "Ecosystem Interactions with Land Use Change" (R. DeFries, G. Asner, & R. Houghton, Eds.). AGU Chapman Monograph Series, Washington, DC.

Folke, C., Pritchard, L. Jr., Berkes, F., Colding, J., & Svedin, U. (1998). "The Problem of Fit Between Ecosystems and Institutions." IHDP Working Paper. URL: http://www.ihdp.uni-bonn.de/html/publications/workingpaper/wp02m.htm.

Funtowicz, S.O., & Ravetz, J.R. (1993). Science for the postnormal age. *Futures* **25**, 739–755.

Groffman, P.M., & Pace, M.L. (1998). Synthesis: What kind of discipline is this anyhow? *In* "Successes, Limitations, and Frontiers in Ecosystem Science" (M.L. Pace, & P.M. Groffman, Eds.). Springer-Verlag, New York, pp. 473–481.

Gunderson, L., & Holling, C., Eds. (2002). "Panarchy: Understanding Transformations in Human and Natural Systems." Island Press, Washington, DC.

Gunderson, L., Holling, C., & Light, S. (1995). "Barriers and Bridges to the Renewal of Ecosystems and Institutions." Columbia University Press, New York.

Hardesty, J., Adams, J., Gordon, D., & Provencher, L. (2000). Simulating management with models. *Conserv. Biol. Pract.* **1**, 26–31.

Harris, M. (1998). "Lament for an Ocean: The Collapse of the Atlantic Fishery—A True Crime Story." McClelland & Stewart, Toronto.

Hays, G., Broderick, A., Godley, B., Luschi, P., & Nichois, W. (2003). Satellite telemetry suggests high levels of fishing-induced mortality in marine turtles. *Mar. Ecol. Progr. Ser.* **262**, 305–309.

Hilborn, R., & Mangel, M. (1997). "The Ecological Detective: Confronting Models with Data." Princeton University Press, Princeton, NJ.

Hilborn, R., & Walters, C.J. (1992). "Quantitative Fisheries Stock Assessment: Choice, Dynamics, and Uncertainty." Chapman & Hall, New York.

Holling, C.S. (1978). "Adaptive Environmental Assessment and Management." Blackburn Press, Caldwell, NJ.

Holling, C.S. (1986). The resilience of terrestrial ecosystems: Local surprise and global change. *In* "Sustainable Development of the Biosphere" (W.C. Clark, & R.E. Munn, Eds.). Cambridge University Press, Cambridge, pp. 292–317.

Holling, C.S., & Meffe, G.K. (1996). Command and control, and the pathology of natural-resource management. *Conserv. Biol.* **10**, 328–337.

Hutchings, J.A., Walters, C., & Haedrich, R.L. (1997). Is scientific inquiry incompatible with government information control? *Can. J. Fish. Aquat. Sci.* **54**, 1198–1210.

Jackson, J.B.C., Kirby, M.X., Berger, W.H., Bjorndal, K.A., Botsford, L.W., Bourque, B.J., Bradbury, R.H., Cooke, R., Erlandson, J., Estes, J.A., Hughes, T.P., Kidwell, S., Lange, C.B., Lenihan, H.S., Pandolfi, J.M., Peterson, C.H., Steneck, R.S., Tegner, M.J., &

Warner, R.R. (2001). Historical overfishing and the recent collapse of coastal ecosystems. *Science* **293**, 629–638.

Janssen, M.A. (2003). "Complexity and Ecosystem Management: The Theory and Practice of Multiagent Systems." Edward Elgar, Northampton, MA.

Kay, J.J., Regier, H.A., Boyle, M., & Francis, G. (1999). An ecosystem approach for sustainability: Addressing the challenge of complexity. *Futures* **31**, 721–742.

Larkin, P.A. (1977). An epitaph for the concept of maximum sustained yield. Transactions of the American Fisheries Society 106: 1–11.

Lee, K. (1993). "Compass and Gyroscope: Integrating Science and Politics for the Environment." Island Press, Washington, DC.

Levin, S.A. (1999). "Fragile Dominion: Complexity and the Commons." Perseus Publishing, Reading, MA.

McNeill, J.R. (2000). "Something New Under the Sun: An Environmental History of the Twentieth-Century World." Norton, New York.

Michael, D.N. (1973). "On Learning to Plan and Planning to Learn." Jossey-Bass Publishers, San Francisco.

Nakicenovic, N., & Swart, R. (2000). "Emissions Scenarios." Cambridge University Press, Cambridge.

Ostrom, E. (1990). "Governing the Commons: The Evolution of Institutions for Collective Action." Cambridge University Press, New York.

Peterson, G.D. (2000). Political ecology and ecological resilience: An integration of human and ecological dynamics. *Ecol. Econ.* **35**, 323–336.

Peterson, G.D. (2002). Forest dynamics in the Southeastern United States: Managing multiple stable states. *In* "Resilience and the behavior of large-scale ecosystems" (L. Gunderson & L. Pritchard Jr. Eds.). Island Press, Washington, DC, pp. 227–246.

Peterson, G.D., Carpenter, S.R., & Brock, W.A. (2003). Model uncertainty and the management of multistate ecosystems: A rational route to collapse. *Ecology* **84**, 1403–1411.

Peterson, G.D., Beard, D., Beisner, B., Bennett, E., Carpenter, S., Cumming, G., Dent, L., & Havlicek, T. (2003). Assessing future ecosystem services: A case study of the northern highland lake district, Wisconsin. *Conserv. Ecol.* **7**, 1. URL: http://www.consecol.org/vol7/iss3/art1.

Peterson, G.D., Cumming, G., & Carpenter, S.R. (2003). Scenario planning: A tool for conservation in an uncertain world. *Conserv. Biol.* **17**, 358–366.

Raskin, P., Gallopin, G., Gutman, P., Hammond, A., & Swart, R. (1998). "Bending the Curve: Toward Global Sustainability," PoleStar Series 8. Stockholm Environment Institute, Sweden.

Ricker, W.E. (1975). Computation and interpretation of biological statistics of fish populations. Bulletin 191, Fisheries Research Board of Canada, Ottawa, Ontario.

Ripple, W.J., & Beschta, R. (2003). Wolf reintroduction, predation risk, and cottonwood recovery in Yellowstone National Park. *For. Ecol. Manage.* **184**, 299–313.

Röling, N.G., & Wagemakers, M.A.E. (1998). "Facilitating Sustainable Agriculture: Participatory Learning and Adaptive Management in Times of Environmental Uncertainty." Cambridge University Press, New York.

Sala, O.E., Chapin, F.S., Armesto, J.J., Berlow, E., Bloomfield, J., Dirzo, R., Huber-Sanwald, E., Huenneke, L.F., Jackson, R.B., Kinzig, A., Leemans, R., Lodge, D.M.,

Mooney, H.A., Oesterheld, M., Poff, N.L., Sykes, M.T., Walker, B.H., Walker, M., & Wall, D.H. (2000). Biodiversity: Global biodiversity scenarios for the year 2100. *Science* **287**, 1770–1774.

Schaefer, M.B. (1954). Some aspects of the dynamics of populations important to the management of commercial marine fisheries. Bulletin of the Inter-American tropical tuna commission 1:25–26.

Scott, J.C. (1998). "Seeing Like a State: How Certain Schemes to Improve the Human Condition Have Failed." Yale University Press, New Haven, CT.

Senge, P. (1990). "The Fifth Discipline: The Art and Practice of the Learning Organization." Doubleday, New York.

van der Heijden, K. (1996). "Scenarios: the Art of Strategic Conversation." John Wiley & Sons, New York.

van Notten, P.W.F., Rotmans, J., van Asselt, M.B.A., & Rothman, D.S. (2003). An updated scenario typology. *Futures* **35**, 423–443.

Wack, P. (1985). Scenarios: Shooting the rapids. *Harv. Bus. Rev.* **63**, 139–150.

Walker, B., Carpenter, S., Anderies, J., Abel, N., Cumming, G., Janssen, M., Lebel, L., Norberg, J., Peterson, G.D., & Pritchard, R. (2002). Resilience management in social–ecological systems: A working hypothesis for a participatory approach. *Conserv. Ecol.* **6,** 14. URL: http://www.consecol.org/vol6/iss1/art14.

Walters, C.J. (1986). "Adaptive Management of Renewable Resources." McGraw-Hill, New York.

Walters, C.J. (1997). Challenges in adaptive management of riparian and coastal ecosystems. *Conserv. Ecol.* **1,** 1. URL: http://www.consecol.org/vol1/iss2/art1.

Walters, C.J., Gunderson, L., & Holling, C.S. (1992). Experimental policies for water management in the Everglades. *Ecol. Appl.* **2**, 189–202.

Walters, C.J., & Holling, C.S. (1990). Large-scale management experiments and learning by doing. *Ecology* **71**, 2060–2068.

Walters, C.J., Korman, J., Stevens, L.E., & Gold, B. (2000). Ecosystem modeling for evaluation of adaptive management policies in the Grand Canyon. *Conserv. Ecol.* **4,** 1. URL: http://www.consecol.org/vol4/iss2/art1.

Walters, C.J., & Maguire, J.J. (1996). Lessons for stock assessment from the northern cod collapse. *Rev. Fish Biol. Fish.* **6**, 125–137.

Waltner-Toews, D., Kay, J.J., Neudoerffer, C., & Gitau, T. (2003). Perspective changes everything: Managing ecosystems from the inside out. *Front. Ecol. Environ.* **1**, 23–30.

Westley, F. (1995). Governing design: The management of social systems and ecosystems management. *In* "Barriers and Bridges to the Renewal of Ecosystems and Institutions" (L. Gunderson, C.S. Holling, & S.S. Light, Eds.). Columbia University Press, New York, pp. 489–532.

Westley, F., Carpenter, S.R., Brock, W.A., Holling, C.S., & Gunderson, L.H. (2002). Why systems of people and nature are not just social and ecological systems. *In* "Panarchy: understanding transformations in human and natural systems" (L. Gunderson & C. Holling, Eds.). Island Press, Washington, DC, pp. 103–119.

Wollenberg, E., Edmunds, D., & Buck, L. (2000). Using scenarios to make decisions about the future: Anticipatory learning for the adaptive comanagement of community forests. *Lands. Urb. Plan.* **47**, 65–77.

17 | IS ECOSYSTEM MANAGEMENT A POSTMODERN SCIENCE?

Kevin de Laplante

17.1 | INTRODUCTION

The chapters by Allen and his colleagues (Chapter 15) and by Peterson (Chapter 16) present several challenges to readers of this volume. For some, the theoretical framework for ecosystem management endorsed by the authors—a variant of what may be called the "ecosystem approach to ecosystem management" (Crober 1999, Kay *et al.* 1999)—will be unfamiliar, and there may be questions about its scientific credentials and its relation to more widely known forms of ecosystem theory. For others, there may be uncertainty about the point or motivation for the more philosophical elements of the chapter by Allen and his colleagues and perhaps some confusion over these authors' blanket rejection of *modernist realism* and their willingness to embrace the *postmodern* label. This commentary is directed primarily at readers new to the ecosystem theory and management literature who are looking for an overview that will help them understand and evaluate the scientific and philosophical issues raised in these chapters.

This chapter is divided into four sections in addition to this introduction. Section 2 surveys the historical development of ecosystem ecology and how the ecosystem approach to ecosystem management relates to this history. I compare and contrast the conception of ecosystem theory endorsed by Allen and his colleagues and by Peterson with the classical tradition of ecosystem ecology developed by Eugene and Howard Odum in the 1950s and 1960s. Section 3 takes a closer look at *post-normal science,* a term that Allen and his colleagues and Peterson borrow from Silvio Funtowicz and Jerry Ravetz (1991, 1992, 1993) to characterize their conception of ecosystem management science. I review the origins and motivation for the concept of post-normal science and comment on the relationship of post-normal science to postmodern philosophies of science that endorse a global antirealist attitude to scientific knowledge. Section 4 focuses more narrowly on Chapter 15. I offer an interpretation of the main thesis and some critical comments on the argument strategy. Section 5 considers the issue of whether changes in the science and philosophy of ecosystem management constitute a gradual evolutionary development or a more radical paradigm shift.

17.2 | ECOSYSTEM ECOLOGY: CONCEPTUAL AND HISTORICAL BACKGROUND

17.2.1 | Ambiguities of the Ecosystem Concept

One primary obstacle to understanding the ecosystem literature is that the key term, *ecosystem,* has multiple meanings and uses. *Ecosystem* can be used to refer to an *object* of scientific study, to *theories* of the nature of such objects, or to a *methodology* for doing science. Failure to distinguish these senses can easily result in confusion.

17.2.1.1 | Ecosystem as Object

The ecosystem as object or entity is commonly described as a community of organisms with its physical environment, or the total set of biotic elements and physico-chemical processes within a particular spatial region. This concept appears in its modern form in the work of Tansley (1935). Ecologists of all theoretical orientations have found the ecosystem concept useful for one purpose or another (Cherrett 1989). However, it is most commonly used in theoretical contexts when there is some intention to study the *interrelations* of the biotic and abiotic elements in the ecosystem or to treat the ecosystem as a dynamical *unit.*

17.2.1.2 | Ecosystem as Theory

Ecosystem theory does not emerge until almost 10 years after Tansley. The seminal paper is Raymond Lindeman's (1942) "The Trophic–Dynamic Aspect of Ecology," in which he reconceptualized Elton's theory of trophic dynamics and community structure as a theory of biogeochemical cycling driven by a one-way flow of solar energy. The focus of ecosystem theory is the description and explanation of the flow of matter and energy in ecosystems and how these flows influence and are influenced by population and community processes (although a strong emphasis on the latter moves us closer to the domain of population and community ecology).

Ecosystem theory comes in a variety of forms. If the focus of study is primarily the flow of matter (i.e. carbon, nitrogen, phosphorus, water, etc.), then it is common to use the term *biogeochemistry* to describe this branch of ecosystem theory (e.g., see Likens *et al.* 1977). If the focus is more narrowly on balances and ratios of chemical elements in ecosystems, then it is called *ecological stoichiometry* (e.g., see Sterner & Elser 2002). If the focus is on the flow of energy and the constraints imposed by the laws of thermodynamics on this flow, then the theorist is doing *ecological energetics* (e.g., see Weigert 1988).

Finally, if the theorist is primarily interested in studying the organizational and developmental properties of ecosystems taken as wholes, then it is common to describe theories of such properties as belonging to the field of *systems ecology*. Systems ecologists use a variety of formal techniques to describe these structural and dynamical properties (e.g., see Halfon 1979, Ulanowicz 1986, and Jorgensen 2002). Note, however, that the term *systems ecology* also has multiple meanings in the literature (McIntosh 1985, pp. 221–241).

The differences among the various forms of ecosystem theory are a matter of degree, not of kind; in any particular application of ecosystem theory there is often a good deal of overlap among the different forms.

17.2.1.3 | Ecosystem as Method

When I type “an ecosystem approach to” in my Google search engine I get more than 600,000 hits. Browsing the first few pages of hits I see the phrase modifying terms such as conservation, human health, fisheries, urban settlements, understanding cities, public education, management, human well-being, the integrity of the Great Lakes, and family therapy. In this context the term *ecosystem* clearly connotes a certain *style* or *philosophy of research* rather than a specific object of investigation or a theory of the behaviour of such objects.

What is this style of research? It will vary from application to application, but it typically involves distinguishing a focal system of investigation (a fishery, a business, a family unit, a forest, whatever) and situating it explicitly within some broader environmental context. To take an *ecosystem approach* is to assume that at least some of the properties and behaviours of the focal system will depend on interactions or relations between the focal system and its surrounding environment and that a proper understanding of these properties and behaviours will require bringing these system-environment relations explicitly within the field of investigation.

The analytic tools brought to bear on such investigations will vary from field to field and will range from informal word or picture models to formal network descriptions and quantitative mathematical models. What such investigations have in common is a distrust of more narrowly framed approaches, a belief that anchoring investigations on phenomena operating at only a single spatio-temporal scale risks missing or obscuring the influence of processes operating at smaller and larger spatiotemporal scales on the focal-level dynamics.

Within the literature on biological and ecological complexity the term *hierarchy theory* is commonly used to refer to a set of theoretical principles for conducting research on complex systems from a multiscalar perspective. One of our lead authors is a prominent contributor to this literature (Allen & Starr 1982, O'Neill *et al.* 1986, Ahl & Allen 1996). Although descriptions of the compositional hierarchy of natural systems are often taken to have ontological importance, hierarchy theory is arguably best understood in epistemological or methodological terms as a framework for *investigating* complex natural and social systems; you might view hierarchy theory as a set of prescriptions for how to implement an "ecosystem approach to X".

Thus, we see that the term *ecosystem* has a range of meanings that need to be kept distinct. One can be a reductionist population ecologist but still endorse the concept of an ecosystem as an ecological entity (e.g., see Colinvaux 1978). One can be an ecosystem theorist yet question the utility of the ecosystem as a suitable object of scientific study (e.g., see O'Neill 2001, pp. 5–9). And one can endorse an ecosystem approach to some phenomenon or area of inquiry without necessarily referring either to ecosystems as objects or to traditional ecosystem theory (e.g., see Forget & Lebel 2002).

17.2.2 | The Classical Tradition of Ecosystem Ecology

How should we understand the theoretical perspective that Allen and his colleagues and Peterson bring to their understanding of ecosystem ecology? With all the references to complexity, self-organization, adaptive cycles, and the like, the type of ecosystem theory described in these chapters may strike vs as terribly new and cutting edge. There is some truth to this assessment; recent developments in complexity theory (e.g., nonlinear dynamics and catastrophe theory and Holling's adaptive cycle model, which is meant to apply to a broad class of complex adaptive systems) play a strong role in their version of ecosystem theory. But it is just as accurate, and I think ultimately more informative, to view this style of ecosystem theory as a throwback to the classical tradition of ecosystem ecology of the 1950s and 1960s; you could describe it as positively *old fashioned*.

This will take some explanation. In the 1950s and 1960s, ecosystem theory—as developed and articulated by the most influential ecologists of that generation, Eugene and Howard Odum—dominated the discourse of theoretical ecology. The story of ecosystem ecology and its rise to prominence during this period has been well told by others (Hagen 1992, Golley 1993). For present purposes I wish only to draw attention to three important elements of the conception of ecosystem science promoted by the Odums (for these see Odum 1971, Craige 2001, and Odum & Odum 2001).

First, their conception of ecosystem science emphasized all three of the senses of the ecosystem concept outlined previously. They believed that ecosystems had a genuine existence and ought to be regarded as the fundamental *objects* of investigation for the ecological sciences; they developed an elaborate *theory* of the structure, function, and temporal dynamics of ecosystems; and they advocated an ecosystem approach (or, as they might say, a "systems approach") to the *methodology* for framing and answering questions about ecological phenomena.

Second, they embraced a radical conception of the nature and domain of ecological science. The Odums argued that ecology was just as much a social science as a natural science; after all, if ecology is the science of organism-environment relations and humans are organisms, then ecology is necessarily also a science of human-environment relations. Indeed, they argued that it was best viewed as a *new kind of science*, a science of synthesis and integration that transcended the orthodox disciplinary boundaries that separated the physical, biological, and social sciences.

Third, they believed that the natural processes of ecosystem organization and development offered a suite of useful models for environmentally responsible and sustainable social and economic organization in the human sphere. The writings of Howard Odum in particular were seminal for the new fields of ecological economics and ecological engineering.

The Odums developed this conception of ecosystem ecology in the context of the emerging environmental movement and growing concern over issues of long-term sustainability. It is clear, I think, that the conception of ecosystem management science endorsed by Allen and his colleagues and by Peterson shares all three of these elements of the classical tradition of ecosystem ecology, and I suspect it is similarly motivated. In this sense it is a throwback to an earlier, more ambitious, and more radical conception of ecological science than is dominant today. You might call them proponents of a "new" classical tradition of ecosystem ecology.

17.2.3 | The Rise, Fall, and Reemergence of the Classical Tradition

The beginning of the 1960s was a time of great optimism for ecosystem ecology, but by the early 1970s attitudes towards the classical tradition of ecosystem ecology and ecosystem theory had changed dramatically. New ecology textbooks published in the 1970s downplayed both the ecosystem concept and the ecosystem theory, emphasized evolutionary and population-community processes as the dominant framework for theoretical ecology, and took pains to distance ecological science qua science from environmentalist concerns and environmental problem solving (Hagen 1992).

What accounts for this change in attitude towards the classical tradition of ecosystem ecology? Given the associations I have drawn between the classical tradition and the conception of ecosystem ecology advocated by Allen and his colleagues and by Peterson, the question is important. If the classical tradition has somehow been discredited, are the more contemporary versions open to the same objections? In the following I survey the rise, fall, and reemergence of this tradition and address this question against this historical background.

17.2.3.1 | Criticisms of Classical Ecosystem Theory

There are at least three important factors that contributed to the decline of classical ecosystem ecology.

A. Perceived Incompatibility with Evolutionary Theory

The most important theoretical criticism of ecosystem ecology came from young ecologists committed to the increasingly popular view that neo-Darwinian evolutionary theory should serve as the central unifying framework for all of the biological sciences. They were concerned that classical ecosystem theory either ignored the principles of neo-Darwinian evolutionary theory or misused them (Hagen 1992, pp. 146–165).

Eugene Odum famously argued, for example, that coevolution operated to promote mutualistic relationships within communities, that such relationships promoted overall community stability, and that more stable communities would out-compete less stable communities, resulting in the progressive development of ecosystems towards states of greater biomass, species diversity, and stability (Odum 1969). The primary mechanism of ecosystem development was *group selection.*

But group selectionist explanations in the biological sciences were dealt a devastating blow by George Williams in his 1966 book *Adaptation and Natural Selection.* Williams argued that natural selection functions primarily to select for genes responsible for phenotypic traits that confer a survival advantage on *individual organisms;* only rarely (if ever) does it function at the level of populations or groups. Williams's arguments were widely endorsed by evolutionary biologists and the younger generation of evolutionary ecologists, and they helped undermine support for classical ecosystem theory.

B. Association with Discredited Clementsian Models of Ecological Succession

The classical Clementsian model of succession assumed an orderly transition from one community type to another through mechanisms operating at the level of the whole community. Odum's account of ecological succession employed thermodynamic, information-theoretic, and evolutionary concepts foreign to Clements's account, but it endorsed in broad outline the phenomenology of successional development that Clements had laid out.

However, the Clementsian phenomenology of succession eventually came under severe attack. It was found, for example, that both early- and late-successional species can be present continuously within a community throughout the succession process, a discovery that conflicted with the Clementsian model of community development as a sequence of virtually wholesale species replacements (Drury & Nisbet 1973).

The models that later came to dominate mainstream accounts of ecological succession posited mechanisms located in the properties and dynamics of individual organisms and populations rather than in the whole communities or ecosystems (Glenn-Lewin, Peet, & Veblen 1992). The critique of Clementsianism and the perceived success of these newer models of ecological succession did much to draw the mainstream of theoretical ecology from the holism that was so closely associated with classical ecosystem theory.

C. Perceived Failure of International Biological Program Modelling Projects

In the 1950s, ecosystem ecology was essentially a post–World War II cold war science. The nuclear threat posed by the Soviet Union prompted the US government, through the arm of the Atomic Energy Commission, to initiate a series of studies to investigate the effects of radiation exposure on biological and agricultural systems. These studies naturally lent themselves to an ecosystem approach. Through the mid-1960s and early 1970s, large-scale ecosystem modelling projects were the main component of the US contribution to the International Biological Program, a multinational venture aimed at understanding the biological basis of productivity and human welfare. Formal, whole-system modelling methods played a large role in these projects.

Retrospective analyses generally agree, however, that these large-scale ecosystem projects met with only limited success, both in terms of contributing to a general theoretical understanding of ecological processes and in terms of generating predictively useful models (Hagen 1992, pp. 164–180; Golley 1993, pp. 131–140). The perceived failure of these modelling efforts further contributed to a diminishing of the status of systems-oriented ecosystem science.

17.2.3.2 | Ecosystem Ecology After the Fall

In response to the criticisms noted previously and the changing culture of professional ecology, ecosystem ecology as a unified discipline broke apart. A tightly coupled combination of ecosystem energetics and biogeochemistry retained its mainstream respectability and remains the dominant tradition of ecosystem science elaborated upon in textbooks on the subject (Aber & Melillo 2001, Chapin, Matson, & Mooney 2002), but the more abstract systems approaches were effectively marginalized.

Systems ecology continued to be developed by a handful of researchers in a handful of institutions through the 1980s and 1990s (most notably at the University of Georgia and the University of Florida, the home-bases of Eugene and Howard Odum, respectively) but largely on the outskirts of mainstream ecological theory. Many systems ecologists shifted their attention to problems of applied ecosystem management and analysis of human-environment interactions, where audiences were more receptive to their concepts and methods.

This mostly continues to be the case. In the United States, at least, the more theoretical forms of systems ecology are still a minority tradition within the mainstream of theoretical ecology, but they appear to have found a home in several applied and hybrid ecological disciplines: conservation biology/ecology, ecosystem health and management, sustainability theory, ecological economics, etc. (e.g., see Costanza, Norton, & Haskell 1992; Pimentel, Westra, & Noss 2000; Gunderson & Holling 2002).

17.2.3.3 | How the "New" Classical Tradition Differs from the Old

The conception of ecosystem ecology that Allen and his colleagues and Peterson support is similar to the classical tradition in its emphasis on synthesis and integration, in the breadth of its domain, and in its conviction that there are general principles of ecosystem growth, development, and self-organization that may function as useful models for sustainable management of socioecological systems. It differs, however, in several important respects.

First, like most of contemporary ecology, it rejects the classical Clementsian phenomenology of succession and consequently is not tied to any conception of ecosystem dynamics as deterministically progressing towards a unique state of stable equilibrium. It accepts that stochastic processes (both internal and external to the system) and sensitivity to initial conditions make it impossible to predict ecosystem behaviours with precision or certainty. The focus of the new approaches is not on precise prediction of future ecosystem behaviours but on developing a global understanding of the different stability regimes that an ecosystem might occupy and on the key ecosystem processes that influence transitions between such regimes (Gunderson & Holling 2002, pp. 25–62).

Second, although it is committed to the reality of self-organizing processes that confer a degree of unity and order at higher levels of organization, it is not committed to group selectionist mechanisms to explain such processes. It is more common to see self-organization explained in terms of processes that are either non-Darwinian in nature (e.g., as described in thermodynamic and network theories of self-organization; see Jorgensen 2002) or consequences of natural selection acting at the level of individuals (Holland 1995, Levin 1999). These are the sorts of processes posited in the version of ecosystem theory that Allen and his colleagues and Peterson endorse (Peterson 2000; Allen, Tainter & Hoekstra 2003, pp. 320–379). It should be noted, however, that multilevel selection theory has regained a certain theoretical and philosophical respectability in recent years (Sterelny 1996), and calls have been made for its renewed application to community and ecosystem ecology (Wilson 1997).

Thus, we see that at least two of the principle objections to classical ecosystem ecology do not apply to contemporary ecosystem theories. There remain, I believe, important questions for ecosystem theorists to consider regarding the degree to which specific ecosystem hypotheses are subject to confirmation or disconfirmation by empirical data (Sagoff 2003). But this is a different objection from the claim that these ecosystem theories are inconsistent with accepted biological science. If this is the case, it is not obviously so.

17.3 | *POST-NORMAL* SCIENCE

The third objection to ecosystem ecology explained previously involved the charge that classical methods of modelling ecosystems did not yield predictively useful models. Allen and his colleagues and Peterson accept that accurate prediction is not a primary goal either of ecosystem management practice or of the ecosystem theories that they support. They argue, however, that ecosystem science—or more precisely, the application of ecosystem science to management contexts—should not be discredited because of this and invoke the concept of post-normal science to help make this point.

17.3.1 | Origins of the Term

The term *post-normal science* was coined by Funtowicz and Ravetz (1991, 1992, 1993), although antecedents of the philosophy and sociology of science associated with their use of the term can be found in Ravetz's 1971 *Scientific Knowledge and Its Social Problems*. This impressive but largely ignored work was influenced by Kuhn's 1962 *The Structure of Scientific Revolutions*, but it owes a greater debt to the neo-Marxist tradition of critical science studies (e.g., see Bernal 1942 and Rose & Rose 1969) and Ravetz's involvement in the British radical science movement (Sardar 2000).

Neo-Marxist critiques of science focus on the role that scientific practices and scientific authority play in legitimizing and reinforcing political, economic, and other institutional frameworks. Ravetz was particularly concerned with what he perceived as the vulnerability of industrialized science to corruption, that is, the production of shoddy, entrepreneurial, and dirty science driven by a reckless and short-sighted pursuit of vested power interests and technological development. He argued that more attention should be placed on the problem of ensuring the quality of both the practice and the products of science. To this end, Ravetz developed the notion of science as *craft work*, emphasizing virtues (quality) of the scientific enterprise that relate the practice of science, at all its stages, to explicit social goals and values. Ravetz argued further that improving the quality of science requires abandoning the Enlightenment ideology of science as an objective method of inquiry whose ultimate goal is the discovery of "true facts" about an independently existing external world.

The concept of *post-normal science* is a continuation and extension of this work. Funtowicz and Ravetz give a simple description of the contexts in which orthodox scientific methods and standards are likely to be inapplicable or result in poor-quality science, namely, contexts in which the behaviours of systems are too complex to predict with any certainty and are difficult to control, in which the risks associated with different outcomes associated with different courses of action are high, in which values associated with these outcomes are in dispute, and in which there are pressing time constraints on policy decision making. Ecosystem management is one area in which such conditions may routinely occur, but it is not the only area (e.g., consider medical and agricultural biotechnology, ecotoxicology and public health, the AIDS epidemic, etc.).

The contrast of *post-normal science* with the Kuhnian notion of *normal science* was intentional. Funtowicz and Ravetz characterize the dominant ideology of science—an ideology that values quantitative over qualitative description, prediction and control over uncertainty and uncontrollability, facts over values, etc.—as *normal* in the Kuhnian sense of being the framework in which the *puzzle-solving* activities of science are typically conducted (Kuhn 1962, pp. 35–42). They argue, however, that when dealing with systems and situations involving high risk and high uncertainty, this ideology is inadequate to even frame the puzzles that need to be solved, much less to solve them. Hence, science in this mode is not *normal;* the standard problem-solving procedures that are effective in low-risk, low-uncertainty situations simply are not adequate to address the challenges of scientific practice in situations of high risk and high uncertainty.

Funtowicz and Ravetz also wanted to challenge the notion of *normal science* as a context in which the exemplars for problem-solving strategies are determined solely by scientific experts. In post-normal contexts there may be several parties that have a stake in the outcomes of policy decisions, including ordinary citizens with no institutional accreditation or specialized scientific training. The policy formation process must therefore involve a framework for integrating the concerns and values of this diverse community of stakeholders.

As Chapter 16 shows, a large part of the practice of post-normal science involves collaboration with what Funtowicz and Ravetz call "extended peer communities." This may seem to involve a radical dilution of scientific authority, but recall that we are describing here contexts in which, in a real sense, there are no scientific authorities (or perhaps better, in which there is a plurality of authorities on the various empirical, sociopolitical, and normative dimensions of the problem context).

17.3.2 | Does *Post-normal* Imply *Postmodern*?

Allen and his colleagues describe themselves as committed to a *postmodern* conception of ecosystem science and management and, indeed, of scientific practice. Precisely what such a view entails is not altogether clear, but they are explicit about one aspect: they are *global antirealists* about scientific knowledge. The question I wish to consider is this: does accepting the postnormal character of ecosystem management entail such a view of science as a whole?

I need to define a few terms. I use the term *scientific realism* to refer to the position that (1) one of the proper goals of science and scientists is to describe the world not just as it appears in experience but also as it is in reality independent of experience; (2) it makes sense to describe scientific theories as "true" or "approximately true" descriptions of the world in the sense that the entities and processes posited by a theory have some relationship of reference, correspondence, or verisimilitude to entities and processes in the external world; and (3) we are sometimes warranted in judging scientific theories to be true or approximately true. *Antirealist* positions come in many forms, but they generally deny one or more of these claims.

It is important to distinguish global and local arguments for scientific realism and antirealism. Global arguments entail conclusions for the practice and status of science as a whole; local arguments apply only to specific sciences or theories within a science. One can consistently adopt a realist interpretation of some local branches of science (say, solid state physics) and an antirealist interpretation of other local branches (say, social psychology). However, so-called postmodern philosophies of science tend to endorse some form of global antirealism (Sardar 2000, Brown 2001). In their rejection of the utility of the concept of *truth* in scientific practice, Allen and his colleagues have at least this much claim to the label *postmodern*.

Now, Allen and his colleagues are clearly influenced by the writings of Funtowicz and Ravetz, who are themselves global antirealists about scientific knowledge (I interpret Ravetz 1971 as an extended argument for such a position). However, it seems clear to me that a mere commitment to viewing ecosystem management as a post-normal science does not by itself entail a global antirealist view of scientific knowledge.

Grant for the sake of argument that science in the post-normal mode is at least *prima facie* hostile to realism; if as a scientist one is primarily concerned with discerning truths about an objectively existing world, then one is likely not doing post-normal science. The primary goals of ecosystem management science in the post-normal mode are better characterized by terms such as *sustainability, integrity, health, security, safety,* and *welfare,* not by the accumulation of facts about an independently existing world (although various forms of fact-gathering will be a necessary component of any practice that aims at such goals).

But the conditions under which solid state physics is practiced are generally *not* the conditions of high-risk, radical uncertainty and contested values that characterize the post-normal mode (at least not obviously so; one could contest this claim). There may be good arguments for global antirealism about science that would include the knowledge claims of solid state physics, but the most common arguments that Funtowicz and Ravetz (and Allen and his colleagues and Peterson) give for treating ecosystem management as a post-normal science typically are not global arguments of this kind. In short, it is at least consistent to interpret a policy-oriented, value-laden, issue-driven field such as ecosystem management in antirealist terms and to interpret other areas of science, including areas of ecological science, in realist terms.

I make this point because I believe there is obvious merit in the concept of post-normal science when applied to contexts such as ecosystem management, but I am concerned that it may be too-summarily dismissed by ecologists who, rightly or wrongly, feel they cannot swallow the broader antirealist philosophy of science and postmodern rhetoric that is commonly associated with it.

17.4 | THE "PARADIGM OF NARRATIVE": DEFENDING THE HOLLING FIGURE-EIGHT

In this section I take a closer look at the thesis and argument strategy of Chapter 15 by Allen et al.

Peterson in Chapter 16 illustrates well the extent to which ecosystem management theory has crossed into the realm of the social sciences.

The point of scenario planning, for example, is to help inform current management practices by considering, through exercises in hypothetical storytelling, a variety of possible futures that include important uncertainties, not to make predictions about which future is most likely to occur.

Now, a defender of a more orthodox conception of ecological science might suggest that this is all well and good, but it is not *ecology*. Ecology is supposed to be a natural science; this is something else, something fuzzier, more speculative. Take that Holling figure-eight diagram, for instance. It is a pretty picture, but what does it actually represent? What are the dimensions of the axes? How can we quantitatively measure these dimensions? It is easy for a critic to reject such pictures as little more than graphical storytelling and consequently dismiss the whole theory of ecosystem development that the picture is intended to capture. This is the kind of criticism that Allen and his colleagues describe in the concluding section of their chapter.

The best way of reading Chapter 15 is, I think, as a defence of the ecosystem approach to ecosystem management against this sort of objection. I interpret their response as going something like this:

> You say the Holling figure-eight is just a narrative storytelling device, and because of this it should not be taken seriously as science. We say that *all science* involves narrative storytelling. Indeed, we say that narrative storytelling is fundamental to how mathematical models in any discipline, even the hardest of hard sciences, acquire specific, meaningful content. So this distinction—between disciplines that do and disciplines that do not employ narrative devices to give their theories meaningful content and practical applicability—cannot be used to distinguish social from natural science, soft from hard science, or immature from mature science. Hence, this distinction cannot be used to dismiss narrative constructions such as Holling's adaptive cycle model of ecosystem development as immature or nonscientific.

The rather abstract description of the epistemology of modelling in Chapter 15 is intended as support for this argument; they are trying to defend a conception of ecosystem theory and ecosystem management science by challenging some entrenched views that they believe bias the opinions of their critics.

The major weakness of this argumentative strategy, in my opinion, is that if one is inclined to be critical of ecosystem science from the perspective of orthodox standards of scientific methodology, then one's criticism is unlikely to turn on a point as rarefied as whether physics and ecology share the same epistemology of mathematical modelling.

One's criticism is more likely to be based on worries about the content or testability or usefulness of the theories, not the second-order question of how such theories acquire *a semantics*.

In addition, I think the general conclusion of the argument that Allen and his colleagues want to make—that all science, indeed all mathematical modelling, involves narrative in an essential way—will be viewed by many philosophers as antecedently more plausible than many of the premises they wish to invoke in that argument. I do not think many philosophers or ecologists will be persuaded of the truth of the conclusion in virtue of agreeing with the positions of Rosen and Pattee on the epistemology of modelling. This is not to say that Rosen and Pattee do not have important or true things to say on many issues concerning the science and philosophy of complexity (indeed, I am an admirer of their work on these topics). But there are inevitable problems with comprehension when appealing to these sources. Rosen's language is sufficiently idiosyncratic to often confuse the uninitiated and when combined with Pattee's terminology becomes even more so. In addition, there are substantive issues with these sources that will not go unchallenged. For example, Allen and his colleagues appeal to Pattee's views on *duality* and *complementarity* in mathematical modelling, which are themselves strongly informed by the writings of Niels Bohr and John von Neumann on the role of the observer in quantum measurement. But the complementarity interpretation of quantum mechanics is only one of a range of possible approaches to the problem of quantum measurement and frankly is not a wildly popular one among contemporary philosophers of physics (e.g., see van Fraassen 1991, pp. 241–272, and Sklar 1992, pp. 172–185). If accepting Pattee's views on the role of language and the observer in mathematical modelling requires accepting a general epistemology based, in part if not in whole, on this interpretation of quantum mechanics (as I suspect it does for Pattee), then this will be viewed as a serious liability by many.

On the other hand, the conclusion itself is more than plausible. Many philosophers have written on the ubiquity of metaphor and narrative in scientific explanation (Hesse 1980, Roth 1989, Harre 1990), and some have written specifically on the role of the human interpreter in the semantics and epistemology of modelling (e.g., see van Fraassen 1997). Allen and his colleagues may not be impressed with the approaches of these authors, but my point is a strategic one: if they are interested in communicating a novel philosophy of science to a broad audience, they might do well to become more acquainted with the resources that are out there and the common ground that might be established by drawing on those resources.

Chapter 15 is rich and provocative, and my comments have only addressed their argument in the broadest terms. Much more can be said about their specific views on the nature of mechanism, reductionism, explanation, truth, and construction in science, but space restrictions do not allow a more thorough review here.

17.5 | THEORY CHANGE IN ECOSYSTEM ECOLOGY: GRADUAL DEVELOPMENT OR PARADIGM SHIFT?

The current volume is concerned with the history of theory change in ecology. Given the historical outline and analysis developed here, how should we regard the "new" classical tradition of ecosystem ecology? Is it simply a rehash of the classical tradition, an evolutionary descendent from it, or an utterly new species of ecosystem theory?

I have identified some areas in which the newer forms of ecosystem theory differ from the older framework of the Odums, most notably on the phenomenology of succession and ecosystem change, on the theoretical basis for ecosystem growth and development, and on broader conceptions of the aims and goals of an effective science of ecosystem management. But my analysis has placed greater emphasis on the *continuity* that exists between the newer and the older forms of systems-oriented ecosystem ecology, especially with respect to agreement on the existence and significance of self-organizing processes operative at community and ecosystem levels of organization, the need for a multiscalar ecosystem approach to ecosystem management, and a commitment to a conception of the nature and domain of ecological science that spans both the natural and the social sciences and that breaks down traditional disciplinary boundaries. If forced to choose among biological metaphors, I would describe the "new" classical tradition of ecosystem ecology as a somewhat better-adapted evolutionary descendent of the old, but certainly not a new species of ecosystem theory.

Allen and his colleagues and Peterson, on the other hand, emphasize *discontinuities* between the older and the newer forms of ecosystem science and are comfortable describing their work as representative of a new paradigm for ecosystem science. But their use of this term attaches to different aspects of ecosystem science than my analysis has focused on. Peterson emphasizes the shift from MVP analysis to adaptive management that has occurred in (at least some branches of) the conservation sciences and the attendant growth of a post-normal management philosophy. The language of paradigms and paradigm shifts is sufficiently vague and unanchored in contemporary scientific discourse that it is, in my view, an arbitrary choice to view the shift from MVP to adaptive management as a paradigm shift rather than a process of gradual evolution within a broader conservation framework.

What is more interesting is the conceptual shift in philosophy that invites the terms *post-normal* or *postmodern*. Allen and his colleagues focus on this conceptual shift, identifying discontinuities at deeper levels of metaphysical, epistemological, and normative commitment: mechanistic versus formal/final causation; reductionism versus holism; objective, value-free science versus a science of quality; causal-mechanical explanation versus narrative explanation. These are real differences that a Kuhnian would recognize as different paradigms for the practice of science.

But it seems to me that one can practice ecosystem management using the tools of ecosystem ecology and the methodology of adaptive management without necessarily privileging the categories on the right over the categories on the left or even accepting this dichotomous way of characterizing the space of philosophical possibilities. I view Allen and his colleagues as defending a radical alternative *philosophy* of science but not necessarily a radical alternative *science*. In this respect I do not see the changes in the science of ecosystems and ecosystem management described in Chapters 14 and 15 as the development of a radically new scientific paradigm for ecosystem ecology, although I admit the philosophy of science that has been stimulated by these changes has a revolutionary flavour to it.

17.6 | CONCLUSION

Ecosystem ecology is a diverse and influential branch of ecological science. The particular style of ecosystem science that the contributing authors endorse elicits strong responses in both critics and advocates. It harkens back to the classical tradition of ecosystem science that viewed ecology as a new kind of science, a science of synthesis and integration that transcends orthodox boundaries between the natural and the social sciences and between the pure and the applied sciences.

I have tried to provide a conceptual framework and a historical overview of this tradition that connects it to contemporary developments in the science and practice of ecosystem management. Although ecosystem science has changed since the Odums, my analysis has emphasized the continuities that exist between the newer forms of ecosystem theory and the classical tradition of ecosystem ecology.

The "new" classical tradition of ecosystem ecology has stimulated the development of philosophies of science that have been described as *post-normal* or *postmodern*. I argue that, although the concept of post-normal science, as developed by Funtowicz and Ravetz, is a potentially useful framework for understanding the distinctiveness and challenges of ecosystem management, endorsing the concept of post-normal science in this context does not commit one to the global antirealist philosophies of science associated with postmodernism.

ACKNOWLEDGEMENTS

I am grateful to Kim Cuddington, Gregory Mikkelson, and an anonymous reviewer for helpful comments on an earlier draft of this chapter.

REFERENCES

Aber, J.D., & Melillo, J.M. (2001). "Terrestrial Ecosystems," 2nd Edition. Academic Press, New York.

Ahl, V., & Allen, T.F.H. (1996). "Hierarchy Theory: A Vision, Vocabulary, and Epistemology." Columbia University Press, New York.

Allen, T.F.H., & Starr, T. (1982). "Hierarchy: Perspectives for Ecological Complexity." University of Chicago Press, Chicago.

Allen, T.F.H., Tainter, J.A., & Hoekstra, T.W. (2003). "Supply-Side Sustainability". Columbia University Press, New York.

Bernal, J.D. (1942). "The Social Function of Science." Routledge, London.

Brown, J.R. (2001). "Who Rules in Science: An Opinionated Guide to the Wars." Harvard University Press, Cambridge, MA.

Chapin, F.S. III, Matson, P.A., & Mooney, H.A. (2002). "Principles of Terrestrial Ecosystem Ecology." Springer-Verlag, New York.

Cherrett, J.M. (1989). Key concepts: The results of a survey of our members' opinions. *In* "Ecological Concepts: The Contribution of Ecology to an Understanding of the Natural World" (J.M. Cherrett, Ed.). Blackwell Science, Oxford, pp. 1–16.

Colinvaux, P. (1978). "Why Big Fierce Animals are Rare." Princeton University Press, Princeton, N.J.

Costanza, R., Norton, B.G., & Haskell, B.D., Eds. (1992). "Ecosystem Health: New Goals for Environmental Management." Island Press, Washington, DC.

Craige, B.J. (2001). "Eugene Odum: Ecosystem Ecologist and Environmentalist." University of Georgia Press, Athens, GA.

Crober, A. (1999). "The Ecosystem Approach To Ecosystem Management," Senior Honours Thesis, Department of Geography, Faculty of Environmental Studies. University of Waterloo, Canada.

Drury, W.H., & Nisbet, I. (1973). Succession. *J. Arnold Arbor.* **54**, 331–368.

Forget, G., & Lebel, J. (2002). An ecosystem approach to human health. *Int. J. Occupat. Environ. Health* **7**, 3–38.

Funtowicz, S.O., & Ravetz, J.R. (1991). A new scientific methodology for global environmental issues. *In* "Ecological Economics: The Science and Management of Sustainability" (R. Costanza, Ed.). Columbia University Press, New York, pp. 137–152.

Funtowicz, S.O., & Ravetz, J.R. (1992). Three types of risk assessment and the emergence of postnormal science. *In* "Social Theories of Risk" (D. Golding & S. Krimsky, Eds.). Praeger, Westport, CT, pp. 251–273.

Funtowicz, S.O., & Ravetz, J.R. (1993). Science for the postnormal age. *Futures* **25**, 739–55.

Glenn-Lewin, D.C., Peet, R.K., & Veblen, T.T. (1992). "Plant Succession: Theory and Prediction." Chapman & Hall, New York.

Golley, F.B. (1993). "A History of the Ecosystem Concept in Ecology: More Than the Sum of the Parts." Yale University Press, London.

Gunderson, L., & Holling, C.S., Eds. (2002). "Panarchy: Understanding Transformations in Systems of Humans and Nature." Island Press, Washington, DC.

Hagen, J.B. (1992). "An Entangled Bank: The Origins of Ecosystem Ecology." Rutgers University Press, New Brunswick, NJ.

Halfon, E. (1979). "Theoretical Systems Ecology: Advances and Case Studies." Academic Press, New York.

Harre, R. (1990). Some narrative conventions of scientific discourse. *In* "Narrative and Culture" (C. Nash, Ed.). Routledge, London, pp. 81–101.

Hesse, M. (1980). "Revolutions and Reconstructions in the Philosophy of Science." Indiana University Press, New York.

Holland, J. (1995). "Hidden Order: How Adaptation Builds Complexity." Addison-Wesley, Reading, MA.

Jorgensen, S.E. (2002). "Integration of Ecosystem Theories: A Pattern," 3rd Edition. Kluwer, Dordrecht, The Netherlands.

Kay, J., Regier, H., Boyle, M., & Francis, G. (1999). An ecosystem approach for sustainability: Addressing the challenge of complexity. *Futures* **31**, 721–742.

Kuhn, T.S. (1962). "The Structure of Scientific Revolutions." University of Chicago Press, Chicago.

Levin, S.A. (1999). "Fragile Dominion: Complexity and the Commons." Perseus Publishing, Cambridge, MA.

Likens, G.E., Bormann, F.H., Pierce, R.S., Eaton, J.S., & Johnson, N.M. (1977). "Biogeochemistry of a Forested Ecosystem." Springer-Verlag, New York.

Lindeman, R.L. (1942). The trophic–dynamic aspect of ecology. *Ecology* **23**, 399–418.

McIntosh, R.P. (1985). "The Background of Ecology: Concept and Theory." Cambridge University Press, New York.

Odum, E.P. (1969). The strategy of ecosystem development. *Science* **164**, 262–270.

Odum, H.T. (1971). "Environment, Power, and Society." Wiley-Interchange, New York.

Odum, H.T., & Odum, E.C. (2001). "The Prosperous Way Down: Principles and Policies." University Press of Colorado, Boulder, CO.

O'Neill, R.V. (2001). Is it time to bury the ecosystem concept? (with full military honors of course!) *Ecology* **82**, 3275–3284.

O'Neill, R.V., DeAngelis, D.L., Waide, J.B., & Allen, T.F.H. (1986). "A Hierarchical Concept of Ecosystems." Princeton University Press, Princeton, NJ.

Peterson, G.D. (2000). Scaling ecological dynamics: Self-organization, hierarchical structure, and ecological resilience. *Climate Change* **44**, 291–309.

Pimentel, D., Westra, L., & Noss, R.F., Eds. (2000). "Ecological Integrity: Integrating Environment, Conservation, and Health." Island Press, Washington, DC.

Ravetz, J.R. (1971). "Scientific Knowledge and Its Social Problems." Oxford University Press, New York.

Rose, H., & Rose, S. (1969). "Science and Society." Penguin, London.

Roth, P.A. (1989). How narratives explain. *Soc. Res.* **56**, 449–478.

Sagoff, M. (2003). The plaza and the pendulum. *Biol. Phil.* **18**, 529–552.

Sardar, Z. (2000). "Thomas Kuhn and the Science Wars." Totem Books, New York.

Sklar, L. (1992). "Philosophy of Physics." Westview Press, San Francisco.

Sterelny, K. (1996). Return of the group. *Phil. Sci.* **63**, 562–584.

Sterner, R.W., & Elser, J.J. (2002). "Ecological Stoichiometry: The Biology of Elements from Molecules to the Biosphere." Princeton University Press, Princeton, NJ.

Tansley, A.G. (1935). The use and abuse of vegetational concepts and terms. *Ecology* **16**, 284–307.

Ulanowicz, R.E. (1986). "Growth and Development: Ecosystems Phenomenology." Springer-Verlag, New York.

van Fraassen, B.C. (1991). "Quantum Mechanics: An Empiricist View." Clarendon Press, Oxford.

van Fraassen, B.C. (1997). Structure and perspective: Philosophical perplexity and paradox. *In* "Logic and Scientific Methods" (M.L. Dalla Chiara, Ed.). Kluwer, Dordrecht, The Netherlands, pp. 511–530.

Weigert, R.G. (1988). The past, present, and future of ecological energetics. *In* "Concepts of Ecosystem Ecology: A Comparative View" (L.R. Pomeroy & J.J. Alberts, Eds.). Springer-Verlag, New York, pp. 29–55.

Williams, G.C. (1966). "Adaptation and Natural Selection: A Critique of Some Current Evolutionary Thought." Princeton University Press, Princeton, NJ.

Wilson, D.S. (1997). Biological communities as functionally organized units. *Ecology* **78**, 2018–2024.

VII | CONCLUSION

18 | KUHNIAN PARADIGMS LOST: EMBRACING THE PLURALISM OF ECOLOGICAL THEORY

Kim Cuddington and Beatrix E. Beisner

This collection was motivated by a desire to explore the historical roots of some major ecological theories and by curiosity as to whether the various subdisciplines to which the work belonged, or even ecology as a whole, had undergone some kind of *paradigm shift*. We had noted an increasing usage of Kuhnian language in the ecological literature and a concurrent disregard for the historical origins of new work. We wondered whether these two trends were related. That is, were younger scientists ignorant of the history of their subdiscipline because these sciences had undergone a paradigm shift that rendered older research irrelevant? To answer this question, we invited both scientists and philosophers to comment on theory development in ecology.

Although this collection is by no means an exhaustive survey of scientists' and philosophers' thoughts regarding theory change and the relative importance of historical developments, we believe the sample pool is wide enough to provide an indication of the diversity of opinion on these topics. As in many collections of the type, the multiple voices sometimes make the whole seem an arbitrary aggregation rather than a multifocus image. Our contributions range from highly ordered and informative analyses of equations and their derivations to anecdotal descriptions of postmodern philosophy (and the type of contribution did not depend on whether the contributor was a scientist or a philosopher). Accordingly, any attempt to derive an overall conclusion regarding our questions about theory development based on a reading of these contributions will likely be more reflective of our personal position than anything else. However, because our main objective was to stimulate reflection on these topics, we unabashedly offer our conclusions based on a reading of this work and invite the reader to energetically disagree. We begin with a brief review of Kuhn and other philosophies of theory change in the sciences, then move to an analysis of the different contributions. We will to attempt to explain what we believe to be the main lessons from this exercise: that Kuhnian paradigm shifts rarely occur in ecology and that theory development is best described as an evolutionary process, which can lead to a multiplicity of approaches. The take-home message for scientists is that to be ignorant of history is to be ignorant about current scientific developments.

18.1 | KUHN AND BEYOND

Kuhn (1962) was not the first to describe growth and change in scientific knowledge over time. His novel contributions are twofold. First, he suggested not only that scientists support particular concepts or theories but also that theories, methods, and standards of evaluation are accepted in an almost indivisible lump. A *paradigm*, then, is a particular way of doing science in a given subdiscipline. This claim is not as simple as suggesting that scientists' hypotheses derive directly from their theory. Rather, Kuhn is stating that the standards for evaluation of those hypotheses are transmitted with the theory. Second, Kuhn claimed that the definitions and standards of one paradigm could not be directly translated into the language of another paradigm. Because of this *incommensurability*, Kuhn suggests that it is not an accurate reading of history to claim that anomalous data and disconfirming instances forced scientists to make rational shifts to a new and better conceptual framework. Instead, individuals have to shift their allegiance from one paradigm to another in a kind of gestalt revolution. This claim has been interpreted as indicating that science is irrational, or at least culturally influenced.

The paradigm shift model may not be historically accurate, even for major theoretical developments (Büttner, Renn, & Schemmel 2003; Wilkins 1996). Moreover, the philosophy of science has progressed beyond Kuhn, and there are other models of theory change (Lakatos 1970, Hacking 1983, Hull 1988). It may be that theory change is a more gradual process and that Kuhn, in his description of the history of science, has juxtaposed the beginning and end point of a change in theory and thus made the process seem nonrational and sudden (Laudan 1984). A more careful analysis may reveal patterns that are not well accounted for by a Kuhnian model (Culp & Kitcher 1989). One can also argue that an intimate understanding of a research field is required to appreciate gradual development and shifts accompanying an evolutionary model of theory change. For example, to examine the rise of cladistics, Hull joined this research community to better acquaint himself with the discipline. In his analysis, Hull (1988) claimed that theory change in cladistics is more accurately depicted as an evolutionary process. However, even as an outside commentator, Castle demonstrated that in the development of macropalaeontological models there is an intellectual heritage from ecology and a slow evolution of these ideas to fit their new subject matter (Castle 2001).

18.2 | PARADIGM SHIFTS IN ECOLOGICAL THEORY?

Overall, our contributors seem to identify theory changes in ecology as evolutionary outgrowths of previous work. Such a finding does not preclude the occurrence of Kuhnian paradigm shifts in our field. It may be that most of the changes described herein fall into the boundaries of *normal science* as described by Kuhn. This seems a reasonable reading of Kuhn as well: certainly paradigm shifts could not possibly occur as frequently as they are claimed to in ecology (see Fig. 1.1 from Chapter 1). Therefore, scientists should view the frequent use of Kuhnian language to describe theory change as merely rhetoric, or perhaps loose use of language. However, Holt, Day, and Allen *et al.* seem to identify some genuine paradigms within ecology in their description of the merger of community ecology and evolutionary biology, the emergence of evolutionary ecology, and the application of ecosystem ecology to management, respectively. Whether revolutionary paradigm shifts have occurred seems a different and unanswered question. Tellingly, our authors use a detailed historical analysis to reach conclusions about the promise of future directions. More interestingly, many of them point to pluralism as a best bet for scientific progress.

We find perhaps the strongest disagreement on theory change in Part I on population ecology. De Roos and Persson fundamentally disagree with Hastings' suggestion that scientists will and should continue to use unstructured models in the foreseeable future. De Roos and Persson suggest that these models are outdated and of no practical utility on their own. Using such top-down models, they argue, is not a particularly useful way of describing populations: their use is a symptom of an old paradigm on the way out. Odenbaugh, in his commentary, is optimistic about current advances in the formulation and analysis of structured models, but like Hastings he advances a pluralist approach to modelling populations, although perhaps more cautiously. Unstructured models have been given a fair trial, he suggests, and although still useful, we should concentrate on newer approaches. Interestingly, Odenbaugh also suggests that the assumptions of unstructured population models (for example, identical individuals), have more to do with mathematical tractability than with a conceptual framework for modelling populations. This theme is revisited in Part II on epidemiological ecology.

Heesterbeek, like De Roos and Persson, equates particular model assumptions with a scientific paradigm. He describes the use of mass-action assumptions as the initial paradigm for modelling disease transmission. However, he suggests that this assumption, like the identical individuals assumption for population dynamics, was adopted for mathematical convenience rather than because it expressed real beliefs about the systems under consideration. Explanatory power must also play a role, however, because later in the chapter he refers to mass-action as "one of the most powerful and most useful old metaphors of ecology" that is still useful today. Similar claims could be made of the assumptions of unstructured population models described by Hastings.

Heesterbeek also identifies an early divergence between the *a priori* mathematical approach of Ross and McKendrick and the statistical approaches of Brownlee, Greenwood, and co-workers that could mark the emergence of separate paradigms in theoretical epidemiology. Once again, we find a divergence of methods and questions based on methodological differences. Castle, in his commentary on community ecology, remarks on a similar divergence. Koopman also notes this separation between mathematical and statistical approaches in epidemiology but suggests that a rapprochement is at hand in the simultaneous pursuit of many modelling approaches. In a theme common to Hastings, Odenbaugh, Keeling, Holt, and Day, he suggests that relating the different techniques will be a fruitful area of investigation.

Keeling, in his description of modern advances in epidemiology, also advocates a kind of pluralism in that he suggests that the mass-action assumption is retained in local-scale interactions for some model formulations. By illustrating the relationships among different approaches, Keeling seems to imply that there has been a steady growth and divergence of methods rather than an abrupt shift. Moreover, there are at least two directions a scientist can travel in the effort to relax mass-action assumptions: computationally intensive simulations models and analytically tractable approximation techniques. Here, too, Keeling favours pluralism.

A common theme in these two sections is that some methodological assumption is identified as key to defining paradigm, or perhaps better, a research program. Hastings and Odenbaugh describe the identical individuals assumption as the key driver of early population models, but note that this assumption was likely made for analytical tractability rather than because it expressed a conceptual belief regarding real populations. Heesterbeek and Keeling describe the mass-action assumption in models of disease transmission in the same way. Moreover, Odenbaugh and Koopman stress the importance of advances in computation, as does McCann in his description of community ecology, and Keeling emphasizes the role of advances in mathematical techniques in precipitating changes in the questions that ecologists can ask. For example, as these authors describe, advances in computer technology and the development of moment-closure and pair-approximation techniques are permitting the growth of spatially explicit models. For the first time, theoretical ecologists are able to make predictions about the importance of local ecological interactions. Does this imply that previously theorists believed such interactions were unimportant? It seems unlikely. Often we feel that the roles of such methodological commitments and the effects of methodological advances are ignored in both histories and philosophies of science. Some accounts would have us believe that, until recently, all theoretical ecologists believed natural systems were always found in equilibrium states (Botkin 1990). However, readings of original literature in population ecology demonstrate this is simply untrue (Cuddington 2001). It seems clear that, no less than experimental science, theory in ecology can be driven by methodological considerations as much as by conceptual commitments.

In their discussion of community ecology, Ives and McCann concur that there have been at least three major theoretical positions regarding the diversity-stability debate. Both also identify these positions with the different methodologies from which they were derived. Ives describes a naïve historical analysis of this theory change as intuition being overturned by mathematical modelling, which was, in turn, deposed by experimental data. However, he suggests that a deeper analysis of how the meaning of *stability* has changed illustrates how these different perspectives are complementary rather than disparate.

Although McCann does not explicitly comment on the issue, it does seem for him that the trajectory of the diversity-stability debate does not really constitute a paradigm shift either. He notes that May (1974) was aware that the results of his early analysis did not match findings from empirical data. May suggested that factors not included in his analysis such as patterns of interaction strengths and compartmentalization of food webs may play a role in the stabilization of natural systems. On a related topic, McCann points out how the area of food web research emerged more or less directly from May's early work.

Castle, in his analysis of community ecology, suggests that an outsider to the field may be disturbed by the apparent lack of a common core. Theory change has involved debates about conceptual issues, building models, and the phenomena themselves. Someone with a Kuhnian perspective may suggest that this is an infant subdiscipline, which lacks a common paradigm. Castle finds it suggestive, however, that Ives and McCann both reach the conclusion that the current need is for more data-driven models. Castle claims that the debates in community ecology are not evidence of an immature science without a cohesive foundation. Community ecology has reached a point at which there is greater conceptual agreement than might, at first glance, be obvious.

In the section on evolutionary ecology, Holt identifies the potential integration of evolutionary biology and community ecology as a new paradigm for ecology. He notes that these sciences, despite some early positions, have been truly distinct disciplines for at least 50 years, and certainly since Odum's influential work in 1971. This separation has hinged on the argument that evolutionary and ecological mechanisms operate on different timescales. In his description of community ecology, he identifies several fruitful directions such as the use of metabolic-based models and community module models. And interestingly, he notes that "several community ecologists have championed such pluralist approaches to the development of community theory". Once again, a pluralist position, at least with respect to modelling, is explicitly stated.

Day, in his description of the links between population ecology and evolutionary biology, picks up on the rejection of the assumption of separation of timescales as central to the formation of the field of evolutionary ecology. This assumption, unlike that of identical individuals in population ecology or mass-action in disease transmission, seems to reflect an ontological commitment in ecology and evolutionary biology rather than a methodological convenience, although it is certainly that as well. Ironically, Day notes that much work in evolutionary ecology has again made the assumption that the ecological and evolutionary dynamics proceed at different speeds. Therefore, with respect to theory change, the formation of the field of evolutionary ecology seems to have involved a big conceptual shift, although this shift would have united two subdisciplines and therefore seems somewhat different than a classical Kuhnian paradigm shift. In contrast, Day suggests that a new and fruitful area of investigation, adaptive dynamics, is a direct descendent of game theory, not a fundamentally new approach. Both the actual conceptual change that occurred in the formation of evolutionary ecology and the potential shift that Holt promotes that would produce an evolutionary community ecology seem likely candidates for paradigms in the Kuhnian sense. However, as Day describes, the formulation of new models simultaneously describing evolutionary and ecological interactions followed an incremental and logical sequence, not a sudden change.

Sterelny, in his commentary on these two chapters, suggests that a synthesis between community and evolutionary ecology, as suggested by Holt, would be a good candidate for a paradigm shift if it were to occur. However, it is still not clear if local communities are simply aggregates of species or cohesive units with important emergent properties. If, for example, diversity is a causally important characteristic of communities that drives stability, then there is a fundamental sense in which the community as a whole determines ecological processes. Therefore, the connection between ecology and evolutionary biology would be at best indirect because communities are not evolutionary units.

Allen and his colleagues, and Peterson to a lesser extent, claim that a postmodern, narrative-driven science of ecosystem management constitutes a new paradigm for ecology; one that members of the older, modernist, realist school may attack without understanding. This is the strongest statement in our collection that a new way of doing science in ecology has emerged and, furthermore, has done so on a trajectory similar to that described by Kuhn. We agree that the emergence of the ecosystem as an object of scientific study is a likely candidate for a new paradigm. The slow growth of this position, however, as traced by de Laplante, leads us to believe that this perspective emerged from a route best described as evolutionary rather than revolutionary.

In their analyses, both Allen and his colleagues and de Laplante trace the origin of ecosystem ecology back to Tansley (1935). De Laplante suggests that the definition of paradigm, as it is used in ecology, is sufficiently vague that we could equally describe the move from Odum to Allen *et al.* in terms of the scientific changes, as either a paradigm shift or an evolutionary outgrowth. Odum also stressed the interdisciplinary nature of ecosystem management and the need to include humans as part of the managed system. From a philosophical point of view, Allen and his colleagues' antirealism could seem to constitute a real shift in the way ecologists do science. However, it is not clear either to de Laplante or to us that global antirealism is a necessary component of the type of ecosystem management described by Allen *et al.* Moreover, few scientists of our acquaintance believe they are discovering the *truth* about systems; instead, they describe their enterprise as "telling stories about nature". Even if they are not global antirealists in the same sense as Allen and his colleagues, most ecologists at least do not adhere to the strong realist position described by these authors. Again, we suggest that ecologists are a heterogeneous bunch and pluralism seems to be the preferred approach even with regard to philosophical commitments. Some of us are realists; many of us believe select parts of our practice, such as model building, are more similar to narrative than to a quest for truth; and some, like Allen and his colleagues, are global antirealists. Once again, we note that a pluralist approach seems to be working for both our metascientific commitments and for our scientific frameworks.

18.3 | CONCLUDING REMARKS

As we noted earlier, it is unlikely that the findings of this collection can be definitive on the question of whether there are paradigm shifts in ecology. As de Laplante remarks, the use of the term *paradigm shift* in ecology is so varied as to imply anything from revolutionary shift to slow evolutionary growth. The answer to the question of what ecologists mean when they use the phrase *paradigm shift* is likely to be: anything. On the other hand, perhaps this is understandable because the answer to the question of how theory changes in ecology seems to be that it changes in a variety of ways. In many cases, new theoretical directions are outgrowths of older material. For example, it seems to us that structured population models emerged directly from older models. In some cases, a merger between different areas occurs, as in the formation of evolutionary ecology. In other areas, such as ecosystem ecology, perhaps a truly new idea created a new subdiscipline. Equally interesting is the point that new conceptual commitments do not always drive theory change. Methodological advances can also precipitate change. Most importantly, however, the contributors have illustrated that a historical viewpoint improves our ability to evaluate the importance of older developments and to anticipate fruitful new directions.

REFERENCES

Botkin, D.B. (1990). "Discordant Harmonies: A New Ecology for the Twenty-First Century." Oxford University Press, New York.

Büttner, J., Renn, J., & Schemmel, M. (2003). Exploring the limits of classical physics: Plank, Einstein, and the structure of a scientific revolution. *Stud. Hist. Phil. Mod. Phys.* **34**, 37–59.

Castle, D. (2001). A gradualist theory of discovery in ecology. *Biol. Phil.* **16**, 547–571.

Cuddington, K. (2001). The "balance of nature" metaphor and equilibrium in population ecology. *Biol. Phil.* **16**, 463–479.

Culp, S., & Kitcher, P. (1989). Theory structure and theory change in contemporary molecular biology. *Brit. J. Phil. Sci.* **40**, 459–483.

Hacking, I. (1983). "Representing and Intervening: Introductory Topics in the Philosophy of Natural Science." Cambridge University Press, Cambridge.

Hull, D. (1988). "Science as a Process: An Evolutionary Account of the Social and Conceptual Development of Science". University of Chicago Press, Chicago.

Kuhn, T. (1962). "The Structure of Scientific Revolutions." University of Chicago Press, Chicago.

Lakatos, I. (1970). Falsification and the methodology of scientific research programs. In "Criticism and the Growth of Knowledge" (I. Lakatos & A. Musgrave, Eds.). Cambridge University Press, Cambridge, pp. 91–196.

Laudan, L. (1984). "Science and Values." University of California Press, Berkeley, CA.

May, R.M. (1974). "Stability and Complexity in Model Ecosystems," 2nd Edition. Princeton University Press, Princeton, NJ.

Odum, E.P. (1971). "Fundamentals of Ecology," 3rd edition. Saunders, Philadelphia.

Tansley, A.G. (1935). The use and abuse of vegetational concepts and terms. *Ecology* **16**, 284–307.

Wilkins, A.S. (1996). Are there "Kuhnian" revolutions in biology? *Bioessays* **18**, 695–696.

INDEX